About the author

Tony Worsley is Professor of Public Health at the University of Wollongong, New South Wales. Until recently, he was Senior Research Advisor at the Victorian Health Promotion Foundation, and before that, Professor of Public Health Nutrition and Head of the School of ... and Nutrition Sciences, Deakin University. He has been a member of the ... of Sciences' Nutrition Committee, Co-Executive Editor ... App ... *Journal of Behavioral Nutrition and Physical Activity*.

He has held senior academic appointments in several universities and CSIRO. These include the University of Adelaide where he was Professor of Public Health, the CSIRO Division of Human Nutrition (Head of the Food Policy Research Unit), the Australian National University's National Centre for Epidemiology and Population Health, and Otago University (New Zealand) where he was professor in Social Nutrition.

He has wide experience in the evaluation of public health nutrition programs, and in the promotion and maintenance of behaviour change. His current research involves several overlapping areas including: behavioural and nutritional epidemiology, studies of products at the food–drug interface, food and nutrition policy research, health and nutrition promotion. He has published widely in scientific and professional journals. He has authored several books including *Public Health Nutrition* (with M. Lawrence), *Food People and Health*, *The Food System*, *The Use and Abuse of Vitamins* and the *Body Owner's Manual*.

Recent research projects include: the development of dietary approaches to stop hypertension; examination of the adoption of plant-based foods by consumers and industry; consumers' attitudes towards children's foods at school, and baby boomers' future food and health needs.

Nutrition Promotion

Theories and methods, systems and settings

TONY WORSLEY

CABI is a trading name of CAB International

CABI Head Office
Nosworthy Way
Wallingford
Oxfordshire OX10 8DE
UK

Tel: +44 (0)1491 832111
Fax: +44 (0)1491 833508
E-mail: cabi@cabi.org
Website: www.cabi.org
 CABI North American Office
875 Massachusetts Avenue
7th Floor
Cambridge, MA 02139
USA

Tel: +1 617 395 4056
Fax: +1 617 354 6875
E-mail: cabi-nao@cabi.org

A catalogue record for this book is available from the British Library, London, UK.

Library of Congress Cataloging-in-Publication Data

Worsley, Tony.
  Nutrition promotion : theories and methods, systems and settings / Tony Worsley.
  p. ; cm.
  Includes bibliographical references and index.
  ISBN 978-1-84593-463-7 (alk. paper)
  1. Nutrition. 2. Health promotion. 3. Medicine, Preventive. I. Title.
  [DNLM: 1. Health Promotion—methods. 2. Nutrition Physiology. QU 145 W931n 2008]

  RA784.W677 2008
  613.2—dc22

2008024376

ISBN-13: 978 1 84593 463 7

Contents

Figures and Tables

Figures

Tables

Preface

This book is intended to be an introduction to an important branch of public health, nutrition promotion. The onset of the current metabolic disease epidemic and the challenge to maintain ecological sustainability have brought about renewed interest in this area of applied research and practice. The field began over a century ago with North American initiatives in nutrition education and home economics. Although it has come closer to the broader subdiscipline of health promotion during the past quarter of a century, nutrition promotion remains a distinctive area. It applies many of the principles and practical skills of health promotion and education to food and health problems.

I had three aims in writing this book. The first was quite straightforward: to draw attention to nutrition promotion as an important aspect of public health which occupies the overlapping area between the health and food and other societal systems. The subdiscipline requires deep knowledge of health but also a similar level of knowledge of the workings of the various food sectors and of society and government, especially in the ways they impact on the population's daily activities. Nutrition promotion is an urgent requirement in societies that are experiencing 'affluenza', which threatens both personal health and sustainability.

My second aim was to provide a balanced account of the theories and methods used in the area. There are many ways to practise nutrition promotion, but all of them have to face the test of their appropriateness for the population groups and contexts in which they are practised. Both holistic and reductionist methods can be useful, depending on the settings in which population groups are found. So, while it is currently fashionable to emphasise environmental and policy theories, the social and individual influences on population food consumption also need to be considered.

The generation of a spirit of optimism was my third aim. It is true that humanity faces some huge problems in feeding and maintaining itself in a healthy and sustainable manner, but that is no reason to become pessimistic and to adopt the hopeless attitude that nutrition promotion does not work. People who hold this view are often very erudite in their own specialities but usually have not read the literature, which includes numerous examples of success. So the book includes accounts of practical and theoretical methods, evaluation techniques and a fair number of examples of projects which show us that nutrition promotion is highly useful and feasible.

Who is this book for?

When writing the book I had several groups of likely readers in mind, both traditional and new. Among the established professions, the book should be useful for home economists, who for over a century have educated generations about food and nutrition and their relevance to daily life, along with nutritionists and dietitians working primarily in the community but also in health services, local government and educational settings, primary and secondary teachers, maternal and child health nurses, general practitioners, and specialist physicians such as cardiologists. Perhaps one major change that has occurred in the past quarter of a century has been what might be termed the 'democratisation' of nutrition—the broadening of interest in food and health issues along with the realisation that the genesis and resolution of them are to be found in many parts of society, not only in the traditional therapeutic and educational settings. There has been a particular renewal of interest in food in public health, and in several countries the new sub-speciality of public health nutrition is blossoming. However, many others besides public health professionals have a stake in the promotion of nutrition—for example, town and land planners, media professionals, policy and regulatory specialists, allied health professionals like physiotherapists, social workers and speech therapists, local government officers, agriculturalists, and food manufac-turers and retailers among many others. In short, this book should be of use to those who attempt in some way to promote healthy sustainable food consumption in the community.

The book is arranged in two parts. Chapters 1 to 7 deal with the rationale, prin-ciples, challenges, theories and methods of nutrition promotion, and Chapters 8 to 13 examine the particular challenges and examples presented by various popula-tion systems and settings. There have probably been thousands of studies reported in the published and 'grey' literatures which represent only a small fraction of those actually conducted, so the choice of examples has been difficult. I have tried to provide accounts of projects which are either classical in the ways they influenced later thinking or those which in some way are ground breaking (at least in my view).

I know that I have omitted many excellent projects, but I hope I have motivated readers to explore this fascinating literature for themselves.

Tony Worsley
Melbourne,
March 2008

1

Introduction:
THE AIMS OF NUTRITION PROMOTION

Introduction

For aeons, humans have passed on their knowledge and skills concerning food production and preparation from parent to child. For example, hunter-gatherer societies had extensive botanical knowledge, particularly about the safety of plant foods, which they taught their children (Diamond 1997). As society passed through agricultural and industrial phases, the relevance of these folkways and the role of the family in food education were weakened. Towards the end of the nineteenth century in North America and Western Europe, nutrition science began to develop, encouraged by public-minded individuals and capitalist and military interests who were concerned, for a variety of reasons, about the widespread malnutrition experienced by the working classes. Educators found that malnourished children undertaking compulsory primary education were too malnourished to learn, employers reported low productivity and widespread injuries among manual labourers, and during World War I the military had to reject many conscripts because of severe nutritional deficiencies. So, for these and other reasons—some humane and others less so—the emerging science of nutrition was used to remedy and prevent widespread under-nutrition. Nutrition promotion and communication, then, have been around for a long time as both academic and public health disciplines.

The ways in which nutrition science has been promoted and the emphases which have been applied during the past century have varied according to prevailing social and political conditions. For example, in the early twentieth century in industrialised countries, protein-energy malnutrition was a widespread problem so nutritionists responded by promoting diets which contained large amounts of energy and saturated fats. Today, in the midst of an obesity epidemic, such a strategy seems bizarre. Nutrition promotion is very dependent upon the prevailing social, economic and

1

epidemiological orthodoxies. Major differences exist between societies which are in different phases of economic transition. Rich, post-industrialised societies suffer from the effects of excessive energy and salt consumption, leading to non-communicable diseases like heart disease, cancers and non-insulin-dependent diabetes, and obesity. Economically poorer societies suffer from a wider range of diseases such as infectious and parasitic diseases, as well as non-communicable diseases. These different economic and disease 'climates' require different types of nutrition and health-promotion responses.

The nature of public health

Nutrition promotion and communication are part of public health. This means that they are more concerned with the prevention and amelioration of diseases among populations than with the treatment of sick individuals. While nutrition promotion often takes place in clinical settings (like doctors' offices), it does not focus on the treatment of individual patients. That is the important work of clinical nutritionists and dietitians. Instead, nutrition promoters also work with groups of apparently well people, or with individuals who have major influence over the population (e.g. newspaper editors, urban planners, government departments) or who control the content and distribution of the food supply (such as food manufacturers, food retailers and regulatory authorities). (Of course, many dietitians and clinicians often promote nutrition—but nutrition promotion goes beyond the confines of the clinic and includes the efforts of many people in addition to clinicians.)

Nutrition promoters focus on populations of people. While the term 'population' can refer to all the humans living in the world or in a country, typically public health workers attend to more specifically defined sub-populations such as men and women, children, young people, the elderly, low-income groups, and so on. The 'four pillars of public health and sociology' are gender, age, ethnicity and socio-economic status because these factors have major influence on health and society, and have to be taken into account in most forms of health promotion.

Various demographic criteria are used to define these sub-populations such as sex (male or female) and age (e.g. people over 60 years of age). Socioeconomic status is a key criterion. People from low socioeconomic status backgrounds tend to be more obese and have worse health than those from higher status strata, so they may require different health and nutrition promotion strategies. Indicators of socioeconomic status include education (e.g. tertiary educated versus non-tertiary educated), income (e.g. those in the top 25 per cent of income earners) and occupation. Employment status (e.g. employed full time, part time, unemployed), and ethnicity (e.g. Indigenous and/or speakers of a minority language) are also important demographic factors.

Life stage is another important concept. For example, pregnant women have quite different nutrition and health needs than non-pregnant women of similar age. Similarly, growing children have quite different nutritional needs from adults.

Public health programs can be conducted at several levels in society and in different ways. Table 1.1 shows how the public health continuum might be applied to nutrition promotion. It should be clear from the examples that many different groups of people and professions can be involved: parents, teachers, home economists, dietitians, nurses, GPs, gardeners, greengrocers and many others. All of them can promote healthy food and nutrition in some way.

TABLE 1.1 THE PUBLIC HEALTH CONTINUUM

Public health programs at the community on regional level	Community development	Group work	Disease prevention prevention and health promotion with individuals and families
For example: Local food policy Changing suppliers of foods Healthy workplace policy	*For example:* Local community groups become a lobbying force (e.g. against promotion of obesogenic food to children) Lay outreach workers Community involved in food-mapping activities (e.g. about access to nutritious foods)	*For example:* Health education activities such as cooking classes Community food gardening groups	*For example:* Work with individuals in their own home to advise on skills such as cooking and budgeting Advice in clinic from GP or dietitian 'Healthy Start' programs for mothers and infants

Source: Caraher, M. 2007, personal communication.

Marginalisation

A major preoccupation of present-day public health is marginalisation. Groups of people often live at the margins of mainstream society and may experience several disadvantages which prevent them from accessing the resources required for optimal health. Financially poor people, people with major physical and mental disabilities, people with little education, new migrants or refugees, people from minority ethnic or language groups and many retired people can often be marginalised in that they may find it difficult to gain access to a healthy diet or to other health resources such as employment, housing, safety and health services. Marginalised groups, then, are often a key focus of nutrition and health promotion.

This interest in marginalised groups is derived from a distinctive set of public health values and beliefs. Public health, as the name implies, is about the promotion of the public good in the health arena. It assumes that society is made up not only of individuals and their families and friends, but also institutions like health services, community organisations and business coalitions. This is a controversial viewpoint which is certainly not shared by everyone in neo-liberal 'individualist' societies. It proposes that, in addition to individuals' interests and goals, we should also pursue community goals in which everyone has a stake. These include the protection of the environment and the care of children, sick and infirm people. Consequently, it proposes that government and good governance are important for the maintenance of the population's health and well-being.

Associated with this idea of the public good are some important values (or guiding principles) which public health workers use to judge the outcomes of their efforts. These are outlined below.

Equity

This is the belief that everyone in society should be given access to sufficient resources which they need to attain optimal health. It is not the same as 'equality', which often means we treat everyone in the same way (e.g. everyone receives identical levels of health care). Instead, equity recognises that people are not all the same, that some have greater needs than others and that they should have as much help from society as they require. It is much easier for well-off people to gain access to health resources like medical care, and so attain good health, than it is for poor people. Equity suggests that poorer, less well-resourced people should receive more help from the community than their better-off peers.

There is much debate in many countries about the desirability and affordability of equity goals. Some societies are more willing to remedy social inequalities than others. In addition, right- and left-wing politicians differ not only in their acceptance of the equity principle but also in the ways they believe equity may be achieved. For example, conservative thinkers see small business as an excellent way to spread wealth around the community, thus increasing equity, while left-wing proponents see greater roles for the state (e.g. through taxation policies which redistribute wealth).

Efficacy

This is related to concepts of efficiency and effectiveness. Resources for public health and nutrition promotion are usually very limited, so it is important that they be expended in ways that bring about the most positive outcomes for particular population groups. An efficacious nutrition promotion program is one that spends resources in ways that achieve optimal outcomes with little wasted money and time (i.e. more 'bang for the buck'). Of course, such efficacy needs to be balanced with long-term outcomes. The US Women Infants and Children (WIC) programs gain $10 for every $1 invested, but the gains occur ten or more years later! Efficacy is often thought of as 'quick wins', and we can lose sight of the value of long-term investment as opposed to immediate costs. The danger here is that we may have an inverse care law (Caraher 2007, personal communication). We might run fruit and vegetable interventions which increase their mean consumption across the population, but at the same time create an even bigger gap between rich and poor (since the well-off are more able to respond to interventions).

It is essential that promotion programs deal with problems which are *important* to specific population groups. This is where efficacy differs from effectiveness. Effectiveness is about the attainment of desirable goals within a normally operating

---

BOX 1.1 DEFINITIONS

- **Effectiveness**: a measure of the benefit resulting from an intervention for a given health problem under usual conditions of clinical care for a particular group; this form of evaluation considers both the *efficacy* of an intervention and its acceptance by those to whom it is offered, answering the question: 'Does the practice do more good than harm to people to whom it is offered?'
- **Efficacy**: a measure of the benefit resulting from an intervention for a given health problem under the ideal conditions of an investigation; it answers the question: 'Does the practice do more good than harm to people who fully comply with the recommendations?'

Source: Buckingham et al., The Evidence-based Medicine Toolkit, www.ebm.med.ualberta.ca

social system. To illustrate the difference, a school nutrition-promotion program conducted by university researchers may be efficacious in increasing children's consumption of fruit. However, once the intervention stops (because the monies supporting the intervention are spent), the children's eating habits may soon worsen. In contrast, an effective school program is implemented as a normal part of school life. For example, Japanese primary schools do not allow children to eat high-energy snacks but they routinely serve highly nutritious lunches. This routine system is straightforward to operate, and has been sustained for many years.

Democracy, good governance and participation

Current public health has a strong democratic flavour which emphasises respect for individuals and community groups, and encourages their voluntary participation in decisions which affect their lives. Public health emphasises cooperation among community groups and supports 'good governance' at local, national and international levels. This means that governments should be agencies which promote the health of all groups in the population 'without fear or favour'—that is, they should not be corrupt. This view has recently been promoted by the Nuffield Foundation (2007). The range of strategies used by public health practitioners is shown in Box 1.2.

BOX 1.2 THE INTERVENTION LADDER

- Eliminate choice—isolate infectious people.
- Restrict choice—remove unhealthy ingredients from food.
- Guide choice through disincentives—tax cigarettes.
- Guide choice through incentives—tax breaks on bicycles.
- Guide choice through changing default policy—offer salad instead of chips as a side dish.
- Enable choice—help to change people's behaviour—free fruit in schools, bike lanes.
- Provide information—campaigns—2–5 a day.
- Do nothing or monitor the situation.

Source: http://www.nuffieldbioethics.org

Social determinants or symptoms? Food poverty or poverty?

It is pretty clear that some people are far worse off than others. A key dilemma for nutrition promoters is: Do we aim to prevent the *food poverty* of these marginalised groups or do we act to prevent their *poverty*? The latter is a far more difficult task than the former. Although people from different social economic backgrounds often have differing nutritional and health needs, a key aim of public health is to remove or modify the social determinants of their ill-health and poor dietary patterns. It is not just that obese people need better dietary and physical habits; underlying factors like poverty and related social conditions also need to be tackled. There is a danger that simple, nutritionally based approaches may actually make social inequities worse. For example, provision of breakfast programs for malnourished children may stigmatise them and make their lives more miserable.

What's different about food?

Martin Caraher at the City University, London has provided a superb answer to this fundamental question which really justifies the need for nutrition promotion:

Food is different to other areas of health concern like drug use, smoking and other lifestyle behaviours. It is not about addiction (though craving for chocolate might feel addictive!). Instead it is a necessity both in our private lives (our fridges are stacked with it) and in public life (food is served at most social events). Food projects do have some similarities with other community and voluntary sector initiatives such as healthy living centres, stop-smoking groups, drug projects, credit unions or advice centres. However, there are important differences. Food is a more complicated issue for most individuals and households than credit or clothing. Food choice and management is a daily habit, as well as being part of our self and family identity. It is deeply embedded in cultural, social and religious beliefs and practice. Food is private in that it is stored and consumed in the domestic domain, but it is also communal (shopping and eating) and therefore is a public good, because few people in affluent societies today grow or rear their own food. Access to food—that is, the shops and markets people can reach, what they buy or how much—is governed by decisions in which few ordinary citizens play any part. Initiatives to change factors within the complex business of obtaining, preparing or consuming food will inevitably be varied in nature and outcomes. (Caraher 2007, personal communication)

What is nutrition promotion?

Nutrition promotion isn't simply educating people about nutrients. It is partly about nutrition, but it is far more. Nutrition science arose when it was recognised that certain biologically active compounds were essential for life (hence vita-amines). And, in its narrowest sense, nutrition science is about the functions of nutrients in the body. However, a broader definition of nutrition also encompasses the influences that ensure our bodies have appropriate types and amounts of nutrients. These include physiological, psychological, cultural, social and economic influences on food consumption (discussed in Chapters 5 and 6). So nutrition science is fundamentally about the ways food influences personal and population health.

Nutrition promotion is much more aligned with public health than with education—though it uses educational techniques a lot. This is because, like public health, it shares three core strategies (van der Maesen and Nijhuis 2000) which aim to:

- improve social conditions that threaten health (e.g. the occurrence of poverty);
- prevent social conditions that threaten health (e.g. poverty, gender and ethnic discrimination);
- neutralise existing social conditions that cause ill-health (e.g. high unemployment rates might be reduced through skills training schemes).

These three approaches involve different sorts of interventions, only some of which may involve nutrition education. These may range from doing something in 'the here and now' (like teaching new mothers about how to access healthy foods) to long-term planning (e.g. planning to grow more vegetables close to cities to reduce fossil fuel used in transportation).

One of the big strategies for the future is the 'New Nutrition' (Cannon and Leitzmann 2005). It focuses on the issue of food sustainability (and on the adequate distribution of healthy food in global society). While most of the world's population is adequately fed (though at least a quarter are not), this has been at great ecological cost. About 40 per cent of the human footprint is related to the food sector. The New Nutrition aims to take policy actions so that food is produced in ways which are sustainable in the long term.

Nutrition promotion, then, is about ensuring that all people in society are well fed (according to their nutritional and cultural needs) through the use of 'upstream' and downstream strategies (see next section). It is about food as well as nutrition. For example, if poor people are forced to eat in a day centre or buy lower quality foods, they are consuming in ways which are not deemed normal by society and so suffer some inequity. In Ireland, the national Anti-Poverty Strategy uses normative expectations of foods and meals (rather than nutrients) as part of its measure of 'consistent' poverty. So in Ireland it is expected that people should be able to have a meal with meat, fish or chicken every second day, have a roast or its equivalent once a week, and not go without a substantial meal for two weeks. People should have the time and money to make these normative choices (Dowler and Finer 2003).

In summary, nutrition promotion is about the promotion of food and nutrition knowledge among food consumers as well as the modification of the food production and distribution sectors so that they foster the optimal health of the population.

Nutrition education or nutrition promotion? Upstream and downstream strategies

Nutrition promotion and nutrition education are often confused. Nutrition education is about the provision of knowledge and skills for food consumers so that they can perform healthier eating and drinking behaviours. It may also include communications which are designed to motivate them to consume a healthier diet (e.g. through admonitions such as 'Eating too much saturated fat may cause heart disease'). In economic parlance, the education approach is called a 'downstream' or 'demand-side' strategy as it aims to alter the demand from consumers for various food and beverage products (e.g. avoidance of saturated fats reduces the demand for traditional meat products). The main criticisms of this approach are that it is difficult to change many people's food behaviours merely by informing them about nutrition or even the consequences of a poor diet, and also that it places all the emphasis for population health on individuals who usually have little direct influence over the composition and availability of foods. However, such criticism is rather sweeping, and there is little doubt that increased knowledge (of an appropriate sort) can influence behaviour change.

In contrast, nutrition promotion, while including nutrition education and communication with consumers, also tries to influence the composition and availability of foods and beverages. That is, it tries to change the food supply ('upstream' or 'supply-side' strategies). For example, governments could raise taxes on high-fat or high-energy foods to reduce their consumption, or they might subsidise the production and sale of 'healthy' foods in order to make them more attractive to consumers and producers. At the local community level, nutrition promoters may set up community (or school) fruit and vegetable gardens so that fresh produce becomes more available, or they may form food banks and buying cooperatives to make healthier foods more available to low-income families. Such food supply change strategies involve a lot of education and communication; however, in contrast to broad nutrition education, the communications are more likely to be targeted to individuals who occupy strategic roles in food production and distribution. For example, education of product or retail managers about nutrition may help them to either alter their products (e.g. produce lower salt bread) or to promote healthier products (e.g. through signage in supermarkets pointing out the salt content of comparable products (such as different brands of baked beans). So the distinction between demand- and supply-side approaches, while useful in theory, is often blurred in practice.

The goals of nutrition education

Typically, nutrition education has been aimed at specific population groups such as pregnant women and new mothers, and schoolchildren. This has now extended to other groups such as pre-school aged children attending long day care centres. The aims of nutrition education, outlined by Gussow and Contento (1984) and Johnson and Johnson (1985a), are the same as those of education in general. Essentially, they are to provide learners with sufficient knowledge and skills so that they can make wise decisions (in this case, food choices) which benefit rather than harm them. In doing this, learners need to assimilate some basic cognitive frameworks or principles so that they can understand and make sense of the world. For example, in nutrition the concepts of energy and energy balance are fundamental to understanding diverse phenomena such as children's growth, and body weight maintenance. Psychologists call these frameworks *schema*. Nutrition educators have thought a lot about the key principles in nutrition. They include:

- energy and energy balance;
- anti-oxidants and free radicals;
- vitamins and minerals as essential 'enzymatic' (catalytic) factors (e.g. folate, homocysteine and inflammatory processes);

- proteins, and growth and repair concepts;
- nutrient sufficiency and excess;
- saturated fats, serum cholesterol and heart disease.

These are very important concepts for the professional nutritionist, especially for those charged with individual patient care or who monitor population nutrition status. Unfortunately, most lay people have to contend with additional issues in their daily lives, such as:

- what sorts of foods (variety) they should eat;
- how often they should consume foods from particular food groups (e.g. 'Do I need to eat fruit every day or every week or less often?');
- whether some foods are better choices than others, and if so what they are; and
- how energy intakes and energy outputs are kept in balance.

These are examples of the cognitive aspects of knowledge. Such knowledge is generally of two forms: *declarative knowledge* —about *what is* (e.g. that lemons contain large amounts of vitamin C); and *procedural knowledge* —*how to* knowledge (often expressed as skills).

A major criticism of traditional nutrition education is that it focuses too much on inappropriate declarative knowledge. Most adults, for example, probably find the vitamin C content of lemons irrelevant to their life concerns, although parents of young children may find it important. Of rather more relevance for most consumers is procedural knowledge, since they often want to know *how* to lose weight, *how* to eat in an environmentally friendly way, *how* to make vegetables more appetising, or *how* to run a school canteen at a profit by selling a range of 'healthier' foods. The delivery and acquisition of such knowledge is highly context-dependent (e.g. are the vegetables being cooked by a chef for restaurant customers or by an older man with few cooking skills for himself?).

In addition to the acquisition of cognitive knowledge and behavioural skills, another implicit aim of education is motivational—that is, to make a topic such as 'healthy eating' attractive to learners so that they apply it in their daily lives. An unpleasant example of how not to do this can be seen in the aversion of many older New Zealanders and Australians to milk. When they see it, they often think of the curdled milk they were forced to drink during their school days—milk that had been delivered and left to stand outside the classroom for hours on hot summer days.

Much early learning is part of socialisation. Children are expected to assimilate socially desirable forms of behaviour—for example, how to use a knife and fork properly. Skemp (1979) calls this function of education 'schooling'; it is done more for society's benefit than for the individual's. Without such socialisation, the person will be stigmatised or rejected by others. However, the main aim of education is to enable the person to live well and happily—and healthily. To some degree, this means enabling learners to make conscious decisions about events which affect their well-being—for example, to know that the smell of a hot bread shop is enticing but also that it is likely to make them eat food unnecessarily and so perhaps should be avoided. A major purpose of education is to put the person in charge of their life instead of those unconscious forces (like advertising, smells of food, vending machines and convenience food outlets) which are placed in their way by people who are usually uninterested in their personal welfare.

Consumers or citizens?

We often refer to people as 'food consumers' because they consume food. However, this term can have a pejorative connotation in that it can refer to the passivity of consumers who blindly accept decisions made by governments and industry in food matters. So when a new product is advertised, many people go ahead and purchase the product without thinking too much about it. The main aim of education in a democratic society, however, is to produce *citizens*: thinking consumers who are capable of considering the pros and cons of any particular action (e.g. the purchase of new or unfamiliar food products).

This view of education has profound ramifications for nutrition promotion, and for the content of curricula and other learning programs. As well as teaching learners about the nutrient content of foods and the health consequences of their consumption, a citizen education program will provide critical analysis of the groups who are trying to influence food purchasing and consumption. It will provide skills to enable learners to cope with advertising and marketing campaigns, and will encourage them to see the consequences of particular courses of action. It is not enough to know about nutrients; people need to know about the forces which attempt to influence their food consumption so they can decide whether to go along with them or not.

Why nutrition promotion?

Foods and beverages are essential for health, but they are produced by the food system, not by the health system. Therefore, nutrition promoters have to work in both systems and often in others, such as the education system. Both systems are complex and demand detailed knowledge and skills from practitioners.

Historically, nutrition promotion was preceded by nutrition education, which emerged as a distinct field of practice in the early twentieth century. Its practitioners tended to be medical and nutrition scientists with strong biological science orientations who often worked closely with the food production industry, the emphasis being on nutrients and nutrient deficiencies. Perhaps this 'nutriocentric' orientation led to the notion that 'there are no such things as good or bad foods, only good or poor diets'. This view effectively stops any criticism of any individual food product since almost every food contains a nutrient which can be construed to be 'good' or 'health-promoting'. In reality, however, food consumers have to decide whether eating too much or too little of certain foods is in their health interests—so they have to evaluate foods as more or less good for their health.

Much early nutrition 'promotion' consisted of educational messages exhorting individuals to change their ways by providing them with science-based knowledge (e.g. 'Eat more butter because it's got a lot of energy in it'). That is, the onus for change was placed on the consumer. Little was done to alter the *supply* of healthy foods. Few exhortations were aimed at food manufacturers and distributors.

In contrast, the health promotion tradition comes from a mixing of at least two groups of professionals. The first comprises health educators such as school teachers, home economists, maternal and child health nurses who inform and counsel individuals (often large numbers of them) about the best ways to behave to foster their health and their families' health—for example, how to feed babies, the use of sunscreen, information about hygiene practices, and so on. The second group consists of community development workers who actively bring people together in groups so that they can deal with local issues such as ways to deal with factories that cause air pollution, or how to gain better local health services for young children, adolescents or people with dementia. Both these groupings have more holistic orientations than traditional nutrition educators.

The nutrition education and health promotion traditions are merging to form a new discipline known as nutrition promotion, which is the application of health promotion principles and methods to population food and health problems.

What is a healthy diet?

Nutrition promotion is about the promotion of healthy eating and drinking habits—in short, a healthy diet. Its purpose is to change people's food consumption behaviours. The behavioural basis of nutrition promotion (and public health nutrition) should not be overlooked. There is little point in nutritionists identifying the poor nutrition status of a population if that knowledge is not applied to bringing about changes in the population's eating and drinking habits. People maintain their nutrition and health status through eating and drinking!

This raises an important issue: what is a 'healthy diet'? The simple answer is: 'a diet which makes people healthy'. But what does 'healthy' mean? Most people's definition of 'health' is fairly loose and nebulous. The World Health Organization's (WHO) definition of health is:

> *Health is a state of complete physical, mental and social well-being and not merely the absence of disease or infirmity.* (WHO 1946)

That sounds clear enough, but how do we define these forms of well-being? For some, physical well-being will mean being able to run 10 kilometres every day; however, this wouldn't apply to many people—for example, those in wheelchairs. People vary and people's expectations about what they should be able to do vary a lot. Other forms of well-being exist in addition to those mentioned by the WHO and are considered to be very important in some cultures. For example, the Maori definition of health adds the concept of *spiritual well-being* to the WHO definition. It is worth noting that 'health' for most people is a necessary condition which enables them to enjoy 'flourishing lives'. It is not a major goal for most people—except when they are ill.

The definition of health has a lot to do with economic development and with cultural expectations. For example, in a heavy manufacturing economy most people need to perform a lot of physical work (lifting heavy weights, walking a lot) in order to earn their living. In a post-industrial society, however, there may be a greater emphasis on the ability to withstand work stress generated by 'just in time' management systems. In some cultures (e.g. Polynesian cultures), health is associated with being big and strong, whereas in others, such as Chinese and Anglo cultures, being slim and youthful-looking is often equated with health, especially women's health. So words like *health* and *healthy* are problematic. Broadly, being healthy means being able to do most of the activities expected by one's culture. That is, health is culturally defined to a large degree. Health is not an end in itself. It is a condition that is necessary for people to pursue 'good' lives.

Humans are omnivores, and they can survive and reproduce on a variety of diets. These range from those which include a lot of animal products (like the diets of the traditional Masai) to the majority of traditional agricultural cuisines which are largely plant based with a little meat or fish—for example, the traditional Okinawan diet. There is increasing evidence that most traditional cuisines, like the Mediterranean, Andean, Hangchow and Okinawan cuisines, are composed of similar animal and plant components (Trichopoulou et al. 2003). For example, all of them include green vegetables, though this may vary from spinach to seaweed. As

Wahlqvist (1995) notes, traditional diets are 'eco-nutritional'—humans eat foods in order to gain energy and body defence mechanisms from animals and plants. For example, the flowers on vegetables like broccoli and berries like blackberries are the parts of the plant most exposed to solar radiation (and thus free radicals), and so are rich in anti-oxidants to protect against the effects of that radiation. By eating such foods, humans acquire these defences.

Research with long-lived groups (eg. *Food habits in later life* [Wahlquist 1995] and SENECA studies [Haveman-Nies et al. 2003]) and observations of healthy and unhealthy populations suggest that optimal health is associated with the consumption of a fairly wide variety of foods. Although in theory we might be able to live on milk, bread and cabbage (if boredom didn't stop us eating these foods after a few days), in practice different traditional cultural groups around the world seem to share broadly similar food patterns composed of around a dozen groups of foods. Wahlqvist's comparisons of the features of some traditional agrarian cuisines with a typical 'fast food diet' are shown in Table 1.2. While the details of the foods differ a lot, botanically the parts of plants consumed are similar (such as leaves, flowers, stems, roots, seeds and fruit). In most agricultural societies, meat was consumed sparingly, though in some fish was commonly eaten.

As a further illustration, Trichopoulou's list (Trichopoulou et al. 2003) of nine important components of the Mediterranean diet is shown in Figure 1.1. Apart from the Mediterranean emphasis on olive oil, these components can be seen in many other cuisines which have nourished agrarian peoples during the past 10 thousand years. The sad irony here is that traditional diets (along with their ecological advantages) are being deserted by young people in the countries in which they were developed.

Many food taxonomies (or food group systems) have been developed. Although they differ in detail, they are substantially similar. The most common dietary guideline recommends people to choose foods from each of these major food groups ('eat a wide variety of foods' is the first dietary guideline in the *Dietary Guidelines for Australians*) (NHMRC 2003a). Inherent in the notion of eating a variety of foods is the concept of 'balance': that we should eat more or less of certain groups of foods—for example, more cereal foods than meats.

There are several caveats to this variety generalisation, however. First, life stage influences people's dietary needs. Growing children, for example, have quite different requirements from adults, requiring more protein-rich foods. Second, there have been two massive developments during the past century which affect human food consumption: the huge growth in the numbers of people surviving for two or more decades beyond 60 years of age; and the deterioration of the ecosystems which support food production.

TABLE 1.2 COMPARISONS OF THE MAIN CONSTITUENTS OF MAJOR CUISINES

Cuisine characteristics	Okinawan	Rural Greece	French	Yankee New England	Current Western fast food diet
Vegetables Root	High	High			Low
Green leafy	Very high	High			Low
Legumes	High	High			Very low
Fruit	Low	High	High		Low
Wholegrain cereals	High	High	High		Low
	Rice	Wheat	Wheat Barley		Refined wheat flour
Refined carbohydrate Added sugar	>Zero	>Zero	>Zero	>Zero?	High and increasing
Meat	Low	Low	Low	High	High
Offal	High	High	High	High	>Zero
Fish	High but small amounts	High but small amounts	High see Friday fish	High	Low 10 per cent/week
Added fat	Zero	High Olive oil	High	High	High
Sweet treat foods	Low	Low	High	High	High
Meals eaten in company	Always	Always	Always	Always	Varies Informal
Festival foods	Many	Many	Many	Sunday	None
Alcohol with food	Yes	Yes	Yes	No	No
Herbs and spices	Yes	Yes	Yes	Yes	No
Fermented foods	Yes	Yes	Yes	No Salted	No
Eating on demand vs rules about food consumption	Formal	Formal	Formal	Formal	Informal Age of plenty

Source: Wahlqvist 2004, personal communication.

Longevity is increasing rapidly in most countries. Even a demographically 'young' country like Indonesia has over 25 million people (out of over 215 million) who are over 60 years of age. In Western countries, longevity has been increasing by about one year every three years. Unfortunately, many of these additional years will be spent facing illness and disability. In human history, this demographic shift is truly revolutionary. There has never been a whole generation of people who have lived beyond 60 years to an advanced age.

FIGURE 1.1 THE MEDITERRANEAN DIET PYRAMID

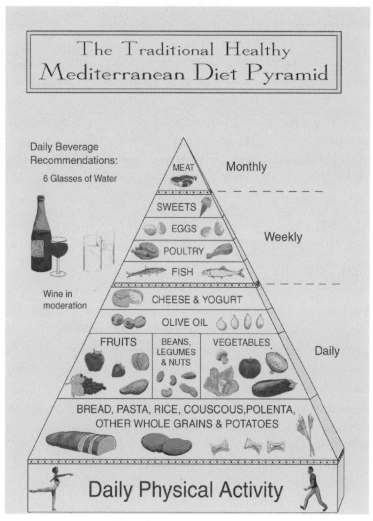

© 2000 Oldways Preservation and Exchange Trust

This increased population longevity poses major nutritional challenges. Elderly people tend to absorb nutrients less well than younger people, but little is known about their actual nutritional requirements except that they require more nutrient-dense diets than other adults. People over 60 years suffer most from the burden of disease. For example, about one in five Australians aged over 60 years either has type 2 diabetes or is 'pre-diabetic'—about to develop the disease. These people present the greatest challenges and opportunities for nutrition promotion. Fortunately, dietary and lifestyle change can slow down or reverse such disease processes. Consequently, there are major opportunities for nutrition promotion among older age groups.

A second 'caveat' to the variety generalisation concerns the 'ecological crisis'. The deterioration of the ecosystems supporting food production is perhaps the major challenge facing nutrition promotion. As long ago as 1984, Gussow and Contento observed that sustainable food production was an essential aim of nutrition education. We need to find ways to enable everyone on earth, now and in the future, to consume a healthy variety of foods. This is one of the main themes of the 'New Nutrition Project' and its founding charter, the Giessen Declaration (see Chapter 4). In the coming years, much of nutrition and nutrition and nutrition promotion will be 'eco-nutrition'.

Should everyone get the same types and amounts of foods?

This question refers to the equity principle, which is central to public health and nutrition promotion. If everyone had the same needs, the answer would be a simple 'yes'; however, people do not have the same nutritional needs. Some groups have different requirements to others, such as pregnant and lactating women, children and adolescents, people in manual occupations (including sportspeople), older people and many individuals who suffer from illness and disease. So the issue is whether people have access to the foods which will meet their nutritional needs. Clearly principles of natural justice and equity demand that they should have such access. Despite this, many do not, so nutrition promoters have to ensure that this is rectified. British nutritionists met the needs of various groups in Britain in World War II through a sophisticated food rationing scheme. People were allowed to buy only foods identified by nutritionists as appropriate for their age, sex and physical workloads. As a result, the general health of the UK population dramatically improved despite the war!

Which resources are required for optimal population nutrition?

The simple answer is 'a varied food supply'. However, much depends on the state of economic development of a society. In hunter-gatherer societies with low population densities food resources were often rich, varying throughout the year. Most tribes appear to have been able to find sufficient food through only a few hours' work each day (Blainey 1975). In agricultural societies, which have dominated the past 5 to 10 thousand years, greater organisation of food production is necessary, especially distribution and financial systems such as roads, rivers, canals and monetary and legal systems. Problems of food storage and distribution occur in these societies, including poor harvests, failure of storage systems (e.g. destruction of stored foods by vermin), interruption or absence of transport systems by weather conditions (e.g. floods) or lawlessness. The high population densities also increase the risk of water-borne parasites which compete with humans for nutrients (from within). These problems are increased in heavy industrial societies, where longer chains of supply increase the need for greater food safety precautions during distribution and consumption.

As population densities of societies have grown, they have brought with them increased needs for physical and organisational infrastructures which can ensure that food is available to everyone according to their requirements. This has meant the greater involvement of governments and the development of food and nutrition policies. A simple example of such involvement was the operation of the parish relief schemes in Medieval England. When grain supplies were interrupted by poor harvests, parish bailiffs were required to open the storage granaries to release grain on to the market (Laslett 2000).

It is quite clear that at least 800 million people in the world do not have sufficient food to meet their daily needs. This is a major challenge for nutrition promoters and food policy-makers which is going to increase as the world population nudges up to 9 billion by 2050. Three key resources will be required: highly efficient and sustainable production, distribution and administrative systems. Our current systems tend to be highly inefficient in ecological terms, destroying arable land, polluting water resources and producing greenhouse gases on a vast scale.

What influences the amounts and types of food consumed by the population?

On the supply side, the activities of farmers, fishers, horticulturalists, food manufacturers, distributors, wholesalers and retailers affect the availability of foods. The degrees of concentration of industry ownership, import tariffs and tax policies, and production subsidies and levies together affect the prices of foods and thus their availability across the population.

The activities of food marketers and advertisers on behalf of manufacturers help stimulate demand for particular groups of foods, usually in the form of brand advertising. For example, horticultural products like potatoes and cherries are infrequently advertised because the fragmented nature of the industry prevents growers from being able to afford expensive advertising campaigns. On the other hand, products like confectionery and takeaway foods are usually manufactured by large international companies which can afford advertising to gain and maintain market share.

Local planning policies are also important since they can allow or prevent the location of large supermarkets on the outskirts of cities, thus promoting or denying financially disadvantaged people access to their lower-priced products.

Governments, if they are strong, relatively free of corruption and well organised, can have a great influence over the food supply. For example, governments can tax food products that are not in the public interest and subsidise the production and marketing of foods that are health-promoting. Unfortunately, prevailing neo-liberal political philosophies prevent full use of government powers in many countries (Pusey 2003). (The neoliberal view is that government is best if it is 'small' and plays as little a role in economic life as possible.)

On the demand side — which is largely about the activities of consumers — there are several influential factors, the principal one being the level and distribution of affluence of the community. If the community is poverty stricken, many people are unlikely to be able to afford sufficient quantities and variety of foods. The level and distribution of affluence depends on economic development, which in turn is related to the education of the community — especially the education of girls and women. Communities which have substantial average incomes but which have poor distribution of wealth tend to have large pockets of poverty and associated malnutrition and poor health. Harlem, part of New York City, is a classic example of such an island of poverty within a highly affluent nation. While a certain basic level of wealth is essential for health, there does not appear to be a direct relationship between national wealth and health and nutrition status. Many high-income countries have greater health and nutrition problems than less affluent countries.

A further set of factors affects the demand for food. In the main, these relate to general education and to special food and nutrition education. General education creates the demand for investment in the future (e.g. pension schemes) and sophisticated forms of demand for such things as mass media, books and information products. Food and nutrition education increases the ability of people to demand and enjoy healthier products and, theoretically, to be less vulnerable to nutrition scams and quackery. Such education also allows the general population to take more control over their food consumption and it also enables people in special roles, such as doctors, teachers and food quality managers, to apply nutrition principles in their work such as the selection of nutritious foodstuffs for catering or food manufacturing. Table 1.3 summarises some of the nutrition and equity issues associated with different types of food supply.

The stakeholders in nutrition promotion

Nutrition educators and promoters must respond to the needs of several groups of people, so-called *stakeholders*. For example, in primary school food and nutrition education there are several stakeholder groups, each with their different needs, expectations and agendas, including:

- *The learners* — Primary children's need to learn is great, but their learning has to be framed according to their cognitive ability, their culture (e.g. Are they allowed to eat pork? Is their family vegetarian?) and their interests and beliefs about food, eating and health.
- *The parents* — Parents have expectations about what children should be eating and what they should learn about food and other aspects of the world. They may be nervous about their child participating in practical activities like cooking (e.g. they may not want their child near hot stoves and boiling water). They may have strong beliefs about certain foods, such as particular meats, and about the respective roles of boys and girls in food shopping, preparation and consumption — often girls and women are expected to perform 'domestic' chores like cooking and families may differ in the freedom they give their children to make their own food choices. More subtly, parents may have different opinions about aspects of health — for example, they may see normal female development like the laying down of fat around the hips as 'obesity' and so they may be highly critical of their daughters' supposed over-eating. Or they may take the opposite view, denying that their obese children in fact have a problem.
- *Nutrition scientists* — Such experts often have expectations about the ways 'their' discipline should be taught. Many of them assume that everyone should be just as interested in nutrition as they are. Alas, this is usually not so!

TABLE 1.3 NUTRITION AND EQUITY ISSUES ASSOCIATED WITH DIFFERENT TYPES OF ECONOMIC DEVELOPMENT

Society level	Food supply	Equity issues	Nutrition problems
Food shortages—acute and chronic	Low food production Poor distribution infrastructure Poor governance	Substantial proportion of population has poor access to sufficient food Orphans of parents killed by infectious diseases and violence have little access to foods in present or future	Starvation High prevalence of nutrient deficiencies Parasitic and infectious diseases prevalent High infant and maternal mortality
Unbalanced food supply in poverty	Low food production Poor distribution infrastructure Poor governance	Substantial proportion of population dependent on a few staple foods (e.g. rice, maize)	High prevalence of nutrient deficiencies Parasitic and infectious diseases prevalent Nutrition status dependent on harvest and other conditions
Unbalanced food supply in affluence	High food production High-energy foods cheaper than fresh foods Good distribution Lack of food and nutrition policy	Substantial proportion of population affected by energy excesses Some groups with poor access to healthy foods	Obesity and degenerative diseases Micronutrient deficiencies possible
Balanced food supply in affluence	Fresh foods cheaper and more available than high-energy foods	All population has access to fresh foods	Main problems related to special needs of life stage

Source: Popkin, B., Horton, S. and Kim, S., The Nutrition Transition and Prevention of Diet-related Chronic diseases in Asia and the Pacific, *Food and Nutrition Bulletin Special Supplement* (December 2001), Asian Development Bank http://www.adb.org/.

Scientists are usually preoccupied with measurement and accuracy—for example, that there are 95 milligrams and not 94 milligrams of vitamin C in 100 grams of a certain food. In daily life this does not matter; what the consumer wants to know is: 'Is there too much or too little of it, and should I eat it?' Most scientists shy away from answering such specific questions, perhaps because of their need for certainty.

- *The food industry*—This industry has a vested interest in children's food activities. For example, several supermarket chains have operated school information campaigns, teaching children about nutrition while quietly encouraging children and their families to shop at their chain. They find it difficult if the curriculum includes criticism of their products or product categories. (For example, how many schools teach children how to follow a vegetarian dietary option?) They have a vested interest in ensuring that children develop into consumers of their products. Marketing programs conducted by Coca-Cola, Pepsi, McDonald's and other fast food companies are widespread in schools in many countries.

- *Governments (national, state and local)*—Governments represent society in general. They have a large stake in nutrition promotion. For example, several governments are now actively involved in promoting healthy eating and school meal reform because they want to avoid the future costs to tax-payers associated with children's obesity. A good example is the New South Wales 'Fresh Taste' school canteen program. Local government in particular may be responsible for the regulation of food safety in schools and pre-school centres.

- *The local community*—Community members such as retired persons and community developers, and *health organisations* such as the National Heart Foundation, may also try to influence children at school as well as other groups. The problem here is that unless the curriculum is independently designed in the child's best interests, the learning opportunities to which children are exposed may be skewed towards one agency's particular agenda while other, equally valid, views may be ignored.

This brief analysis of the primary school setting could be repeated for other settings. There are always stakeholders and vested interests, but they are not always obvious. Nutrition promoters need to be able to identify and deal with them appropriately. The success of nutrition promotion depends on the identification and cooperation of key stakeholders.

Tensions: Bottom-up or top-down?

Nutrition promotion involves interactions with people who have differing interests. It is a tension-filled area. To make matters more interesting, there are a number of rifts between the professionals themselves. They are difficult to describe but they relate to how nutrition promotion should be done. We can call this the 'bottom-up–top-down' dichotomy. The bottom-up approach emphasises participation and control of promotion activities by local people. Its focus is on finding practical ways to solve locally identified problems. So practicality, specificity and attention to local contexts are the hallmarks of this approach. In contrast, the top-down approach is more general. It is often posited in terms of general theoretical principles and control from knowledgeable people 'above' the people to whom the intervention is aimed. The emphasis is on understanding how general approaches work—on solving the ideal, general problem.

In the past, considerable heat was generated as the two approaches were considered to be incompatible. However, both approaches can work well depending on the context. For example, local food policies are unlikely to succeed if central or state governments do not provide the funds required for local projects. Similarly, edicts 'from on high' are unlikely to be enacted unless there are vibrant local organisations with on-the-ground capabilities. It is very important that the two approaches are integrated. Too often in the past, experts have issued top-down edicts without consulting the people who will be most affected.

This dichotomy will be apparent throughout this book. Typically, the experimental top-down 'scientific' approach, exemplified in many university-based nutrition promotion projects, stands in stark contrast to the uncontrolled local initiatives which appear, with some uncertainty, to solve local food and health issues. The Ottawa Charter (together with subsequent modifications such as the Jakarta Declaration) represents a standoff between the strategies favoured by these two approaches. For example, the emphasis on education and the acquisition of skills and knowledge by individuals in 'Developing Personal Skills' is a traditional approach favoured by top-down theorists who believe that expert knowledge can help people make their lives healthier. On the other hand, 'Strengthening Community Action' clearly emphasises the importance of bottom-up approaches.

Although the Ottawa Charter was proposed as a way of unifying ways to promote health and emerged from the health system, it is just as relevant to issues which emerge from the food system. The point is that people's health can be

BOX 1.3 THE OTTAWA CHARTER ON HEALTH PROMOTION

The Charter includes five broad health promotion strategies.

The Charter recognises that for public policy to have the best influence on health, action is needed from a range of sectors, not just the traditional 'health systems'. The Charter also recognises that the capacity of individuals to alter their behaviour is greatly influenced by social and cultural factors.

The five strategies are:

1. Building healthy public policy

Policy decisions made to improve the health of individuals can be made at all levels of government and by non-government organisations. The objective should be to make the healthier choice the easier choice for all people, policy-makers as well as individuals. Some examples of building healthy public policy are the compulsory wearing of hats in school playgrounds to protect against skin cancer; non-smoking areas in restaurants and smoking bans in workplaces and shopping centres; and school food policies which control the types of foods sold in school canteens.

2. Creating supportive environments

People's social and physical environments influence the population's health and food behaviours. This strategy encourages the creation and modification of environments and settings so that they foster healthy behaviours. Health behaviour change works at its best when people take care of one another, support each other's efforts and accept that positive health behaviour is the normal pattern for living. Examples include smoking bans in the workplace; school canteens that promote healthy foods; and the 'baby-friendly hospital' initiative which actively promotes breastfeeding and discourages the use of milk formula preparations by new mothers in hospitals.

3. Strengthening community action

This strategy aims to empower communities to use their resources to work together to set goals, determine priorities and plan and implement strategies. Many local community groups, such as child-minding centres, schools, health and medical centres, industry groups, shopping centres, the media, interest groups and local governments, have learned to work together to achieve common goals. Examples include local government food policies which foster the supply of healthy foods; food banks; and the Parents' Jury which monitors the marketing and supply of children's food products.

Box 1.3 continued

4. Developing personal skills

The development of personal skills to enable people to have better control over their lives is a key part of health promotion. Such skills may be learnt in schools, workplaces, community and recreational groups, vocational programs and through the media. We all need many skills to be healthy—for example, knowing how to shop for and prepare food, knowing how to relax and deal with stress, or how to deal with conflict, or how to engage in physical activities without injuring oneself. Even happiness depends on our acquisition of skills—for example, how to deal with disappointments, or how to control our moods.

5. Reorienting health services

Everyone who is involved in health services needs to work towards a health-care system that promotes health. Reorientation of the health services requires more attention to be given to health research, professional education and training. When people are sick in hospital, or when they visit a GP or other health professional, they are often at a point where they are willing to change the ways in which they live. For example, if doctors provide simple advice to help people quit smoking or change dietary habits, many people will listen and try to change their behaviours. Unfortunately, many health service practitioners do not see health promotion as being part of their role. The 'fee for service' payment system for health professionals actually penalises doctors and other health professionals who spend time advising their patients.

Source: Based on Carey et al. (2003).

improved by its five basic strategies, irrespective of the source of the problems they face.

The Ottawa Charter (see Box 1.3) is quite old and has a very strong rhetorical feel about it (most of us would agree with its provisions to some degree). This should not blind us to its utility. Many thousands of health promotion workers around the world follow one or more of its strategies, whether they are operating children's farms and gardens in cities or conducting mass communication programs. It is a useful reference document which helps health promoters check whether they are using all the strategies available to them.

Challenges for nutrition promotion

There are several challenges for nutrition promotion (and the wider area of public health nutrition) which revolve around methodological problems, as well as the definition of its scope.

The key methodological challenge for nutrition promotion is to unify the advantages of the top-down and bottom-up approaches to produce effective, sustainable programs. We will see examples of this later, especially in relation to recent work-site programs (Chapter 9) in which project designers have sought strong inputs and feedback from workers and other stakeholders in the design and implementation of projects.

At first sight, the focus of nutrition promotion might appear to be straight-forward—it is about promoting the nutrition status of the population. The complexity arises when we ask: 'In relation to which populations and which problems?' The population can be subdivided into several life stages. Traditionally, nutrition promotion has mainly been associated with mothers and children. This emphasis has remained, but has broadened to include fathers and other caregivers as well as to place a greater emphasis on the social structures which influence children's food consumption (such as the family, school food policies and community control over high-energy and high-salt food manufacturers' attempts to sell their products to children).

Increasing numbers of people are living at the other end of the lifespan—old age. It is this group which presents the greatest challenge to nutrition promoters because the prevalence of non-communicable disease, and thus suffering and expenditure, is greatest among this age group. Although many health promotion studies show that even very old people can reduce their risk of disease, or even ameliorate existing diseases, there has been relatively little interest in promoting nutrition and lifestyle change among the elderly.

Children and older people are just two population groups among several which present opportunities for nutrition promotion. Some disadvantaged ethnic minorities (e.g. urban and remote Aboriginal groups in Australia) and other vulnerable groups such as people in low socioeconomic strata experience a high prevalence of non-communicable diseases. Other, less recognised groups have received even less attention. For example, at least one in nine people live with physical or intellectual disabilities which affect their ability to work (ABS 2004; US Census Bureau 2002).

Many people with severe disabilities experience swallowing difficulties which make eating problematic; many die prematurely after suffering from nutrition-related conditions such as obesity, heart disease, diabetes or osteoporosis. In most countries there are no nutrition programs (or any other health promotion programs) for these groups. For most practical purposes, they are invisible. So identifying and defining sub-populations with serious nutrition problems is a key challenge.

To who else do we promote nutrition?

The basic argument for nutrition *education* is that the market economy has destroyed traditional food practices to such an extent that consumers need nutrition knowledge to be able to distinguish suitable foods from less suitable foods. This is probably more true for certain groups of people than for others—for example, adult shoppers have to make buying decisions, while children generally do not. Clearly, nutrition promotion is about influencing populations' nutritional and health status.

However, are 'the masses' the only targets for nutrition promotion? It is only in the past 25 years or so that the value of nutrition education for those who supply food to the general population (farmers, food manufacturers, caterers and retailers) has begun to be recognised. These groups control the food supply—so, *in theory at least*, the more they know about the population's nutrition and health needs, the more able they will be to provide foods that will keep the population healthy. Whether they actually do so depends on the forces which influence them, such as the profit motive, community opinion and government food and health policies, among others. This is an aspect of the supply-side approach to nutrition promotion as distinct to the demand-side approach which tries to influence consumers' food choices directly.

The nature of evidence in public health[1]

Much of the work in nutrition promotion is based on epidemiology, which essentially counts frequencies of deaths and sickness in the population. You should be familiar with terms such as *prevalence, incidence, mortality and morbidity rates, disability,* and *disability adjusted life years* (see Australia's Health on the AIHW website for useful definitions of these and other terms: www.aihw.gov.au).

The nature of evidence is an important area in public health. How do we know, for example, whether exposure to a food constituent actually causes bowel cancer,

---

1 This section adapted from Carey et al. (2003).

or that high saturated fat intake causes heart disease? Evidence and evidence testing are essential for nutrition promoters—they need to know whether their interventions are effective in promoting the population's health and well-being.

Many of us have strong opinions about the main influences on people's eating behaviours. These opinions are often based on our personal experiences. Some of them may be highly cherished—they are our 'pet' theories. They may be true, false or a little of each. Our aim is to understand the influences on people's food consumption behaviours. So it is wise to adopt a sceptical attitude to all propositions, *including our own*! Many scientists trust Karl Popper's falsification approach—they try to test or falsify their favourite beliefs or hypotheses. So if you think that fish oil will prevent depression, you should try to look for evidence which might disprove that assumption. It is a useful approach because it makes us question our own beliefs.

Scientific sources of nutrition evidence

The study of the relationship of diet and nutrition to the distribution of disease in a population is called *nutritional epidemiology*. Nutritional epidemiologists examine whether groups with a higher prevalence of a particular disease differ in their food and nutrient intakes from those with a lower prevalence of the disease. From these studies, they make recommendations about changes to dietary intake and lifestyle. There are several study designs which can inform us. They include the following:

- *Randomised controlled trials (RCTs)*—These are experiments. For example, one group of people is given a treatment (such as a calcium supplement), and another group is not (the control group). An independent person allocates the subjects to either the treatment or control group at random. This person keeps the identity of each subject secret until the end of the trial. Usually, the investigators and the subjects do not know whether they are in the treatment or control group (they are both 'blind' to the treatment allocation). The aim of the study might be to assess the effects of calcium supplementation on systolic blood pressure, so the blood pressure of all the participants in both groups would be measured before and after the treatment, which might last for a period of days or weeks. Comparisons would be made between the control and treatment groups to see whether there was a statistically significant difference between the two groups.
- *Uncontrolled experiments*—These are similar to RCTs except the allocation to treatment or control is not done in a random manner. Usually a characteristic,

such as the serum cholesterol level of one group of people—for example, vegetarians—is compared with that of another group—such as meat eaters. A key problem about this type of study is that the groups may differ in ways other than the characteristic being compared, so it is hard to interpret the results. For example, vegetarians may have lower serum cholesterol levels than meat eaters, but they may also be younger, less likely to drink alcohol and more likely to be female. So is it the vegetarianism, or the age or gender, or the alcohol differences which could be causing vegetarians' lower serum cholesterol levels?

- *Prospective cohort studies*—These are usually long-term trials in which a group of people is followed for years, with measurements of health-related variables being taken at the beginning (baseline) and periodically thereafter. For example, the US Nurses study has followed over 40 000 American nurses for over 20 years. Biomedical, social, economic and food consumption variables have been measured every few years. During this time, some of the nurses have developed heart disease and other conditions, and some have died. This enables the researchers to test their hypotheses about the links between, say, saturated fat intake and heart disease mortality, by comparing the diets of those who died of heart disease with those of the people who did not.

- *Case control studies*—Researchers may compare the characteristics of people who have died of a disease with the characteristics of those who are in good health. For example, we could compare the medical and dietary histories of people who developed bowel cancer with the history of those who did not, and deduce what may have caused the disease. A problem with simple case control studies is that there may be additional factors which have not been considered, and which may distinguish the two groups.

- *Population health surveys*—Samples of large populations may be selected at random (i.e. each person has an equal chance of being selected for the survey) and aspects of their health may be examined. Often, comparisons are made between various groups of people in the sample. For example, in national nutrition surveys comparisons may be made between the food and nutrient intakes of men and women, and members of various age groups. Sampling techniques allow us to estimate the health status of the whole population based on the findings from relatively small samples.

Case control, uncontrolled trials, prospective cohorts and population health surveys are essentially observational studies, unlike RCTs in which the investigators can manipulate the exposure of the subjects to hypothesised 'active' influences or substances.

There are many other sources of evidence—for example, comparisons between the prevalence of heart disease among Japanese living in Japan versus Japanese living in Hawaii with American-style dietary patterns ('ecological observations'). In addition, animal experiments are some of the most powerful ways to examine nutritional processes, though generalisation of their findings to humans may be difficult.

A *meta-analysis* is a statistical (numerical) summary of the findings of a number of studies on the same topic. Some studies might show no effects, others negative effects, and still others may show positive effects of a nutrition intervention on a disease or health outcome. The meta-analysis weighs up all the evidence about these effects in order to conclude whether a particular intervention is likely to bring about a particular outcome (e.g. Do high intakes of saturated fat result in type 2 diabetes?). The Cochrane Collaboration (www.cochraneconsumer.com) is a worldwide network which publishes the results of meta-analyses of health and medical studies. This emphasis on collecting rigorous evidence is sometimes called *evidence-based medicine*. *Systematic reviews* often utilise meta-analysis (see Table 1.4).

TABLE 1.4 NHMRC LEVELS OF EVIDENCE

I	Evidence obtained from a systematic review of all relevant randomised controlled trials.
II	Evidence obtained from at least one properly designed randomised controlled trial.
III–1	Evidence obtained from well-designed pseudo-randomised controlled trials (alternate allocation or some other method).
III–2	Evidence obtained from comparative studies (including systematic reviews of such studies) with concurrent controls and allocation not randomised, cohort studies, case-control studies, or interrupted time series with a control group.
III–3	Evidence obtained from comparative studies with historical control, two or more single-arm studies, or interrupted time series without a parallel control group.
IV	Evidence obtained from case series, either post-test or pre-test/post-test.

Source: NHMRC (1999).

The statistician Bradford Hill (1965) developed a set of criteria which are used to help decide whether disease outcomes can be attributed to exposure to particular influences—that is, whether a factor can be viewed as *causal*. Here is a useful summary of his postulates (from www.childrens-mercy.org/stats/ask/causation.asp):

1. *Strength* — Is the risk so large that we can easily rule out other factors?
2. *Consistency* — Have the results been replicated by different researchers and under different conditions?
3. *Specificity* — Is the exposure associated with a very specific disease as opposed to a wide range of diseases?
4. *Temporality* — Did the exposure precede the disease?
5. *Biological gradient* — Are increasing exposures associated with increasing risks of disease?
6. *Plausibility* — Is there a credible scientific mechanism that can explain the association?
7. *Coherence* — Is the association consistent with the natural history of the disease?
8. *Experimental evidence* — Does a physical intervention show results consistent with the association?
9. *Analogy* — Is there a similar result with which we can draw a relationship?

Sources: Bradford Hill (1965); Kelsey et al. (1996); Gerstman (1998).

All scientific work is incomplete — whether it be observational or experimental. All scientific work is liable to be upset or modified by advancing knowledge. This does not confer upon us a freedom to ignore the knowledge we already have, or to postpone the action that it appears to demand at a given time. Who knows, asked Robert Browning, but that the world may end tonight? True, but on available evidence most of us make ready to commute on the 8.30 next day. (Bradford Hill 1965, p. 300)

Problems of applying evidence-based guidelines to food and nutrition

The levels of evidence criteria (in Table 1.4) originated from studies of drug effects in which double-blind control trials are possible. Unfortunately, nutritional and food phenomena are much more complex and it is often difficult to conduct randomised control trials in real-life settings, so recently steps have begun to be taken to develop sets of evidence-based criteria specifically designed for population nutrition studies. The World Cancer Research Fund and the American Institute for Cancer Research (2007) in reviewing sources of evidence notes that:

The best evidence that aspects of food, nutrition, physical activity and body fatness can modify the risk of cancer does not come from any one type of scientific investigation. It comes from a combination of different types of epidemiological and other studies, supported by evidence of plausible biological mechanisms.

There are many ways to collect evidence in nutrition science. All of them are useful. However, when a few studies suggest a nutritional relationship — for example, between food X and bowel cancer — most scientists will not immediately conclude that food X *causes* bowel cancer (or some other condition). Usually, they will wait for many other studies with differing designs which show the same finding before they will begin to accept it as 'fact'. When you read newspaper articles or listen to TV reports, be open minded but sceptical. Just because the media say something is true, that doesn't mean it is!

Conclusions

In this chapter, we have examined the nature of public health nutrition and nutrition promotion including a brief examination of its history, values, aims and the populations it tries to serve. We have noted that nutrition promotion is a very broad set of activities which uses food and nutrition to promote the population's health and well-being. In the next chapter, we will examine the food system and the players within it, with whom nutrition promoters often need to interact.

Discussion questions

1.1 Discuss the aims of nutrition promotion in relation to the values and the agenda of public health.

1.2 Outline the continuum of public health strategies in relation to nutrition promotion.

1.3 Discuss the concepts of 'healthy foods' and 'healthy diet'.

1.4 Explain why we need nutrition promotion. Be sure to explain the aims of nutrition promotion.

1.5 Explain with your own examples how the Ottawa Charter strategies might be used to promote nutrition in the general population.

1.6 What are the main challenges facing nutrition promotion?

1.7 Compare and contrast, with examples, bottom-up and top-down approaches to nutrition promotion.

2

Nutrition promotion in the food system

Introduction

Systems and settings

People live, eat and drink within a number of social, commercial and physical environments. These environments influence their food behaviours so nutrition promoters need to know how they do so and how they may be used to promote healthy eating. Three main systems have important roles: the food system, health services and the education system. We can call them *macro-systems*. These systems influence specific settings in which people prepare, purchase or consume foods, such as the school canteen, the local supermarket, the family home and so on (see Table 2.1).

This chapter describes the players and activities involved in these macro-systems. The most pervasive and influential system is the global food system, which supplies food to most people. In addition, several non-food systems are also of interest to nutrition promoters, especially the health and education systems, work and leisure settings, the mass media and the family household. People are influenced to varying degrees by these environments, spending more or less time in them. Age and dependency status are important determinants of the settings in which people live—for example, children spend more time in educational institutions than adults.

We can conceptualise the environmental influences on the population in several ways. For example, Hancock and Perkins (1985) summarise traditional views of the environmental influences on health in the Mandela of Health, the individual being surrounded by a number of personal, social and cultural influences (much like an onion). A more recent scheme describes the upstream, midstream and downstream influences on health (Figure 2.1). Other models, which include more detail of environmental influences, are described in Chapter 5.

TABLE 2.1 SELECTED SYSTEMS AND EXAMPLES OF SETTINGS THAT INFLUENCE FOOD CONSUMPTION

Food system	Education system	The household	Employment and workplaces	Health services	Mass media	Government	Trade system
Primary production—selection of foods to be produced	School curricula may include nutrition, cooking and shopping skills (or not)	Gatekeeper functions of parent(s)	Availability of food and food quality at work	Time and facilities for professionals to advise patients	Advertising of food products Marketing of high-energy foods to children via TV and advertorials, and games in children's magazines and on websites.	Local government initiatives, such as community gardens, food safety, early child care, and local food policies, govern local agencies (e.g. child care, hospitals, residential care homes) Meals on Wheels services for infirm elderly	Neo-liberal policies promote economic growth above health and ecological concerns. Self-regulation encourages promotion of unhealthy food products
Secondary production (e.g. use of fats to make cakes; addition of salt to bakery and processed foods)	School food policies and food service; school food may negate teaching in health curriculum	Parental modelling of food behaviours	Workplace policies about food and health promotion	Modelling of food and health behaviours by professionals	Promotion of fads and quackery	Food security measures (e.g. Women Infants Childrens' program	Subsidies may encourage over-production of high fat products

Table 2.1—continued

Food system	Education system	The household	Employment and workplaces	Health services	Mass media	Government	Trade system
Distribution and retailing (e.g. 'two for one' promotion of high-energy products; excessive fossil fuel use in food transport)	Teachers' modelling of food behaviours (healthy or not)	Family income—food security	Breastfeeding facilities for new mothers	Knowledge for professionals about food, nutrition and behaviour change	Articles or programs about food and health	National programs to reduce obesity, promote healthy eating—fruit and vegetable campaigns	Agricultural policies may discourage sale of non-standard foods (e.g. small eggs, fruit)
Fruit and vegetable promotion at point of sale				Fee-for-service impediments to dietary counselling	Internet sites (good and bad) Feedback about dietary behaviours from selected sites	National tax policies may prevent subsidies for health foods, promote *laissez-faire* marketing of obesogenic foods	
Food stamp and other food subsidy schemes (USA)			Time and facilities for physical activity		Viral marketing of products (especially ready-to-drink alcohol-enriched beverages)	State or national government food and nutrition policies (e.g. to promote fruit and vegetables, or lack of them)	

FIGURE 2.1 MAJOR INFLUENCES ON POPULATION HEALTH

Upstream (macro)		Midstream (intermediate)	Downstream (micro)
Global forces	**Determinants of health** Gender Age Education Employment Occupation working conditions Income Housing and area of residence Social capital	**Pyschosocial factors** Demand　　Self-esteem Control　　Coping Perceptions　Anger Social support　Stress Networks　Hostility Attachment　Isolation **Health behaviours** Smoking Diet/nutrition Alcohol Physical activity Self-harm/addiction Preventive health-care use	**Physiological systems** Endocrine Immune Nervous Cardiovascular Musculoskeletal, etc. **Health** Morbidity Self-assessed health **Biological reactions** Hypertension Fibrin production Adrenaline Suppressed immune function Blood lipids Body mass index Glucose intolerance
Government policies Economic Welfare Health Housing Transport Taxation			

Note: There are complex interactions between the categories

Source: Adapted from Turrell et al. (2006).

The food system

The food system is global, very complex and very real, yet for most of the time most affluent city dwellers are hardly aware of it. In Australia, New Zealand, North America, Europe, urban China and India, Japan and similar affluent societies, most — though not all — people have access to a nourishing, safe, plentiful and relatively inexpensive food supply. This is not something that happens automatically, nor can it be guaranteed in the future.

The food system permeates our lives in many ways, not only through the types of foods that are on offer. It creates demand for foods through marketing and advertising, and it has changed the physical environments of countries through shifting patterns of primary production. For example, the rolling wheat paddocks of Western Australia, with barely a fence to be seen, are products of broad land planting, monocultures and harvesting technologies.

When things go wrong with the food system, nutritious foods become unavailable or in short supply and people become ill, starve and die. About one-quarter of the world's population does not have enough food. Even in rich countries, many people do not always have enough to eat. For example, in Victoria, Australia (which has a population of around 5 million) in 2003, Victorian Relief helped to provide a million meals and the Salvation Army met the needs of over 125 000 requests for emergency aid (Burns 2004). Charities report increasing demands for food aid from many people in 'advanced' economies. Projected increases in the prices of staple foods are likely to increase food insecurity in many countries.

The effects of the food system on health and nutrition status are both direct and indirect. Direct effects include under- and over-supply of foods and nutrients which promote health and well-being (or ill-health), and the indirect effects on employment and economic well-being are also major pillars of community health. The food system is all about trade. Foods are commodities which are traded for money. This trade employs large proportions of the world population (possibly up to one-quarter) in production and distribution industries. The income derived from this trade maintains people's livelihoods, and therefore their health. During the economic depression of the late 1980s in New Zealand, for example, small country towns were crippled when their meatworks (and other agricultural ventures) were forced to close. The subsequent health consequences included reports of increased incidence of suicide, depression and domestic violence. Similar effects of economic disruption on community health are seen during periods of prolonged drought (e.g. Sub-Saharan Africa). It is important, then, for nutrition promoters to understand how the present food system works to ensure that it functions to the population's health advantage.

A brief history

The ways in which the modern food system operates are not the only ways humans can produce and consume food. As a species, human beings have experienced at least several phases of food production and consumption (see Table 2.2).

Hunting and gathering

This way of life has existed from the earliest times. It is associated with a nomadic or semi-nomadic existence and with low population densities. A highly varied diet seems to characterise this lifestyle. The chain between food-gathering and consumption is very short—usually most foods are acquired, distributed and

TABLE 2.2 HISTORICAL PHASES OF THE FOOD SYSTEM		
Phase	**Population**	**Nutrition states**
Hunter-gatherers	Nomadic, tribal, low density, tall stature	High, stable
Agricultural	High density, shorter, settled, regional cuisines	Variable, subject to famine and nutrient deficiencies
Early industrial	Increased density, high mortality rates	Social differential, malnutrition common among low socioeconomic status groups
Late industrial	Population stabilised	Improved nutritional status, increased longevity, tall stature
Post-industrial	Affluent food supply dependent on exploitation of ecological resources	Improved but variable Unsustainable food system—unless reformed

consumed within the family or clan. The Aboriginal peoples of Australia, for example, traditionally had extremely varied, stable and nutritious food supplies; famine appears to have been a rare event as many different foods were available in different places at different times of the year (Blainey 1975). Generally, hunter-gathering was associated with varied diets and high nutrition status. Hunter-gatherers were similar in stature to the well-nourished city dwellers of today. Knowledge about foods, where and how to gather or hunt them and how to prepare them (without poisoning the family!) was sophisticated and handed down within tribal and family units from one generation to the next. Food education was part of every child's socialisation, though it often differed for girls and boys.

Agriculture

Between 5000 and 10 000 years ago, certain species of grasses such as barley, wheat and rice became domesticated along with several species of herd animals such as cattle, sheep and goats. This was associated with the end of nomadic life and the rise of agricultural settlements in the great food bowls of Egypt, Asia Minor (western Turkey), Mesopotamia (Iraq), India, China, North, Central and South America and Papua New Guinea. Many of these settlements coalesced into cities or into densely settled regions such as the Highlands of New Guinea and the Amazon rainforests. Huge increases in human numbers occurred in settled areas because the cultivation of crops and the production of domestic animals provided much more food in one area than could be achieved though hunting and gathering. This increase in population was achieved at some cost. The variety of available food was reduced and this gave rise to nutritional deficiencies, especially during times of scarcity and famine when harvests failed.

Jarrod Diamond (1997) has described the domestication of plants and animals and its effects on human health and settlement in entrancing detail. He notes that the presence of large animals on the Eurasian land mass — the ancestors of present-day horses, pigs, sheep and cows — made them attractive propositions for domestication because they could be used to perform work as well as supply large amounts of food and material, such as leather. This domestication brought people into close contact with these animals and their bacteria and viruses jumped to humans, causing diseases such as measles, influenza, mumps, plague and many others. These diseases caused high mortality, particularly among children, but over hundreds of years most people developed immunity to them. They became familiar 'childhood diseases'. That is, the domestication of animals led to Europeans and Asians contracting new diseases and developing relative immunity to them.

When the European Age of Exploration began at the end of the fifteenth century, the effects on the peoples of the Americas, and later other parts of the world, were disastrous. Tens of millions who inhabited North America, the Amazon basin and the Inca and Aztec Empires succumbed to these 'childhood' diseases because they had no immunity to them. It is likely that the Han expansion from China into Southeast Asia, the Pacific and the Indian Ocean had similar effects. So food production was in a very real sense related to the spread of infectious diseases and to the formation of the world as we know it — for example, the current predominance of Caucasians and Afro-Americans in the United States instead of the dominant Native American civilisations of 500 years ago.

The rise of agriculture caused many people to rely on others for their food. This was associated with trade in food. Merchants earned money from buying and selling food, so the chain between food producer and food consumer was established with merchants linking the two. Unlike the present system, the chain in agrarian societies was short — the consumer usually had a good idea where the food came from and who had produced it. Even in countries like present-day Turkey, which have advanced technological forms of production, it is still possible to buy food from street vendors who know exactly where the food was grown and produced. Contrast this with the situation in large urban supermarkets.

The rise of agriculture marks a subtle but very important change in the status of food. Prior to this stage, food had many cultural, social, religious, psychological and physiological properties. It still does, but with the rise of *trade in foods* there is the added notion of *food as a commodity*. Food is a substance that can be bought and sold for money (or bartered for other goods). In agricultural societies, most foods that were bought and sold were raw materials grown by farmers. Most food processing was done in the home or in the village.

The commodity trade in foodstuffs is now worth hundreds of billions of dollars each year. Generally, richer countries buy food commodities from poorer regions like India, South America, Africa and Indo-China, and they also export their own commodities. The increasing globalisation of the food system has made food production and distribution geographically more complex. For example, salmon grown in Chile is shipped to China for processing and then sold in European supermarkets.

Commodity-exporting countries like Brazil, Australia and many poorer countries earn much of their income from the export of mineral and agricultural commodities. Commodities are not worth as much as processed products (e.g. a cake is worth more than the ingredients from which it is made) and the commodity trade is subject to major fluctuations in price levels—most of which are beyond local control. Even rich countries are commodity exporters—for example, many food commodities are produced in the United States and the European Union. However, they are produced through large government subsidies. Exports of these cheap foods tend to disrupt food production in poorer countries by undercutting the prices of locally produced foods.

The great traditional food cultures such as the Western European, Mediterranean, Inca, Meso-American, Japanese (and Okinawan) and Chinese (e.g. Hang Chou) cuisines were all based on agriculture. Plant foods dominated, though fish was included where it was available along with sparse but highly valued servings of meat from domesticated animals such as pigs, cows, goats and sheep. Most domesticated animals were too valuable as sources of work or milk to be bred solely for meat production. Depending on the stability of supply, these cuisines could maintain people in good health into advanced years. However, supply was often threatened by poor harvests, poor storage, excessive taxation, warfare and civil unrest, so that population growth was usually constrained for long periods (McNeill 1998).

Industrialisation

The first Industrial Revolution in the United Kingdom and Western Europe started around the mid-eighteenth century with the invention of steam engines and other labour-displacing machinery, not least of which was the popularisation of iron ploughs which could release more nutrients for crop growth. The revolution led to major increases in population and to increased specialisation in people's work outside the home.

People in urban areas became much more removed from the sources of their food. New transport systems moved foodstuffs over long distances. Australia and New Zealand, Canada and other colonial countries grew in the nineteenth century because they were able to specialise in the production of wool, wheat, meat and other commodities which could be transported cheaply to the expanding British market—half a world away. The development of refrigeration meant that mutton and lamb could be shipped to the British market in good condition. Increased food production came about partly as a result of mechanisation—for example, farm machinery (such as the 'stump plough') and food processing factories such as fruit canneries in fruit growing areas (like Shepparton in Victoria, Australia).

Thus the distance between the farmer and the consumer grew, with food processors, importers and exporters, wholesalers and distributors coming between them. This made it easier for unscrupulous people to adulterate food or to give 'short measures'. As a consequence, members of the public and ethical food manufacturers demanded food regulations and their enforcement by government bodies. Among other things, compositional guidelines were set—for example, that tomato sauce must contain a certain amount of tomato, water and other compounds. Without these regulations, it would be quite possible for tomato sauce to be a red, tomato-tasting liquid with no tomato in it at all!

Nutritionally speaking, the first Industrial Revolution was an immense disaster for people working in cities. For example, in Manchester in the 1820s the average life expectancy of unskilled women was less than 30 years, while that of a high social status 'squire' was barely 40 years (Chadwick 1842; Laslett 2000). Many Western cities (unlike Japanese cities) were filthy, unsewered places, alcoholism was rife, most water was contaminated with effluent, and the diets of the working poor consisted of little more than bread, sugar, lard and tea. Around one-third of British Army conscripts were rejected for duty in World War I because of conditions like rickets and other forms of malnutrition. It has been estimated that about one-quarter of the British population suffered from frank deficiency diseases during the period from the early nineteenth century until World War II. Similar conditions reigned in the labour-hungry industrial cities of the United States, Europe and Russia. Clearly, the industrialised food system did not deliver adequate nutrition for everyone.

Similar conditions apply today in many parts of the economically developing world. In the new economic zones of China, for example, life expectancies are lower than in other parts of the country. In these zones, people suffer from nutrition deficiencies and excessive energy intakes due in part to over-reliance on a narrow fat- and sugar-rich diet. As a result, heart disease and type 2 diabetes are common as they are in other parts of the world which are undergoing the *nutrition transition* (Popkin 2001; James et al. 2001; WHO 2000).

Gradually, however, conditions did improve during the century from the 1850s onwards. This was partly associated with welfare reform, especially during and after World War II, and to a century of advocacy for better water, sanitation, food services and family planning. The Victorian period saw the establishment of state-funded water and sewerage infrastructures in the United Kingdom, Australia and New Zealand, North America, Japan and Western Europe. The home economics movement which started in the United States in the late nineteenth century, with its women-in-the-home focus, was effective in disseminating household management, hygiene and nutrition knowledge and skills.

It was during this period that findings from nutrition science about the prevalence and effects of nutrient deficiencies (funded partly for industrial and military reasons) were acted upon in the form of nutrition education. The Basic Four Food Groups scheme was promoted mainly as a way to encourage the consumption of protein- and energy-rich foods in order to curb protein-energy malnutrition prevailing among industrial workers and their families. This was followed in the 1930s by the Five Food Groups, which forms much of the food folk wisdom of many English-speaking countries.

The present-day global food system

At its simplest, the global food system is about the production, processing, distribution and consumption of food by humans (Heywood and Lund-Adams 1991; see Figure 2.2). Each of these sectors of the food system is in communication with the others, both domestically and internationally.

Within these major sectors, there are sub-sectors—whole groups of industries like horticulture and agriculture in the primary production sector. In turn, each of these sub-sectors is made up of discrete industries. For example, agriculture is made up of wheat farming, beef production, poultry production, rice production and other substantial industries.

The food system is full of specialised parts. Each part is concerned with a narrow set of activities such as the production of certain groups of foods or commodities, like rice or chicken meat. This means that clashes between the various parts of the system are likely—for example, buyers and sellers are likely to disagree over prices. For the system to run smoothly, ways have to be found to balance the conflicting

FIGURE 2.2 SCHEMATIC OF THE MAIN COMPONENTS OF THE GLOBAL FOOD SYSTEM

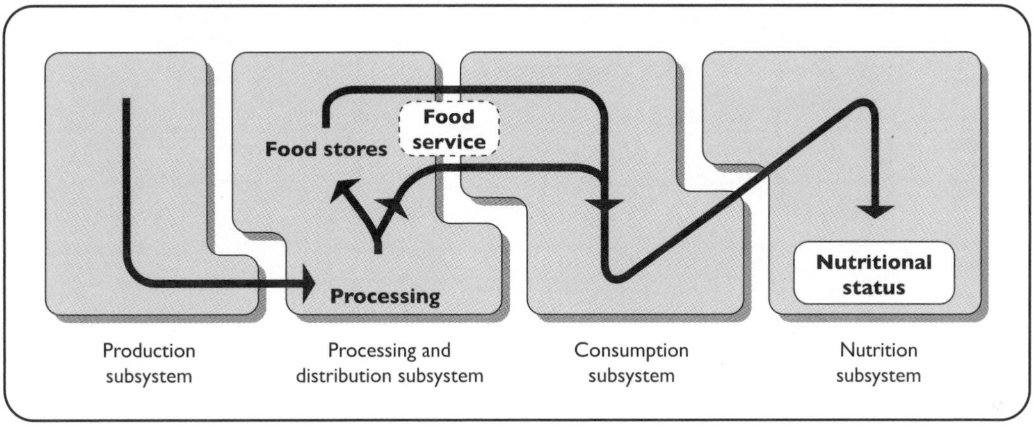

Source: Heywood and Lund-Adams (1991).

interests of the different parts of the system (via food policies). For example, codes of fair trading protect honest manufacturers and consumers from the wrongdoings of unscrupulous traders (such as the sale of adulterated foods). These codes and regulations protect the integrity or 'wholesomeness' of foods supplied to consumers, and so form part of nutrition promotion.

The food system should work in harmony with the physical, biological and social environments on which it depends. For example, the strong value placed on equity in most societies means that the system should feed everyone in ways expected by members of society, and it should be able to do so without threatening the biosphere. At present, it fails to feed about a fifth of the world's population adequately, and it puts the biosphere at risk in many regions of the world through salination, desertification of arable land, water pollution and excessive fossil fuel use. This is particularly acute in fragile ecosystems such as those found in much of Australia. Monocultural agriculture can be a form of 'soil mining' (Diamond 2005), which extracts nutrients in the form of food commodities.

Over-exploitation of ecological systems

The climate change crisis is part of the increasing set of ecological crises which are associated with the worldwide increase in population, the drive for affluence and the use of fossil fuel technologies. Recent food production technologies are partly

FIGURE 2.3 THE COMPLEXITY OF THE PRESENT-DAY FOOD SYSTEM

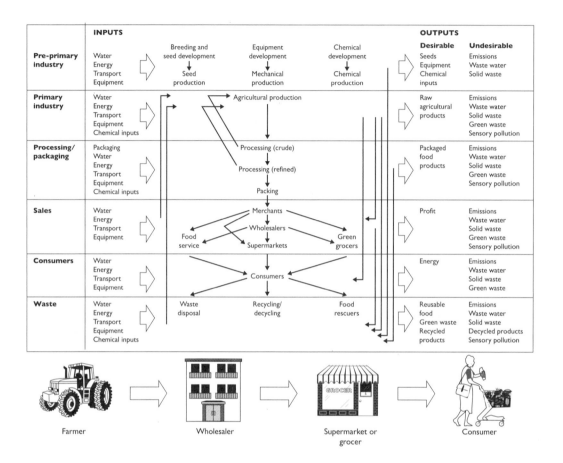

Source: EmilyMorgan@tufts.edu.

responsible for deforestation, salination of soils and desertification and for greenhouse gas emissions. Around 40 per cent of greenhouse gas emissions are linked to the food system, principally through fossil fuel energy use in primary and secondary production, and transport. A key problem is the production of methane, which is a potent greenhouse gas, largely associated with dairying and cattle and sheep meat production. Shortages of water, increases in oil prices, salination and climate change are driving up the costs of food production. The post-World War II era of cheap food commodities, based on cheap energy prices, has ended. Clearly, the food system faces urgent challenges to reduce greenhouse gas emissions, all of which will impinge on food producers and consumers.

The interests of the main players in the food system: Production

Primary producers such as farmers, horticulturists and fishing enterprises tend to be weak economically, though they may be well organised politically. They are 'squeezed' between the economically powerful suppliers of chemicals, seed, fuel and so on, and the food processors and commodity traders who buy their produce. Farmers in developing countries and in commodity-exporting countries who rely on exports of unsubsidised commodities for much of their incomes face severe financial problems caused by subsidised American and European Union agricultural exports.

A profound example of the negative effects on the health of poor farmers of their entry into the world commodity market is to be found in the Punjab. This region feeds much of India with rice and wheat. It is a fertile area which generally has two harvests per year—one of rice, the other of wheat. Before the region's entry into the world commodity market, most of its farmers produced a range of cereals, fruit, vegetables, eggs and meats for their families, and they sold their surplus grain into local markets. This gave them moderately good nutrition status. To take a small example, a chicken which roams the farmyard scavenging for food probably lays a few eggs a week, any one of which would maintain a young girl's iron status. When farmers are induced to use all their land for the production of a commodity crop like wheat or rice, the chickens and home gardens usually disappear. The farmers expect that the more commodity crop they produce, the more money they will have to buy food and other things for their families such as eggs, fruit and vegetables—foods they used to produce for themselves.

Unfortunately, small farmers face two key problems: usually they have to borrow money for seeds and fertilisers, which often comes from money lenders who charge high rates of interest; and second, they become dependent on world prices which are set by powerful, far-away interests. Commodification of farming has led to greater poverty in agricultural regions in many poor countries, which in turn has led to malnutrition among the farmers, their animals and indeed among their crops (since they often cannot afford adequate fertilisers).

A recent phenomenon which puts more pressure on primary producers has been vertical integration of ownership—for example, the ownership of production by supermarket chains (P. McMichael 2007). Manufacturers and retailers have either bought farms of their own, or impose very detailed purchase contracts on farmers which lessens their freedom to manoeuvre. Of course, entry into the world commodity market need not have such dire results. However, to benefit from the world 'free trade' system, farmers need government support in the form of price insurance, industry organisation, administrative support and representation.

Nutrition promoters have to be aware of the basic conditions of trade and related government policies, and need to be organised to influence them. This may be difficult to do, especially if most of the efforts of nutrition promoters are taken up with dealing with the aftermath of poor trade conditions in the form of educating poor people on how to make the best of their minimal resources. Detailed discussion of the health and economic problems generated by the current global food system may be found in P. McMichael (2000) and Lang and Heasman (2004).

Agricultural input suppliers have strong interests in modern agricultural technologies like pesticides, hybrid seeds, biotechnology and machinery. They are large multinational organisations. Modern agriculture could not produce large amounts of food without their products. They are economically powerful and in a good position to maximise their profits from the primary production sector.

They have become involved in a number of controversial issues. One concerns the development of seeds which do not yield viable seeds themselves (Lehmann 1998; www.biotech-monitor.nl/3503.htm). This forces the farmer to reorder seed every year. These seeds have been genetically manipulated so patents may apply to them and prosecutions may be mounted for any unauthorised use of them. A second issue is 'gene harvesting'. Companies search for novel plants—often in poor countries—and attempt to patent one or more genes in the plant, thus forcing users of the plant, who may have used it for generations, to pay for the privilege. A third controversial issue has been the development of genetically modified cereal grains which are unaffected by patented herbicides—this promises farmers the ability to control weeds but it also runs the risk of introducing modified genes into the environment with unknown consequences for society and the biosphere—and, of course, farmers have to buy the seeds and the herbicide from the company!

Nutrition promoters need to examine the evidence around these sorts of issues to decide whether they threaten people's short- and long-term access to healthy food. It is a question of weighing up the risks and benefits to society. For example, companies may promote the potential public health benefits of their practices (e.g. 'genetic modification will feed the world'), though they may not acknowledge the public risks involved or may blur the issues (for instance, that it is not the *amount* of food produced but poor *distribution* which often produces malnutrition).

Secondary producers manufacture primary commodities into thousands of familiar food products such as margarine, bread, sausages and breakfast cereals. Some of these companies are international giants but there are also many small companies which manufacture specialised products and which employ many people. In Australia and New Zealand, small companies account for around 80 per cent of people employed in food manufacturing.

This is a really important sector for nutrition promoters since it designs and makes most of the products consumers buy. Many processed or semi-processed products, such as frozen vegetables and canned fish, are excellent components of healthy diets. Others, like bread and processed meats, are problematic—they contain important nutrients and so are important for health but they also may contain large amounts of nutrients which adversely affect health status (e.g. salt in bread, salt and saturated fats in processed meats). There may be opportunities, then, to reformulate these products or to influence people's consumption of them. Other product categories such as soft drinks are even more problematic but there may be opportunities for nutrition promoters to influence companies' production—for example, through encouragement of smaller package sizes or even substitution by better product categories (e.g. soft drink replaced by bottled water or milk products). Individual nutrition promoters may not be very influential, but organised lobby groups can be effective. One example of the effects of nutrition advocacy has been the recent banning of soft drinks and other high-sugar products from sale in government school canteens in several countries including the United Kingdom and Australia (New South Wales and Victoria); another is the reduction in salt content of many products as a result of the Pick the Tick logo program in Australia and New Zealand (Young and Swinburn 2002) and recent British government and industry initiatives. These problems and opportunities will be discussed in more detail in Chapter 13.

Commodity traders are importers and exporters of unprocessed commodities and processed food products. Many are large multinational companies. The world's wheat trade, for example, is dominated by only a handful of companies. They often switch suppliers according to price and other conditions. Some companies search the world for the lowest priced processed foods and sell them on to retail chains, usually as generic or 'no brand' products. Government quarantine and food regulatory agencies (such as Food Standards Australia and New Zealand and the UK Food Standards Agency) are involved in safety assessments of these products.

Other major food distributors include wholesalers and retailers who arrange the sale of foods to consumers. Over the past 30 years, supermarket chains have grown to dominate much of the food system in Australia and elsewhere (a crucial exception being Japan, where small shops still dominate thanks to government policies). Because of their immense buying power, they can secure favourable deals with manufacturers and sometimes directly with farmers. Some retailers' incomes are sufficient to enable them to offer additional services such as in-store nutrition promotion.

Space in the supermarket is at a premium—many products may be waiting to replace a particular product on the shelves. Shelf space is rented by the manufacturers. If a product does not make a profit for the supermarket within a short time, it will be dropped. Because there are so few retail chains, manufacturers often have to comply closely with the demands of the retailers. If they are denied access to a retailer they may lose access to a large number of shoppers. Manufacturers are often asked to support products during their launch—lowering their prices to the supermarkets to pay for extra advertising, paying for 'two for one' offers or other promotional incentives. Often, 'upsizing' of products like chocolate is initiated by supermarket chains (not manufacturers) to boost sales.

Caterers are of increasing importance because more people are eating away from home in their workplace, in schools, fast food outlets, snack bars, cafes, restaurants, hotels and motels and transport catering facilities such as airports. Around one-third of all meals are now prepared outside the home so food services have an immense role and offer many opportunities for nutrition promotion. These include award schemes like the Heartbeat Awards in the United Kingdom and New Zealand, menu guides to healthy food choices, hospital catering and nutrition codes of conduct. The Australian Heart Foundation has recently extended its Pick the Tick program to cover the nutritional quality of meals served by food service outlets such as McDonald's.

The main advantage of working with food retailers and food service companies is that they have a strong 'service to customer' orientation: if their customers want nutritious foods, they are keen to provide them. We will examine nutrition promotion in supermarket and food service retailing in Chapter 11.

Consumers, the people who buy and consume food (you and me), form a major part of the food system. In a real sense, we are the reason why it exists. If there was no demand for commercial food products (for example, if everyone grew their own food) there would be no food system as we know it.

There is no such thing as 'the typical consumer'. Some people think a lot about the healthiness of foods; some like particular types of food (e.g. meat, 'fine foods'); some have more resources than others. However, more consumers are taking an active role in attempting to ensure that the food system provides products which they want, such as 'low-fat' cheeses and environmentally friendly, ethically produced products. We will look at consumers and their nutrition interests in Chapter 3.

Dealers in information

Several groups transmit information within and beyond the food system.

Advertisers and marketers persuade people within the production and distribution sectors, as well as 'consumers', to buy products and services. They help generate demand for food products. It seems strange that demand has to be generated for food, especially in a world in which one in four people is short of food.

If you walk round a supermarket, it is soon obvious that many of the products are virtually the same—one sweet biscuit is much like another—but the different brands offer different images which appeal to different people. Food products are sold not only to satisfy hunger but to meet 'psychogenic' needs, so called 'emotional benefits' (Chapter 3), such as the need to be accepted by others—'if you buy this type of cola you will be a popular person'. The modern food system sells images and promises along with food. The stark reality is that people can only eat so much food—if companies want to sell more, they have to promise them something *in addition* to a nice taste and 'full tummies' (the satiation of appetite).

Marketing and advertising are frequently regarded by public health workers as 'dubious at best', probably because they often promote products which may contribute to ill-health—for example, fast foods and drinks, tobacco, alcohol and fast cars. However, around 40 per cent of all advertising worldwide is for public-good objectives such as campaigns for safe sex, immunisation and healthy eating. Depending on the nature of the product, marketing can be used for good or ill.

Unfortunately, nutrition promoters are often faced with the negative consequences of food marketing, particularly the over-emphasis on the promotion of high-energy foods and beverages to children (discussed in Chapter 13). Nevertheless, marketers and advertisers can play constructive roles. They frequently have clear ideas about what motivates members of the population and they are often expert communicators. In today's complex society, advertising and marketing can play valuable roles in informing people about the availability of products and services, and helping bring about change. Nutrition promoters can learn much from them, particularly about ways to influence decision-makers which are often used in business-to-business marketing. The application of marketing to the promotion of the public good is called *social marketing*.

Specialised *media* such as technical journals inform professionals of new developments and innovations. These can give companies advantages in their competition for market share. The mass media play several important roles in the system. They act as channels for advertising, they maintain consumer interest in food products through TV and radio programs and articles about food (e.g. recipe pages). In addition, they often probe the causes of food system problems, such as outbreaks of food-borne illness and malnutrition, in the public interest (good examples can be found on *The Guardian* food files website; or Jamie Oliver's School Meal campaign on UK television).

Both specialised and mass media offer splendid opportunities for nutrition promoters. The specialised media allow access to decision-makers in the food system.

Letters and articles and internet communications can be disseminated to present the points of view of nutrition promoters. Similarly, TV and radio program-makers and newspaper and magazine editors are often eager for input about food and nutrition issues (discussed in detail in Chapter 13). Recently, interest has developed in 'ambient media' such as advertising in washrooms for contraceptives or AIDS prevention messages (Turk et al. 2006).

Experts are used in most parts of the food system. They are usually employed by large companies. Food technologists, microbiologists, market researchers and many other specialists are used to generate information which can be used to make decisions or to justify them. The research and development problems on which they work are decided in the interests of the companies. These may or may not coincide with public health interests.

From a public health point of view, the credibility of expertise is crucial. Many people automatically discount the findings or opinions of experts who are employed by commercial interests because they perceive them as being 'bought', or 'paid to toe the company line'. This may be justified, especially when their information is offered in public debate; however, much commercially funded information can be valuable for nutrition promoters, especially technical information about the nutrient content of food products. The main advantage public health workers and nutrition promoters have is that they have credibility in the eyes of the public because their job is to promote the public good. This places a great responsibility on them to ensure that they do not compromise their independence through unwise partnerships with commercial organisations. Partnerships between nutrition promoters and industry can be very valuable but not if the promoter's perceived (or actual) independence is reduced.

Regulators often act as umpires. Because there is often conflict between the various components of the food system (e.g. between buyers and sellers), governments have developed food policies which regulate competing interests. Bureaucrats and various experts usually administer these policies. Similarly, industry organisations like grocery manufacturers' associations employ bureaucrats to balance the competing interests of their members and to advance their joint interests. Ideally, bureaucrats collate and disseminate information upon which decisions can be based.

Government and industry bureaucrats appear to share similar viewpoints. Often they support the 'status quo' and follow government or industry orthodoxies. In the past, at least, they did not hold favourable attitudes towards nutrition promotion (Pusey 1991; Worsley and Murphy 1994), though this may be changing due to the political impact of the 'obesity epidemic'.

In many countries, governments have established regulatory authorities to administer health and nutrition. For example, in Australia and New Zealand, Food Standards Australia and New Zealand (FSANZ) was set up to protect the public's health through the enactment and enforcement of food regulations (food standards). (The manufacture and sale of dietary supplements, however, is regulated through the Therapeutic Goods Administration.) In the United States, the Food and Drug Administration performs similar tasks, as does the Food Safety Authority in the European Community (www.efsa.eu.int/) and the Food Standards Agency in the United Kingdom.

Internationally, the Codex Alimentarius Commission also recommends food standards in order to harmonise trade between member countries. The United States, the European Union and Japan are key members. The World Trade Organisation (WTO) sets the terms for world trade, such as the amounts of subsidies and tariffs which are allowed. There is considerable dispute over the implementation of the WTO's 'free trade' agenda, especially about the difference between its *official* pro-free trade stance and the *actual* protectionist polices of the United States and the European Union.

Other international agencies which influence trade in food commodities and nutrition status are the World Bank, the International Monetary Fund, UNESCO, the Food and Agriculture Organisation and the World Health Organization. The policies of these international bodies are under continuing scrutiny and criticism from a variety of critics (P. McMichael 2007). These agencies play major roles in nutrition promotion around the globe.

Lobby groups are common within the system. They represent specific industries or groups—for example, the margarine manufacturers or the grocery manufacturers. They exist to lobby or advance the interests of their members, usually with respect to government policy. Capital cities like Canberra, Washington and London have large numbers of lobby groups. Nutrition promoters rarely have well-established lobby groups, though sometimes members of public health, dietetic and medical associations combine to mount campaigns for or against specific issues (e.g. for greater regulation of food advertising to children). These alliances, however, tend to be transient without the staying power of the industry lobbies.

Nutritionists, public health and welfare workers and educators have major roles to play in these systems and settings. They complement the activities of the commercial food system. Some would say that they try to fix its failures. This is seen starkly in poor countries and poor areas of rich countries where people have insecure supplies of food and resulting malnutrition. Private charities, other non-government organisations (NGOs) and government-supported schemes provide food to needy people.

Trends and tensions in the food system

There are several trends and tensions in the global food system which are relevant to nutrition promotion. They are summarised in Table 2.3 along with some of their implications.

Other systems which affect population nutrition status

People interact with other environments or systems in addition to the food system. The most relevant are systems in which food is provided to people or in which they learn about food and nutrition. They include:

- the *education system*—including pre-school centres, primary and secondary school and the tertiary system, discussed in Chapter 8;
- *hospital and health services*—from antenatal and post-natal advisory centres to the care of highly dependent patients in hospital and people in residential care, especially severely disabled people (discussed in Chapter 11);
- *people living in institutional care* such as those in prison, in community care and the military (see Chapter 11);
- *workplaces* in which people spend much of their time outside the home (see Chapter 9);
- *community networks and local government*, including myriad activities and groups like sports and leisure centres, scout and guide groups, church groups, neighbourhood groups and environmental associations (examined in Chapter 8);
- the *mass media*, which touch everyone to varying degrees and play major roles in defining normal eating and health (discussed in Chapter 13).

Micro-environments

Within these systems, there are specific settings in which foods and beverages are purchased and consumed. For example, foods are purchased from supermarkets, greengrocers, fast food outlets, vendors at sports venues, vending machines at transport interchanges and from many other places. Similarly, foods and drinks can be consumed in many settings ranging from the street, the schoolyard and McDonald's to the family home. Several aspects of these micro-environments have important influence over food consumption. These include:

- *Access and availability*—The price of food and income levels have a major influence: if foods are too expensive, they are not available for less well-off people. The distance of housing from food outlets, the provision of safe, direct walkways and of public transport, the cost of petrol and the availability of cars are also significant factors. The range of foods provided in shops and restaurants and in workplaces, schools and the family home have influenced the availability of foods. For example, in many remote Aboriginal community stores the range of fresh foods is poor.
- *Familiarity*—the extent to which foods are available in retail outlets (and in other settings) is something members of the local community are used to. Many Sub-Saharan refugees are used to consuming camel meat, but they are unfamiliar with the nutritionally similar beef and lamb on sale in shops in their adopted countries. People tend to eat only foods with which they are familiar.
- *Convenience and proximity*—People tend to consume foods and beverages which are available in convenient places. Thus soft drink vending machines are placed near classrooms and in transport hubs so they are convenient for thirsty people to purchase. Generally, biscuits left in full view of family members will be eaten before biscuits stored out of sight in cupboards.

Food policy—what is it?

Food policies provide the plans and 'rules' which enable systems to provide foods for consumers. In general, food policy has been defined as 'the set of activities and relationships that interact to determine what, how much, by what method and for whom food is produced, distributed and consumed' (Tansey and Worsley 1995). This suggests that the food system can be manipulated to serve the goals of its constituent interest groups. Food policies do not have to be written down on paper—they can be implicit rather than explicit, and they can be implemented at local, state, national or international levels. For example, they may involve informal agreements between producers not to compete against each other, or they may be enshrined legislation, or more likely they exist in hundreds of regulations. Policy is merely a set of arrangements to ensure that certain desired outcomes come about. *Food and nutrition policy* places special emphasis on the provision of healthy food which promotes the population's nutrition status. Food policies can be implemented at local (Grossman and Webb 1991), national (Heywood and Lund-Adams 1991) and international levels.

TABLE 2.3 FOOD SYSTEM TRENDS AND THEIR IMPLICATIONS FOR NUTRITION PROMOTION

Issue	Implications for nutrition promotion
Convenience—consumers want food products which are easy to prepare and consume	Encourages dependence on fast foods and products which may contain large amounts of saturated fats, sugars, salt and energy. Associated with the decline in the population's food preparation skills and the disempowering of consumers (consumer passivity).
The drive for tasty foods	Foods must taste good if they are to be consumed regularly. In the past manufacturers could rely on the addition of salt, sugar and fats to make products appeal to consumers. Today, consumers are becoming more discerning. Nutrition promotion must ensure 'healthy' products taste good.
Value for money	Leads to over-consumption of unhealthy, high-energy products through two for one deals, oversizing, etc.
The drive for health	Mainly in the form of individual's personal health, little encouragement of community health or environmental health. Overemphasis on individual health has encouraged quackery and an absence of strong government actions to further the population's health.
Functional foods	Foods which are made to deliver specific health benefits over and above those of normal foods such as phytosterol margarines, which can reduce serum cholesterol levels. May cause confusion in consumers' minds; may focus too much on a nutrio-centric viewpoint rather than foods as the main source of health and disease; problems are likely to be encountered in substantiating any claims made.
Dietary supplements	These *promise* consumers the ability to control their own health. They include herbal remedies as well as nutrient supplements. Consumer protection measures are required to prevent false unsubstantiated claims; promotion of foods is needed rather than supplements. Clear messages should be given about appropriate supplements for specific population groups.
The quest for slim bodies and beauty	Fostered by mass media and fashion; encourages dieting and quick fixes as well as anorectic eating and poor health. Continuing work required to promote healthier models in media and fashion.
The obesity and metabolic disease epidemic	Mainly due to over-consumption of high-energy foods and physical inactivity. Raises opportunities for novel, low-energy foods and possibly plant-based dietary patterns.

Table 2.3 continued

Issue	Implications for nutrition promotion
The increasing prevalence of obesity and metabolic diseases	Major increasing worldwide problem. Programs needed to encourage food industry to market less energy-dense products and government to regulate sales of high-energy (and salt) foods. Small group and individual level multidisciplinary programs required to enable people to avoid obesity.
Organics and the quest for 'natural' products	Trend to use organic foods could help ecological and employment sustainability. Clear, substantiated production accreditation required.
Ethical food sourcing	Examples include the use of Fairtrade products, which ensure primary producers are fairly paid; avoidance of foods produced by indentured labour; ethical animal production and slaughter. Food shoppers express their political views when they shop.
Vegetarianism	Vegetarianism's major environmental and sustainability advantages should be supported while avoiding possible nutrient deficits. Promotion of healthy vegetarian practices required for specific population groups. Often related to a desire for locally produced foods, animal welfare, environmentalism and 'slow food'. Microbiological and nutritional safety may be an issue.
The drive for pleasure and entertainment	Foods should be enjoyable. Problems occur, however, when the pleasure comes with energy- and salt-dense products.
Climate change and ecological threats	Associated with increasing costs of staple foods. Requires fundamental changes to agriculture, trade and consumption, e.g. reductions in production and use of ruminant animals.
Food insecurity	Increasing numbers of people do not have enough money to purchase basic foods. Associated with poverty, obesity and ill-health. Rising staple food prices will increase personal food insecurity. Nutrition promoters need to establish systems to deal with food insecurity and with factors which threaten the food supply, e.g. drought, civil disturbance, poor land management practices.
Shift to locally produced foods	More consumers prefer foods grown close to their homes (e.g. within a 10 km radius). This may reduce 'food miles' and encourage local employment but may have adverse effects on low paid workers in Africa, Asia and Latin America.

The implementation of food policies: Institutions, instruments and information

From micro to macro

Food and nutrition policies can be implemented at most levels of society, in institutions like child-care centres and primary schools, in local government, in food manufacturing and retailing companies, and at the state and international levels. They differ in detail but they share the aim of helping systems to operate consistently to provide healthy food for all. There are three key factors which enable food and nutrition policies to be implemented: information, institutions and instruments — the 'three is'.

- *Information* — This is essential for decision-making. Food policies need information about levels of imports and exports, the monetary value of commodity production, the safety of foods and their ingredients, the energy content of food, and much more. For example, in order to ensure the sustainability of the food supply, we need accurate information at regular intervals about land degradation, farming practices, the effects of pricing arrangements on land use, fossil fuel use during transportation and so on. Regular monitoring of indices of sustainability will indicate the extent to which particular 'sustainability' policies are working.
- *Institutions* — Food policies cannot be carried out in a vacuum; they need the coordinated efforts of groups of people who can implement them. These groups often belong to special organisations or institutions, or to branches of institutions, which are dedicated to particular policies. They may be specialised agencies which deal only with food issues — such as food regulatory branches of government — or they may be more general institutions such as trade practices commissions and consumer affairs bureaux which attempt to ensure the fairness of trading practices. For example, in Australia the Strategic Inter-governmental Nutrition Alliance (SIGNAL) was set up in the late 1990s to integrate the nutrition activities of the state and federal governments (see Box 2.1).

In addition to Health Departments, Departments of Agriculture and Primary Industry play major roles, as do non-government organisations such as the farmers' organisations, the Heart and Cancer Foundations, food industry organisations and trade unions.

- *Instruments* — In order to implement policies, institutions use instruments just like a gardener uses a spade to dig holes. These may facilitate or inhibit consumers' actions (e.g. compulsory nutrition information on food labels), or they may redistribute goods or services within the system. For example, food stamps and vouchers may be given to people who are unable to buy food. These actions can be taken directly by governments, through legislation, redistributive taxation, sales taxes and tariff policies (e.g. extra taxes on imported foods to protect local products). Alternatively, policies can be implemented indirectly by public and private persuasion — for example, through calls to consumers from the National Heart Foundation to choose low-salt, low-saturated fat foods.

BOX 2.1 WHAT IS SIGNAL?

The Strategic Inter-governmental Nutrition Alliance (SIGNAL) is a sub-committee of the National Public Health Partnership, established to coordinate action to improve the nutritional health of Australians.

SIGNAL is made up of representatives or nominees of: the Australian Department of Health and Ageing; all state/territory government Health Departments; the Australian Institute of Health and Welfare (AIHW); Food Standards Australia and New Zealand (FSANZ); the National Health and Medical Research Council (NHMRC); and the New Zealand Ministry of Health (with observer status). The committee also includes four independent members with expertise in nutrition and public health. The primary goal of SIGNAL is to provide strategic direction for and coordination of national nutrition priorities.

SIGNAL develops annual work plans identifying priority areas within the broad framework of Eat Well Australia—the national public health nutrition strategy. SIGNAL can also respond to topical public health nutrition issues from a government perspective. SIGNAL's role is to:

1. support action on the goals and objectives of Australia's National Health Priority Areas;
2. take a leadership role in building a common approach to public health nutrition across the Commonwealth, state and territory governments, as well as coordinating and supporting national programs;
3. provide expert advice on public health nutrition issues from a government perspective to the National Public Health Partnership and other government agencies;
4. promote better communication about public health nutrition between the health and nutrition professions, industry and governments for disseminating information about public health nutrition; and

5. foster partnerships between the public, non-government and private sectors to advance public health nutrition.

What are SIGNAL's key priority areas?

SIGNAL plays a major role in coordinating the implementation of the national nutrition strategy 'Eat Well Australia: A National Framework for Action in Public Health Nutrition, 2000–2010', which includes an action plan for Aboriginal and Torres Strait Islander peoples. The strategy focuses on increasing the consumption of vegetables and fruit; reducing over-weight and obesity; promoting good nutrition for women and children; and promoting good nutrition for vulnerable and disadvantaged groups. These and other initiatives are designed to address the major nutritional challenges within Australia's National Health Priority Areas, including cardiovascular health, diabetes and cancers.

SIGNAL seeks to improve effectiveness, reduce duplication and achieve economies of scale in developing government programs, campaigns and educational materials. A more co-ordinated approach to workforce development, research, monitoring and evaluation is also being pursued. SIGNAL provides, for the first time at a national level, a government forum for public health nutrition in Australia. It can act as a first 'port of call' for industry groups, professional associations, non-government organisations and consumer groups wishing to work cooperatively with government. Effective links can also be established with scientific experts and others with an interest in health, food and nutrition. This includes clinicians, researchers, educators and the media, and all those with the shared goal of advancing public health nutrition in Australia. As an expert and representative committee established by the National Public Health Partnership, SIGNAL is responsible to that body and through it to the Australian Health Ministers' Advisory Council.

Source: www.nphp.gov.au/workprog/signal.

Criteria for food and nutrition policy

Several criteria can be used to judge the adequacy of food and nutrition policies (Tansey and Worsley 1995). They include the following questions.

How safe is the food?

Foods pose risks to consumers. A good food system will include a number of procedures (e.g. Hazard Analysis Critical Control Points (HACCP) systems in production) and regulatory warning and recall systems which minimise these risks to as many people as possible. Total absolute safety is not possible, but minimisation of risk is.

Unfortunately, foods can be contaminated during production or distribution. Sometimes contaminants can enter the production process—for example, ergot can be introduced during the manufacturing of rye, leading to severe illness and even death. More commonly, microbiological contamination can result in serious food-borne illnesses such as salmonella poisoning or campylobacter infection. Some foods can cause allergic reactions or food sensitivities in a minority of people—for example, some nuts and food colourings can cause anaphylactic shock in some children.

How nutritious or health-promoting is the food?

Food safety is usually considered as a distinct set of issues separate from the nutritional aspects of foods. Until recently, it has been generally assumed that if a wide variety of foods is available, most people will receive adequate nutrition and their health will be maintained. Unfortunately, the over-promotion of high-salt, high-energy foods undermines this assumption (as the obesity epidemic shows).

Specific nutritional problems can occur in particular subpopulations such as pregnant and lactating women, babies and children who are growing rapidly, older persons and people who have poor access to foods. Therefore, it is necessary to monitor the nutrition and health status of these at-risk groups to ensure they have satisfactory nutrition status. For example, the US Pediatric Surveillance Program uses sentinel hospitals' baby height and weight data to monitor the extent of stunting in high-risk populations so that additional food support can be provided where necessary (e.g. via the Women, Infants and Children's Extended Nutrition Program). Most countries (though not Australia) conduct systematic nutritional monitoring of the general population or of specific subpopulations. Usually this is to ensure that the population has access to safe, nutritious foods.

How secure is the food supply?

Food security has been a key goal for all societies during history. To a considerable extent this has been achieved in most countries. However, the mal-distribution of food is a growing problem in many affluent countries and a continuing problem in poorer societies. A good food policy will include monitoring of the availability of food across all levels of society and will devise instruments and procedures to ensure that everyone has enough 'healthy food' to eat at all times and in all circumstances in their lives. Figures 2.3 and 2.4 show data from Canada where the problem of food insecurity has been extensively studied. Large numbers of financially deprived people are affected. The situation in Canada is fairly typical of other Western 'affluent' countries. Box 2.2 outlines the activities of Foodbank Victoria which is typical of many voluntary organisations that ameliorate food insecurity.

Is sufficient food available at a reasonable cost?

A varied, healthy food supply should be available for everyone. This means that, relative to average earnings and expenditures (on rent, transport, schooling, etc.), foods have to be reasonably priced. This is usually achieved through open markets which encourage competition between producers and between food retailers. Consumer protection and fair trading legislation is required to ensure fair trading and prevent the growth of cartels which can inflate the price of food. In some countries, government consumer protection agencies and anti-competitive behaviour agencies monitor the corporate behaviour of food retailers and manufacturers. In others, food subsidies are provided (at least among poorer groups) to ensure the production and distribution of important classes of food (such as fruit or vegetable oils). The task of nutrition promoters is to question the adequacy of current safeguards and to take action to ensure everyone has access to healthy foods. It is remarkable that only Brazil and South Africa have the right to food enshrined in their constitutions. Perhaps this reflects their particular histories, or perhaps it is because the food supply tends to be taken for granted in many developed countries.

Is the food supply equitable and fair?

There are inherent differences in the food and nutrition needs of various population groups like women and men, children and adults and so on. Equity and fairness are two key public health principles which ensure that food should be available to people according to their needs. The Recommended Dietary Intakes (RDIs) and

FIGURE 2.4 NUMBER OF CANADIANS USING FOOD BANKS IN THE MONTH OF MARCH, OVER A 12-YEAR PERIOD; 1989–2004

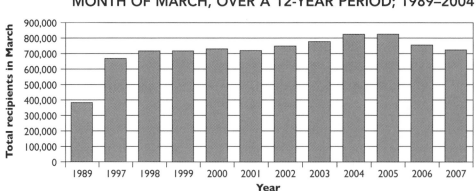

Source: Canadian Association of Food Banks 'Hunger Count' (2007).

FIGURE 2.5 SOURCE OF INCOME AND ODDS OF REPORTING FOOD INSECURITY

Major source of income	Odds (95% CI) of reporting food insecurity	Odds (95% CI) of reporting compromises in food intake
Employment	1.0	1.0
Social assistance	3.1 (2.3–4.0)	3.4 (2.6–4.5)
EI, worker's comp. CTB, support/alimony	1.7 (1.2–2.6)	1.8 (1.1–2.8)
Seniors' benefits	0.9 (0.7–1.5)	1.0 (0.8–1.4)
Other	1.0 (0.7–1.5)	1.1 (0.7–1.6)

Source: Che and Chen, *Health Reports* (2001).

Nutrient Reference Values (NRVs, described in Chapter 3) provide estimates of the nutrition needs of various population groups. A good food policy monitors the extent to which the food supply actually meets the nutrient requirements of these groups. In addition to physiologically based differences between people, there are also major social and cultural differences which food policies have to take into account if they are to achieve fairness and equity. These include recognition of the food prohibitions of religious groups, such as Islamic and Jewish requirements for halal and kosher foods, as well as the demands of other groups, such as vegetarians, for an adequate variety of plant foods. In most countries, market forces usually ensure that these criteria are met.

Sustainability

Foods should be available for everyone on a long-term basis, from generation to generation. This means that food production has to be on sound ecological and economic foundations. A major problem of the present system, in most affluent countries, is that about one-quarter of food production is wasted even before it reaches the consumer. An equivalent amount is wasted by consumers after purchase. This happens for many reasons—for example, poor stock rotation and impulse buying. In several countries, food manufacturers and retailers donate foods to food banks (such as Foodbank Victoria) and other charities for distribution to needy people (see Box 2.2).

BOX 2.2 FOODBANK VICTORIA

Mission: To feed those in need in our community by soliciting and distributing nutritious, surplus food and grocery products through a network of certified, not-for-profit welfare agencies that provide services to people in need throughout the state.

- Foodbank receives regular donations from over 120 food manufacturers.
- Markets partner merchants to donate truckloads of fresh fruit and vegetables each week.
- Foodbank supplies over 350 charitable agencies in Victoria.
- Foodbank levies a service fee based on weight, only partially covering operating costs and income.
- Foodbank Victoria has four full-time and twelve volunteer part-time staff.

Source: www.foodbankaustralia.org.au/foodbank/victoria/profile.asp.

Are commonly available foods enjoyable to eat? Does the food supply promote social relations (conviviality)?

Most of us could live for a long time on a diet of milk, bread and green vegetables. But this would be very boring and unattractive. Few of us would be able to persevere with it for more than a few days. To eat well, and therefore healthily, most people prefer some variety in their diets and often we prefer to eat in company. Eating and drinking act as 'social lubricants' and as a focus for social relations. In other words, enjoyment of food is a key aspect of the healthiness of a food policy. Of course, food variety can be achieved in several ways, such as by eating different foods or by eating the same foods (e.g. potatoes) prepared in different ways. This requires culinary skills and knowledge. Paradoxically, the abundant variety found in affluent countries enables some consumers to eat narrowly all the time (e.g. by eating the same frozen vegetables at main meals).

Ethics

People's views of the ethics of various practices vary; however, there is some consensus across society that animals should be dealt with humanely during rearing, transport and slaughter, and that food workers should not be harmed during their work. Similarly, many people object to food imports from countries which allow poor working conditions among food workers. Others consider the advertising of any product to children as unethical since they are unable to make legal contracts. Lobby groups, such as animal rights activists, trade unions, environmentalists and others, attempt to sway public opinion towards their viewpoints. Consensual ethical standards form the basis of any healthy food policy.

Common sense

A good food system should follow common sense. For example, cardiac patients should not be fed foods high in saturated fats by hospitals, and cabbage grown near city A ought not to be shipped to city B via city C before it is sold to shoppers in city A!

Actions speak louder than words

Policies must be *implemented*! It is not sufficient to provide sets of recommendations and background briefings. People and institutions need to adopt recommendations and put them into practice. Such implementation is the main role of nutrition promoters and food communicators.

Conclusions

The food system is vital to us. It is a complex set of interacting sub-systems which have fairly specific functions. Food policies are ways in which the food system can be made to bring about goals that we all value. The world's population is likely to double during the next 50 years to over 10 billion people. The ways in which we will feed all these people in healthy and sustainable ways present a major and exciting challenge for nutrition promotion. Nutrition promoters need to be familiar with all aspects of the food system—the players, the tensions and trends, and the policies—so that they can implement programs and policies which foster the population's health and well-being. A useful summary of the issues facing the British food system, which are similar to those in other countries, is provided in Box 2.3.

CONSUMER DEMAND	**SAFETY**
Spending a smaller share of incomes on food	Falling incidence of food-borne illness
Increasingly sophisticated purchasing	More reported food allergies
Eating out more	New challenges in a lengthening supply chain

THE FOOD CHAIN

Increasingly dominant retail sector

Increasing variety of places to eat out

Employment concentrated in retail and services rather than farming and fisheries

GLOBAL MARKETS

Increasing commodity prices having the greatest impact on developing countries

Population and income growth underpinning demand for food

Biofuels manufacture set to take a rising share of US and EU grain output

HEALTH

Excessive fat, salt and added sugar in UK diets but not enough fruit and vegetables and oily fish

Obesity increasing

Emergence of diet-related ill health among children

SECURITY

Europe's share of UK food imports rising (imports are rising in other countries too with globalisation)

Resilience of supply chains a key issue

Greater risk of climate change impact on developing countries

Until recently, more UK households have become food secure (this is not the case in other countries)

ENVIRONMENT

Food chain has a high share of greenhouse gas emissions

Concentration of environmental impacts around livestock production

Food travelling further

Source: The Cabinet Office Strategy Unit (2008).

Discussion questions

2.1   Describe, with examples, the ways food is produced, distributed and sold in your state.

2.2   How does the food system contribute to the 'obesity epidemic'? How could food policies be devised to make it less obesogenic?

2.3   Discuss the impact of the ecological movement on the food system. Is it a case of too little, too late?

2.4   Describe, with examples, how policies affect the availability of food in Australia.

2.5   Describe the three main phases in the history of human food production. What key points do we need to learn from this history?

2.6   What are the advantages and disadvantages of the globilisation of the food commodity trade?

3

Food consumers

Introduction

As well as being involved in changing the food supply, nutrition promoters attempt to change the demand for foods by influencing consumers' food behaviours. Therefore, nutrition promoters need to how people think about and use foods and beverages. People's nutrition status depends on their eating and drinking habits. These complex behaviours are influenced by many factors, including their beliefs about foods and beverages, their motivations, their lifestyles and the settings in which they live.

The dependence of nutrition on eating and drinking

Although nutrition science forms an important intellectual basis for nutrition promotion, its potential influence on population nutrition status depends on people's food and beverage behaviours. If people did not buy, grow and prepare foods, and eat and drink them, there would be no point in discussing their nutritional status because they would be dead! Food behaviours are central to nutrition promotion so we need to understand them and the factors which influence them.

People's food and beverage behaviours can be classified in several ways:

- *By type or endpoint* — People have been engaged in the collection or growing of foods (as in horticulture, agriculture and animal husbandry) since time immemorial. Many people sell or purchase foods (e.g. 'do the shopping') while others prepare them (e.g. washing, cutting, cooking foods). And all of us consume foods by eating and drinking.
- *Through cultural taxonomies of foods* — Cultures classify foods in different ways as 'healthy' or 'unhealthy', 'fit to eat' or 'disgusting', 'animal' or 'plant', 'slimming'

or 'fattening', 'dangerous' or 'safe', 'morally good' or 'morally bad', 'ethical' or 'unethical', or along other continua or polarities. People consume foods in order to gain the benefits they perceive are associated with them. For example, Chinese people may eat 'hot' or 'cold' foods to bring about particular states of well-being or health. In this case, 'hot' or 'cold' does not refer to temperature or sensory properties. Psychologists have examined how humans learn to apply these types of criteria during socialisation (Rozin and Vollmecke 1986; Birch 1999). An example of the ways in which Australian women perceive meats is provided in Figure 3.1. In the figure, the closer a property is to the food, the more the women associated the meat with that property.

Most of us are familiar with the Five Food Groups or similar taxonomies used by nutritionists over the years (i.e. fruits, vegetables, cereals, meats, dairy products and their sub-categories). But where do such classifications come from and why do people categorise foods and other things into groups? Clearly, there are so many foods that we have to 'clump' them together in order to refer to them quickly and easily in everyday life. Lay food taxonomies have been studied for over a century by food anthropologists and psychologists. They are important because the properties people attribute to foods usually influence their behaviours. For example, foods which lay people classify as 'children's foods' are less likely to be consumed by adults, especially by young adults. Pilgrim and Kamen (1959) were among the first investigators to look at the ways industrial consumers categorise foods. Their taxonomy based on information from American consumers in the 1950s is shown in Table 3.1.

TABLE 3.1 MID-TWENTIETH CENTURY AMERICAN FOOD TAXONOMY

In terms of types of food	In terms of suitability for people and functions
Meat—solid food—main dish	**Masculine food**
Sweet—desserts	**Children's food**
Fruit	**Healthful food**
Strong flavours—root vegetables, cold vegetables	**Youth food**
Starch—mixed dishes	**Economical food**—tea room
Casserole—creamed	**Common foods only**
Soft—starch	**Luncheon food**
Light main dish—soups	**Southern food** (from South USA)
Hot bread—pork	**Picnic food**
Snacks	

Source: Pilgrim and Kamen (1959).

FIGURE 3.1 WOMEN'S PERCEPTIONS OF THE PROPERTIES OF DIFFERENT MEATS

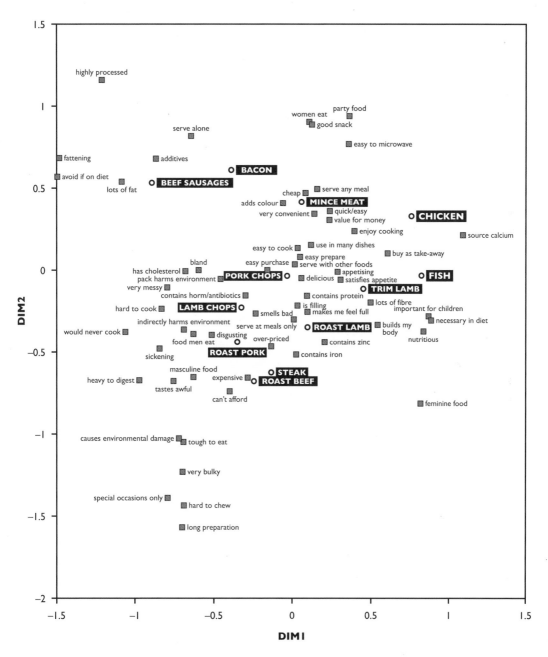

Source: Worsley et al. (1995).

- *According to the goals of those consuming the foods* — It is true that we can 'dispassionately' describe the mechanics of food behaviours such as the number of times a person chews before swallowing. However, eating behaviours are usually part of a series of purposive actions which individuals undertake to achieve personal goals, such as trying to be slim, impressing other people, being physically attractive, caring for the planet (by not eating animals for food), being healthy (whatever that is) or being moral or ethical. Usually, nutritionists are only interested in one goal (health), while most lay people strive to achieve various goals through food consumption (as indicated above).

This motivational approach is central to understanding human food choice, though it is not the only approach. One useful way to understand the purpose of people's food behaviours is provided by the utilitarian philosophy famously espoused by Jeremy Bentham and John Stuart Mill in the eighteenth century (Stephen 1990). They suggested that humans aim to maximise benefits to themselves and to minimise any disadvantages following from their behaviours. This is the basis of many 'attitude-behaviour' theories which have been used in this area. It is worth noting that the notion of 'food choice' is questionable (see Box 3.1).

BOX 3.1 FOOD CHOICE: HOW MUCH CHOICE DO PEOPLE HAVE?

Do you really *choose* your foods? Imagine you are in a large supermarket with over 30 000 products on offer. Such variety! Such choice! Okay, you make your way to the margarine fridge—how many very similar margarines are there? Ten, 20 or 30? Are you going to read the information on each of the 30 packs and weigh up which of them is the best, working your way through each of their attributes: price, weight, cholesterol content, saturated fat content, name of product, colour of pack, country of origin, etc. Most people simply haven't got the time. Instead they use simple 'rules of thumb' (called 'heuristics') like 'I choose the one with the shortest list of ingredients—that way I know it is purer and contains fewer additives', or 'I choose margarine made in Victoria because I want to protect Victorians' jobs'. Often these heuristics are not very scientific or logical, but they get the shopping done quickly.

So you have food choice in theory, but in practice it is limited by the sheer variety of very similar products in today's retail outlet. Of course, there is one very fundamental limit to food choice—if the retail outlet doesn't stock the product, then people are probably not going to buy it or eat it. For example, if you want bilberries (not blueberries) you probably won't find them in an Australian supermarket—you could email the United Kingdom in their summer and perhaps have them airfreighted to Australia (at great cost), but you probably won't bother. So to a large extent what you eat is governed by what is in the retail outlets. Strangely, this tends to be even truer in rural areas than in the cities—the days of farmers growing all their own food are long gone.

Source: Carey et al. (2003).

- *By level of analysis or aggregation* — Food behaviours are complex and are often summarised or 'aggregated' in daily language for ease of referral. A behavioural description such as 'we eat a meal' can be disaggregated into a series of discrete behaviours such as 'looking at the food', 'picking it up' (with a knife and fork, chopsticks, spoon or fingers), putting it into the mouth, chewing it (quickly, slowly, roughly, gently) and swallowing it. This crude approximation does not include the ways in which foods are served — for example, the combinations of liquid, solid and semi-solid foods dictated by particular cultures which might form a meal such as soup, meat and vegetables, and dessert (e.g. Douglas and Nicod 1974).

- *By eating occasion and duration* — This might be the time of day (e.g. breakfast, lunch, dinner, snack, supper), time of year (e.g. types of foods eaten at Ramadan or Christmas), or type of occasion, such as a dinner party, birthday, wedding or funeral. There are many types of culturally ordained occasions (e.g. business lunches) which prescribe or prohibit the types of foods consumed. Some foods are acceptable at certain occasions, others are not. Baked beans, for example, would probably not be served at a wedding! More broadly, during the 40 years between 1961 and 2001 eating and drinking has become frequent throughout the waking hours, instead of being confined to breakfast, lunch, dinner and supper (Future Foundation 2005).

- *By the type of person consuming the food* — In many cultures, certain foods are reserved for men (e.g. meat) or for women (e.g. fruit), or for young or old people. The ways in which foods are served and the order in which people eat often depends on membership of social categories — for example, the men may eat before the women and children, choosing the best foods for themselves (de Garine 1972). In Western societies, some foods are used more by single men, or by single women, or by cohabiting couples (Craig and Truswell 1994; Krammer et al. 1999; Worsley 1988). Food taboos prohibit the use of certain foods by particular categories of people. For example, meat may seen as 'too strong' for children or 'too masculine' for young women. More informally, people on slimming diets may regard foods like chocolate as taboo (forbidden) foods.

Nutritionists often assume that the category of person under consideration is 'the average man or woman' or the 'general population', which is made up of very similar people, like a frame of billiard balls. This 'undifferentiated' view of humans can be useful for understanding some behaviours like the avoidance of particular companies' products during a food recall, but it can blind the nutrition promoter to reality. For example, around 70 per cent of main food shoppers are women, who tend to hold different positions of responsibility for the family food supply than men (Baker and Wardle 2003). They usually have more knowledge about and skills regarding food, and more interest in health than men. Food marketers identify different groups of people with different interests in food (market segmentation; see Box 3.2).

A less obvious but important consideration is that some people in the population are far more influential over the population's nutrition status than members of the general public. These 'influential people' can include government bureaucrats who control the access of food products to the market, marketing and production managers who theoretically can choose more or less nutritious foods to sell or promote, as well as 'opinion leaders' such as doctors, dietitians and journalists. Nutrition promotion has rarely focused on such groups.

This brief survey of the classifications of foods and their associated behaviours clearly shows that there are a multitude of 'eating and drinking behaviours'. Some aspects of them can be publicly observed but other characteristics remain private and have to be inferred through observation. Most people do this on a daily basis when they pose questions like: 'Why on earth is she buying that food?'

Sets of food behaviours

Several major classes of 'goal-oriented' food behaviours, and their associated settings, are likely to affect people's nutrition status.

Growing foods

For farmers, this is a major occupation. However, in today's free market economies, living on a farm may convey few or no nutritional advantages.

Farming families often have worse nutrition status than many urban families simply because they produce only one or two crops (e.g. wheat and rice) and have little money to buy a variety of food products (Conway 1998). Compare this with traditional farms, which produced fruit and vegetables, pigs, chicken and eggs as well as cash commodities.

The production of foods in cities has become less common in recent years. However, peri-urban agriculture accounts for about a quarter of food production in countries like Australia. Foods produced outside the agricultural economy, such as those grown in children's city farms, school gardens and fruit and vegetable gardens, are quite popular and can be consumed by families or bartered between them to improve nutrition status. This can be particularly important in food-insecure communities. However, there are broader and longer term benefits associated with such 'amateur' food production. These include the maintenance of physical and mental health (digging and hoeing, for example, help maintain upper body strength, flexibility and immunity—Savige et al. 2001) and community-building (people socialise when gardening). Gardening also helps children and young people develop sound knowledge of the food system, nutrition and a deeper appreciation of nature. Half a century ago, in Australia and other Western countries, most schools had food gardens. A recent Australian survey showed that 90 per cent of adults think this practice should be revived (Worsley 2006b). In many developing countries, school gardens produce essential foods for children. An important development in several urbanised rich countries such as Japan and the United States has been the development of community-supported agriculture and farmers' markets. Essentially, these movements enable city dwellers to 'adopt' farms which then supply them with fresh produce, by-passing supermarket chains (Friends of the Earth 2006).

Shopping for food

In most countries, this is one of the main classes of essential food behaviours. It provides millions of people with pleasure and millions of others with dismal routine. Shopping behaviours are highly influenced by the situation in which the shopping occurs, and the goals that consumers have in mind are also closely linked to these settings.

Shopping settings are quite familiar: supermarkets, greengrocers, local markets, speciality shops like hot bread shops, cake and pie shops, plus the many food service outlets such as fast food chains, cafes, restaurants, petrol stations, food kiosks in transport interchanges, street food vendors (e.g. sausage sizzles), foods sold at sports venues and the ubiquitous confectionery and beverage vending machines. These can be further subdivided, for example, into large and small supermarkets, supermarkets which serve high socioeconomic status (SES) or low SES communities, and so on. It is worth noting that supermarkets are not an inevitable part of modern urbanised societies; Japanese policies, for example, do not encourage large supermarkets.

The extent to which these settings influence the community's nutrition status varies according to people's reliance on them. Supermarkets tend to have a great influence because, most of the time, many people buy most of their food from them. In contrast, the sales of foods like pies, pasties and high-energy snack foods at sports venues may have relatively little direct effect on population nutrition status since most people (who are not avid sports fans) use them infrequently. The nutritional status of the community surrounding large supermarkets is directly related to the foods sold by them (Cheadle et al. 1993, 1995). This suggests that, most of the time, most people only buy foods from stores which are near them (in the United States and the United Kingdom, the foods bought by people living far from supermarkets tend to have worse nutritional quality than those living near or having car access to them—Sorensen, Hunt et al. 1998; Winkler et al. 2006). Fast food chains like KFC and McDonald's may have a strong but indirect influence on people's nutrition status. We know that more people living in low SES areas (in Western societies) tend to be overweight or obese than those living in high SES areas. Reidpath's group has shown that there are more fast food outlets in the poorer (more obese) areas than in the richer areas of Melbourne (Reidpath et al. 2002).

In many English-speaking countries, about one in three meals is prepared outside the home. This presents nutrition promoters with something of a crisis because many of these foods are eaten in larger portion sizes and are often higher in energy and saturated fats than the home-prepared equivalents (e.g. pasta, chocolate cake) (French 2003; Young and Nestle 2002). However, many cafes and restaurants are happy to promote more nutritious versions of their meals. The United Kingdom and New Zealand Heartbeat and similar campaigns are responses to this situation (see Chapter 13).

Many food products are unfamiliar to consumers—indeed, for many people shopping is something of a mystery. Many questions can go through shoppers' minds—for example: 'Should I make a list before I shop?' 'What happens if I see meat on special—what kind of meal could I make with it?' 'What does 430 mg/100 gm of sodium on a packet of food *mean*? Is it better than a similar product which says 410 mg/100 gm of salt but which is 5c dearer?' 'Should I buy those frozen peas which feel "lumpy" in the packet?' Shoppers do a lot of (fast) thinking and make a lot of decisions during the average shopping trip. They need a lot of knowledge and many skills about buying and using foods. Nutrition promoters need to know about their skills and knowledge needs.

Shopping as decision-making

Shoppers make their purchases in complex and sophisticated ways. They have to take into account time limits, price changes, special offers, stock availability, changes in packaging and much more in addition to the needs and preferences of household members and themselves.

Decision-making is category dependent. For example, nutrition may be an important factor in choosing fruit, meat or margarine but may play no role in the choice of cakes or confectionery. Similarly, some people's purchases may be price dependent but within a price band; other factors, such as nutrition, may play the decisive role. For example, if six margarines appear to be similar in price and quality but one emphasises its low cholesterol properties, it will be more likely to be chosen—assuming that brand loyalty (habit) is not a key factor. Thus claims that nutrition runs first, second or third in importance among consumers' perceptions of product attributes tell us very little. It depends on the context—the category and the distribution of attributes across products. Nevertheless, the widespread interest in nutrition means that it may be used to provide a point of difference between products, favouring those with some link to nutrition.

Decision-making models

In daily life, people often have to do things about which they are not entirely certain within contexts which are constantly changing, such as 'doing the shopping'. Shopping is a series of decisions that have to be made 'on the run' (an *architectural* model) since plans can be changed—for example, a new untried cheese product may be bought. In contrast, models which assume fixed preferences are called 'archeological' models (Payne et al. 1999). Rather than people's preferences for foods being fixed *before* the shopping trip, there is the possibility that they actually form their preferences *during* shopping!

Preparing and cooking foods

Similar considerations apply to the preparation and cooking of foods. Nowadays, food preparation is less and less a family activity. Most parents have jobs which increasingly take up their time, and *time users* like television, computers and team sports tend to reduce further the time available for food preparation. Unfortunately, despite the proliferation of cooking programs on television (which are excellent media for nutrition promoters) we know relatively little about the ways in which most people prepare and cook foods (Caraher et al. 1999). However, the high sales of recipe books and magazines, and occasional surveys, show that food preparation—or at least talking about food—is a major source of pleasure for many.

The high sales of convenience foods suggest that many people simply do not value cooking highly, although most adults think it is a 'good idea' for all boys and girls to learn how to cook at school (Worsley 2006b). Unfortunately, in most schools there is little teaching of these important skills. In part, this is because such traditional female pursuits are under-valued by education authorities who prefer more 'rigorous' academic subjects, and it is also probably due to the fact that cooking can be dangerous (e.g. knives and boiling water are present) which frightens many risk-averse parents. Despite these limitations, many children are interested in cooking and there are some excellent examples of cooking programs for children (see Chapter 9).

Shopping for health or for something else?

There are considerable opportunities for nutrition promotion in food retailing. However, it may be that only well-informed, thinking shoppers are interested in health and nutrition, not the majority of 'passive' consumers (Gabriel and Lang 1995). The majority of shoppers do not think seriously about nutrition every time they go shopping, although about one-third probably do (Scott and Worsley 1994; Wandel 1997; Shine et al. 1997). However, healthy food does not have to be promoted solely on the basis of nutrition. If we know about shoppers' goals—what they want to get out of their shopping trip—then we can promote foods which will meet their needs and wants in ways that support their nutrition status. For example, healthy food choices often save the shopper money—and saving money is a key goal for many shoppers. The majority of consumers prefer healthy foods to unhealthy foods (fat-rich, fast foods)—*so long* as the price, taste and appearance are right.

So what do shoppers want when they shop for food?

The simple answer is that they want *benefits* like good taste, 'value for money' and convenience, and few or no *risks* (bad consequences). This simple demand relates to several social psychological influences which are summarised in the food-related lifestyle model (Grunert et al. 1997).

The food-related lifestyle model

The food-related lifestyle model is probably the simplest but most comprehensive model of consumer food behaviours (other models are described in Chapter 5). It was designed in Denmark by a group of consumer food behaviour and marketing specialists specifically to explain food consumption. It provides a useful checklist of the main influences operating in the daily lives of food consumers (see Figure 3.2).

FIGURE 3.2 THE FOOD-RELATED LIFESTYLE MODEL

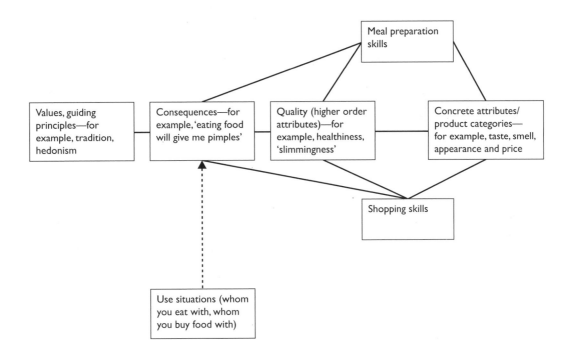

The components of the food lifestyle model include the following:

- *The food's concrete characteristics*, such as appearance, shape, feel, aroma, taste, 'right' price—what the consumer expects. Experienced greengrocers always provide samples of fruit for customers because the characteristics of fruit (e.g. apples) vary a lot. If foods look and taste good, customers are likely to buy them.
- *The positive consequences of buying (and consuming) the food or drink*, which have to outweigh any negative consequences—indeed, certain possible negative consequences like the possibility of becoming ill will completely abort the buying process. Possible benefits include saving money, better health, relatives liking the purchase, friends being impressed when it is served and so on. Nutrition promoters can highlight a variety of positive consequences (e.g. 'it's good for children', 'the family will like it'), not just the health consequences.

One central aspect of consumers' food motivation is the perceived safety of foods. Consumers do not want to be harmed by foods. Usually, most consumers try to reduce the chances of harming themselves. Paul Rozin (1987) has described the 'omnivore's dilemma'—food is necessary for life but it can kill you or make you sick. So consumers are particularly interested in the safety risks of food as well as its benefits. Research from other areas of consumer behaviour strongly suggests that most shoppers check out the labels on products by first looking at the possible dangers involved in its use—for example, whether a washing powder contains caustic compounds (Bettman et al. 1986). The same seems to apply to food products—shoppers are particularly interested in 'negative nutrients' (Scott and Worsley 1994) when they inspect food product labels: ingredients such as 'fats' and 'cholesterol', as well as 'pesticides' and 'artificial chemicals' ('E numbers'). Consumers who suffer from diseases like diabetes (types 1 and 2), and parents of young children, are particularly conscious of possible inherent dangers in foods—especially the dangers posed by 'chemicals'. Wandel (1994) provides a useful description of the food safety concerns of consumers, and Frewer and Miles (2001) provide a detailed discussion of risk perception, communication and trust. Table 3.2 illustrates consumers' key concerns about food and health.

Most consumers, then, try to minimise the risks that foods present while trying to maximise their perceived benefits. Several attitude-behaviour theories such as the theory of planned behaviour (Ajzen 1991) attempt to map out the main factors involved in this process of anticipating the consequences of food choice behaviours. We won't go into these models here except to say that the anticipated consequences of eating foods can be powerful determinants of those behaviours. For example, if you think a food might give you heart disease, you may be unlikely to eat it unless there are tangible benefits. This is often the case when people indulge in high-fat foods like chocolate since the immediate pleasure benefits usually outweigh any long-term benefits of not eating them!

Many theories assume that consumers often reason their way through the anticipated consequences of eating particular foods, figuring out the *likelihood* (some times called the *expectancy*) of possible consequences as well as the *value* (benefits or disadvantages) of those consequences and then 'weighing' them all up to come to a decision about whether or not to undertake the behaviour. Allen and Baines (2002) have shown that this is not the case for 'value-laden' foods like meats—depending on their personal values, people either like or dislike meat, so no reasoning is necessary.

Because of the prominent roles of expectancies and evaluations of the possible consequences, these are called *expectancy-value* theories. There are major limits to the application of these theories, the principal one being that they are

TABLE 3.2 CONSUMERS' RANKINGS OF THE IMPORTANCE OF FOOD AND HEALTH CONCERNS IN SOUTH AUSTRALIA, 1999 AND 1991

Food concerns	1999 importance ranking	1991 importance ranking
Clean handling of food in shops	1	3
The honesty of food labels	2	4
Harmful bacteria in food	3	2
The microbiological safety of imported foods	4	12
The safety of drinking water	5	10
Enforcement of food regulations	6	6
The safety of takeaway foods	7	18
Animal cruelty in food production	8	19
Chemical additives in foods	9	1
TV advertising of junk food to children	10	16
Eating too many fatty foods	11	9
Genetic modification of foods	12	—
Uncertainty about what is in foods	13	15
The cost of basic foods	14	5
Poverty in Australia	15	11
The links between food and cancer	16	8
The irradiation of foods	17	14
The links between food and heart disease	18	13
Importing of foreign food products	19	17
Driftnet fishing	20	7

Source: Worsley (n.d.) unpublished data; see also Worsley and Scott (2000).

more suited to consumption behaviours associated with *novel* foods or food found in unusual situations ('Should I buy that sausage roll from that "dirty looking" stranger?') rather than the *habitual* consumption of daily foods. However, the measurement of expectancies is closely related to the concept of 'risk', which is discussed in Chapter 13.

- *Quality expectations ('higher order attributes')* — Shoppers' expectations of food vary and tend to be highly subjective and abstract, but they are real never- theless. For example, some consumers may look for foods which are 'healthy', others want foods which are 'slimming', still others want 'natural or organic' foods or 'luxurious' foods, and many want 'convenient' foods or foods which are 'good value for money'. Grunert has summarised some of the ways consumers weigh up information such as credence attributes before and after the purchase of food products in his total food quality model (Grunert 2005). He proposes that search processes, experience and credance qualities (perceptions of product characteristics) affect consumers' abilities to judge quality before purchase and their satisfaction after purchase (Table 3.3).

The total food quality model outlines the factors which consumers take into consideration before they purchase the product (their expectations of the product) as well as their post-purchase evaluations, which will influence the likelihood of them repeating the purchase of the product.

- *Shopping and meal preparation scripts or skill* — these are the procedural knowledge (the 'how to' knowledge) that enables people to buy and make enjoyable foods. Scripts that incorporate healthy eating principles are likely to enable children (and adults) to eat more healthily. The assimilation of useful cooking or shopping scripts is one aim of nutrition education. It includes the ability to decode food labels and advertising. The school education process should be expected to develop children's procedural knowledge of shopping and cooking in the form of skills.

  Health, food and nutrition literacy are important concepts. They concern the basic knowledge people require in order to remain healthy in complex societies. Nutbeam (2000) distinguishes three basic levels of health literacy: *functional health literacy* and the communication of information; *interactive health literacy* and the development of personal skills; and *critical health literacy* and both personal and community empowerment. The various individual and community benefits of these forms of literacy are shown in Table 3.4.

  There is increasing interest in related issues of food and nutrition literacy as they are essential for active consumers in a market society (www.food literacy.org).

- *Usage situations* — these are the settings in which foods are bought or consumed. Usage situations have social, temporal and physical aspects. For example, the purchase of meat often occurs in a particular social situation — the butcher's shop. Picture a young shopper who is about to cook a meat dish for her family for the first time. She enters the shop in which there is a 50-year-old man behind a counter wearing an apron — the butcher. There are also two women in a queue behind her. The butcher says: 'What would you like?' Down below the glass counter there are several trays of meats — she

TABLE 3.3 A SUMMARY OF THE TOTAL FOOD QUALITY MODEL

BEFORE PURCHASING: The formation of quality expectations

Cost cues: (e.g. price) and perceived cost cues:	e.g. 'cheap' looking label, 'expensive'
Extrinsic and perceived extrinsic quality cues:	e.g. friends' views, for sale in low cost supermarkets
Intrinsic quality cues and technical product specifications:	e.g. nutrient content, 'feel' of fruit
Quality expectations:	about sensory, health, convenience and processing qualities of the product
Expected purchase motivation fulfilment:	e.g. 'it will taste good'

These factors interact in complex ways and affect intention to buy the product.

AFTER PURCHASING: Quality experience

Consumption of the product	
Household production:	e.g. cook the product
Experienced quality:	e.g. what did it taste like? i.e. sensory, health, convenience, processing experiences of the product
Experienced purchase motive fulfilment:	did it taste good?

Again, these factors interact with each other but the outcomes of the consumer's consideration of these factors will influence future purchasing.

Source: Grunert (2005).

TABLE 3.4 LEVELS OF HEALTH LITERACY

Health literacy level	Content	Individual benefit	Community benefit	Educational examples
Functional: communication of information	Communication of factual information about health and health services	Better knowledge of risks and services and better compliance with recommendations	Greater participation in public health programs (e.g. immunisation)	Communicate via existing channels; make interpersonal contacts and use media
Interactive: development of personal skills	+ skill development in a supportive environment	Able to act more independently, better motivation and self-confidence	Able to influence social norms and interact with social groups	Tailor communications to needs of community; support self-help and other groups; combine communication channnels
Critical: personal and community empowerment	+ information about social economic determinants of health	Greater resilience in adversity	Able to interact with social determinants— community empowerment	Provide technical advice; undertake advocacy; engage in community development

Source: Nutbeam (2000).

hasn't a clue what sorts (cuts) of meat they are—all she knows is that she wants 'beef'. Suppose she says: 'I want some beef'. 'What sort and how much?' he says. It is an embarrassing situation, especially now the older women are looking at each other with knowing looks. Is she likely to return to that butcher? Probably not. It is much easier to pick up some meat in the supermarket, try it out at home and, if things go badly, get rid of the remains before anyone notices.

This is an illustration of one social aspect of a buying situation. Smart butchers lay out prepared meat dishes with large labels on them, together with cooking instructions. They often stand outside their shops inviting shoppers who don't know about 'cuts of meat' inside, reassuring them that all they have to do is to say what sort of meat they want. Social, temporal and physical characteristics of the usage or purchasing situation all conspire to affect the pleasantness or discomfort of the episode.

Marketers often focus on the usage situation because it is so central to product acceptance. Nutrition promoters need to share their focus and ask questions such as: 'Where will the food be bought?' 'Where will it be prepared and consumed?' 'What are the social demands of these situations?'

The home provides a major set of usage situations, whether it is preparing and eating breakfast cereal by oneself, raiding the refrigerator for a snack, eating out for lunch with friends, or eating dinner with other family members. Consumption outside the home is also prominent in many people's lives, especially eating lunch or snacks with workmates.

Clearly, meals, snacks, shopping and consumption situations are phenomena which nutrition promoters should know a lot about. The nutrition status of consumers may be influenced by changing the content, size and timing of habitual meals and snacks in purchasing and consumption situations.

Food lifestyles as settings

Shopping, preparation, cooking and consumption tend to be interwoven into particular lifestyles. People tend to follow habitual behaviours which are interlinked with several aspects of their lives. Box 3.2 on Australian food consumers' lifestyles illustrates this well: food shopping behaviours are associated with people's time constraints, their family type and even their mobile phone ownership. Lifestyle is a 'fuzzy' concept, and people spend their time in broad 'clouds' of habitual behaviours. Kickbusch (1986) defines a healthy lifestyle as 'patterns of behaviour choices made from alternatives that are available to people according to their socio-economic circumstances and to the ease with which they are able to choose certain ones over others'.

Sadly, we know little about food usage situations, especially those in the home. We do know Melbourne people from low SES backgrounds are more likely to sit down together for evening meals than those from higher SES backgrounds

BOX 3.2 AUSTRALIAN FOOD CONSUMERS' LIFESTYLES

1. *Aspiring moderns* (18 per cent of the sample). Mainly 18–44-year-olds who 'eagerly embrace everything that is new in food'; they are interested in health and food relationships (e.g. vitamin B, anti-oxidants). They tend to be good cooks who love their kitchens. This segment is likely to grow in size.

2. *Nutrition-aware sophisticates* (20 per cent). Mainly older white-collar consumers (45–54 years), but they include some younger people who enjoy browsing around boutique grocers. They are keen on best-quality fresh foods such as organic foods and free-range eggs. They will pay more for anything that is produced by 'natural' means. They are attracted to 'real' food rather than new fashions or fads; they are serious about nutrition.

3. *Fashion aficionados* (10 per cent). 'Culture vultures' and frequent diners—the true 'foodies' who live for the pleasure of food, both cafe style and convenience oriented. They are willing to pay for quality, variety and novelty. Not all of these consumers reside in high-density trendy suburbs—quite a few are in the outer suburbs of large metropolitan areas. Most are 55 years and over.

4. *Plain conservatives* (14 per cent). They are driven by routine, convenience and economy. They 'eat to live' and go for 'basic foods'. They tend to be suspicious of new technologies such as 'genetic engineering'—'the old ways are the best ways'. Shopping for them is something to get over and done with—it is not a 'peak experience' as it is for other segments.

5. *Convenience lifestylers* (17 per cent). They are too busy to worry about food—'convenience stores were designed for them'. They are into mobile phones and fast-moving lifestyles. They eat fast and simple meals which are often eaten on the run or skipped. They are likely to be outer suburbanites in the 25–44-year age group. There may be big opportunities to serve this rapidly growing segment with small easy-to-prepare/ready-to-eat snacks through the day.

Source: Carey et al. (2003); adapted from Rudder 1999.

(Campbell and Crawford 2001). Generally, we have little idea about what people do around food in the home apart from anecdotal comments that young children often refuse to eat the same meals as adults or that husbands may veto or discourage their wives' food choices. There have been few studies of meal timing, duration, meal nutrient content and social company despite the growing realisation that these factors can affect satiety (Blundell 1999), mood state (Benton 2002) and diabetes risk (Jenkins et al. 1994).

Drivers of food behaviours

These include needs, wants, goals, motivations and values. Several interrelated concepts have been advanced to explain the factors which drive human behaviours.

Needs and wants

These are marketers' key concepts. Needs are relatively few in number—they are basic requirements for humans, ranging from biological needs like hunger and thirst through to psychological needs for security, social recognition, intimacy and cognitive consistency (or predictability in daily life). These needs have to be satisfied if we are to live. If we go without food for a day or two it soon becomes apparent just how motivating these needs are: most of our behaviour and consciousness will become dominated by food! In contrast, wants are a more subtle concept. There can be many, many wants! Marketers define wants as the ways in which needs can be satisfied. For example, if you are feeling thirsty you could satisfy the thirst (the need) by drinking water, or milk, or beer—or Brand X beer, or Coca-Cola, or Bloggs lemonade.

In materialist societies, manufacturers and marketers ensure there are many ways to satisfy basic needs. They may propose that their product satisfies the need better than competing products (e.g. Beverage X is a 'thirst *quencher*', or food product Y 'fills up the kids' better than alternatives. Often, a product's ability to satisfy a basic need like hunger is associated in advertising with its supposed ability to fulfil another basic need. For example, much cola advertising demonstrates how the product satisfies thirst within a highly social and/or sexual setting—the implication being that drinkers of this cola product will not only satiate their thirst but will also be socially accepted and more likely to satisfy their sexual needs! So-called 'functional foods' are similar. They will satisfy the basic hunger need but also promise to meet health (self-preservation) needs. Wants are very apparent in materialist society, and are related to issues such as *quality expectations* (as specified in the total food quality model).

Consideration of people's diverse needs and wants can be useful in nutrition promotion. For example, the managers of a school canteen in Adelaide realised that, while plain water is a good thirst quencher, it was not seen as a 'sophisticated' beverage by teenagers who were keen to demonstrate their adultness to their peers. However, when it was bottled and labelled as Aqua Vitale, it sold well in the school canteen. The canteen managers had realised that teenagers have strong needs for social recognition which they meet through the foods and beverages they consume. Hence the provision of a product name considered (at the time) to be socially sophisticated.

Motivations

One commonly used model of human needs is Maslow's (1971) hierarchy of needs. While there are numerous exceptions, it provides a useful general model which illustrates the importance of basic human drives like hunger (see Figure 3.3). An even simpler model of human needs is Epstein's cognitive experiential self theory (Epstein 1994), which proposes four innate human needs: the need for social recognition, for intimacy, for control over the immediate environment, and for cognitive consistency.

FIGURE 3.3 MASLOW'S HIERARCHY OF HUMAN NEEDS

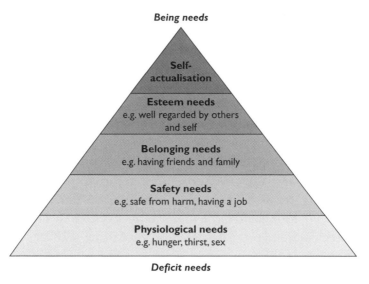

Source: www.ship.edu/~cgboeree/maslow.html.

A useful alternative to this psychodynamic approach is provided by the concept of *life stage* or *lifecycle*. As people progress through life, they tend to occupy various roles and responsibilities which are characteristic of particular age groups (e.g. infants, children, single young adults, cohabitating adults, parents, grandparents/elderly persons). The extent and types of responsibility held at any life stage depend on gender and economic factors—for example, women tend to do the shopping and food preparation at all life stages.

When we look at shopping in this way, nutrition and other food-related qualities can be seen to be mainly instrumental. That is, they help shoppers achieve other goals, many of which are social in nature. Kirk and Gillespie (1990), for example, show that American 'homemakers' interest in nutrition was driven by their desire to be 'good mothers' or to be loved by their family, but in ways which allowed them to hold down full-time jobs.

Steptoe and colleagues (1995) provide a useful summary of food consumers' main motivations (see Box 3.3). These include 'biological' needs (e.g. hunger) as well as 'psychological' motives like 'ethics'. The strength of these motivations differs

BOX 3.3 MOTIVATIONS THAT INFLUENCE DAILY FOOD CHOICE

It is important to me that the food I eat on a typical day:

Health
Contains a lot of vitamins and minerals
Keeps me healthy
Is nutritious
Is high in protein
Is good for my skin/teeth/hair/nails etc.
Is high in fibre and roughage

Mood
Helps me cope with stress
Helps me relax
Keeps me awake/alert
Cheers me up
Makes me feel good

Convenience
Is easy to prepare
Can be cooked very simply
Takes no time to prepare
Can be bought in shops close to where
 I live or work
Is easily available in shops and
 supermarkets

Sensory appeal
Smells nice
Looks nice
Has a pleasant texture
Tastes good

Natural content
Contains no additives
Contains natural ingredients
Contains no artificial ingredients

Price
Is not expensive
Is cheap
Is good value for money

Weight control
Is low in calories
Helps me control my weight
Is low in fat

Familiarity
Is what I usually eat
Is familiar
Is like the food I ate when I was a child

Ethical concern
Comes from countries I approve of politically
Has the country of origin clearly marked
Is packaged in an environmentally friendly
 way

Source: Steptoe et al. 1995.

between people—for example, vegetarians are more likely than others to choose food according to ethical principles, and parents' food selection may be more motivated by considerations of safety (Jussame and Judson 1992). It is also likely that different motivations will come into play in different usage situations—for example, safety concerns are highly apparent during food purchasing, especially of fresh foods. For example, in a study conducted by the author some years ago, almost half of shoppers purchasing fresh oranges were thinking about the possible effects of pesticides which they supposed were on the oranges.

Food and health concerns

As we saw in Table 3.2, shoppers have a wide variety of health concerns about food. Many of them centre on 'me and my family', such as safety concerns and fair trading, but quite a few are about 'others'—other people, animals and the environment (Worsley and Scott 2000; Worsley and Skrzypiec 1998b). Women tend to have more concerns than men, parents more than non-parents, older people more than younger people, and people from lower socioeconomic status backgrounds tend to be more concerned than those from more prosperous circumstances. 'Material and vested interests' are major influences over people's food motivations. For example, if you are poor, you are more likely to be aware of the wide variety of food and health issues than if you are better off. Similarly, if you are a parent, you are more likely to be exposed to children's food and health problems than if you are not.

Identification of the client groups' food motivations, and especially of their food and health concerns, is important for nutrition promotion because it allows the promotion to focus on the key issues facing the group, and thus the development of strategies to deal with them. This 'food issues diagnosis' can be performed quite simply during the needs assessment phase of the nutrition promotion program (see Chapter 6). Findings from food behaviour research can provide useful checklists of likely issues.

Goals

People use foods to achieve many personal and social goals. For example, they may give someone a box of chocolates to express their love or gratitude, or they may prepare vegetables for their baby to ensure that he or she is healthy. That is, many food behaviours are performed to attain an implicit or explicit goal. Sometimes these are well considered behaviours (for example, weighing up all the pros and cons when deciding which restaurant meal to order), but more often than not they are quite unconscious and habitual—like doing the shopping at the local supermarket and buying the usual breakfast cereals, packaged bread, milk and so on.

A good example of a nutrition promotion program which considers consumers' goals is the FoodCents program. This program identified that one of the goals of many financially disadvantaged single mothers was to save money (Foley and Pollard 1998). The program was able to meet this need by demonstrating that healthy foods could be cheaper and tastier than less healthy alternatives. Most of the time, most consumers do not consciously seek health as a goal (unlike nutrition promoters). It is the task of the nutrition promoter to find ways in which healthy foods can be used to attain consumers' non-health goals. This is a strange demand for many health and nutrition professionals but one which is natural for marketers, who always consider the needs and wants of consumers and the ways in which their products might be used to satisfy them.

Nutrition promoters have to ask 'What's in it for me?' from the point of view of the food consumer. For example, a shopper who cares for a young baby wants that infant to grow into a healthy child as easily as possible. Aspects of nutrition to do with baby feeding, weaning, growth and survival are likely to be uppermost in that parent's mind—additives, vitamins, warmth and comfort are likely to be far more salient than cholesterol, fat and heart disease. In addition, female shoppers may also hold responsibility for the feeding of their male partners, who have quite different health, psychological and social needs from themselves, and so they may have to consider them in addition to their own. This approach, then, suggests that nutrition promotion programs should be organised with the needs of the following categories of consumers in mind:

- babies and young infants;
- school-aged children;
- teenagers;
- male spouses/partners,

and in addition:

- single young adults;
- childless couples under 45; and
- retired couples and singles.

Cultural and personal values

There is one final set of behavioural 'drivers' relevant to nutrition promotion. These are cultural and personal values. Children soon acquire them, usually from their parents, during socialisation. At the personal level, values are defined as the 'guiding principles' in one's life (Schwartz 1992). They are very much about what we think is right, appropriate or what 'ought' to be (Feather 1982). Values tell us what is 'good'. For example, most of us think that 'being honest' is a good thing, or perhaps we may believe that achievement and independence are very important states to strive for, or alternatively that harmony and being kind to others are the most important things in life.

Schwartz (1992) has proposed a set of ten key personal values held by people in most large societies. These may be *self-related*, like independence, stimulation and achievement, or *small group-oriented*, like social power, or *community-oriented*, like the values of harmony, universalism, conformity and tradition. All societies foster these values, though to differing extents according to their economic and political histories. It is now becoming clear that various patterns of food use, and food and health concerns, are associated with particular personal values—for example, people who consume vegetarian diets tend to express more egalitarian, harmonious values than meat eaters.

Values influence our food behaviours. Health-conscious people, for example, often place a high premium on security (personal and family safety), whereas many teenagers tend to value stimulation and excitement far more highly, so may not care much about warnings about the dangers of over-consuming fast foods and soft drinks because they are more interested in the sensory and image (sociability) qualities of these foods and drinks. Advertisers latch on to these 'value segments', offering products which 'suit' them (e.g. products with green advantages for those who hold communitarian values, prestige products for those who are more interested in social power, etc.).

Personal values are important because they are the evaluative criteria which enable us to decide whether we like people, objects or performing various behaviours. They are like the 'software of the mind' (Hofstede 2005), enabling us to categorise people and objects (including) foods as 'good' or 'bad' and so worth pursuing or not. The problem, as Schwartz's taxonomy demonstrates, is that there are several possible 'goods' such as social power, achievement, conformity and benevolence, and it may not be apparent which values people hold in any particular situation. This is important for nutrition promotion because groups of people differ in the values they hold, and also in the goals they pursue through food. So if a group of people holds benevolent or universalistic values its members will be concerned about others, including animals and the environment, and so will be more likely to choose foods consistent with this value such as vegetables, organic foods, free-range eggs and so on. An example of a personal values questionnaire is given in Box 3.4.

Bisogni and her colleagues in the United States have drawn attention to a common quandary most of us face in our personal lives (Connors et al. 2001). This is the conflict we may experience between our own values—for example, we may value health but want stimulation and convenience at the same time. People find many ways to resolve such conflicts—perhaps by consuming 'healthy' foods Monday to Friday but splurging on 'tastier' fast foods at weekends. Hopefully nutrition promoters can help resolve such conflicts by helping people find foods that are both healthy *and* convenient *and* tasty.

BOX 3.4 EXAMPLES OF ITEMS FROM A PERSONAL VALUES INVENTORY

What is important in your life?

Listed below are some guiding principles or personal values which many people use in their lives. Please **circle one word** next to each value according to how important it is to you.

	How important?			
Equality (equal opportunity for all)	Not	A little	Quite impt	Very
Inner harmony (peace with myself)	Not	A little	Quite impt	Very
Social power (control over others)	Not	A little	Quite impt	Very
Pleasure (gratification of desires)	Not	A little	Quite impt	Very
Social order (stability of society)	Not	A little	Quite impt	Very
An exciting life (stimulating experiences)	Not	A little	Quite impt	Very
Wealth (material possessions, money)	Not	A little	Quite impt	Very
National security (protection of my nation from enemies)	Not	A little	Quite impt	Very
Self-discipline (self-restraint)	Not	A little	Quite impt	Very

Source: Based on Schwartz (1992).

At the cultural level, values can be seen as 'key themes' in society, around which debate, politics and action rotate. Hofstede (2005) has shown that several value dimensions distinguish countries (and sub-groups within them); these include *egalitarianism, tolerance of difference* (e.g. novelty, strangers, eccentrics), *individualism, masculinity–femininity, short or long-term future orientation* and *power distance* (general acceptance of greater power among the elites). From the point of view of today's food supply, the dominance of materialism (the belief that material things can make you happy) is a key cultural value since it tends to separate consumers who use food to enhance the self (e.g. through the use of slimming foods and socially desirable foods) from those who are more interested in food being used to protect the environment, or to care for animals or underfed children.

Understanding people's values enables nutrition promoters to address issues that are relevant to them and so initiate and maintain communication with them.

In summary, food behaviours are likely to be influenced by the left to right progression of this model:

Values > Motivations > Goals > Behaviours.

Social ideologies and discourses

The values that psychologists measure are the tips of 'attitude-belief-interest-practice icebergs', but there is a lot of stuff 'below' them. People have many beliefs and attitudes about food which are associated with their food behaviours. Their values, needs, wants and motivations are also associated with food behaviours. While there is clear evidence that all these social and psychological factors can influence food choices, we have little knowledge about the complex ways in which they interact, and from a nutrition promotion perspective there may be little to gain from too detailed an analysis of the roles played by the conceptualised variables.

A broader perspective suggests that much of our behaviour is in the form of social discourses. Groups of beliefs and attitudes, values and motives are present along with social relationships and group processes, but all of them are centred on a theme (the discourse) which in various ways is preferred by the group of people in question. For example, vegetarians may be characterised by core beliefs about the sanctity of all animal life (Twigg 1979; Adams 1990), by values of universalism and harmony (Worsley and Skrzypiec 1997), by numerous social practices (not eating meat being only one), and by many beliefs and attitudes which have little to do with meat or vegetables and reach into areas like voting intentions, clothing preferences and egalitarian social relationships (Sims 1978).

We can think of such a social discourse and its associated ideology as being like a rain cloud. It looks soft and woolly and spreads in all directions, but it is very real and can soak a lot of things! There are many social discourses in society about key issues in people's lives such as: equity and gender egalitarianism, nature and the environment, the cult of appearance and the tyranny of slenderness (over a quarter of women are on a slimming diet on any given day, according to Jeffery et al. 1991), purity and cleanliness, and materialism. All of them may relate to people's food choices. For the sake of brevity, only two are described here.

Equity and gender egalitarianism

In many societies like Australia, North America and Western Europe, egalitarianism is a major preoccupation (Hofstede 2005). Many people refute the notion that some people (or animals) are inherently more worthy or powerful than others. Instead, they view the world as having a flat structure: women are equal to men, children to their parents, workers to their managers, animals to humans. This non-hierarchical world-view goes way back in Western culture (and in many non-Western cultures, especially tribal societies). For example, the Levellers in the English Civil War demanded the vote not only for working men but also for women, in part a throwback to the relatively equal status of women in Anglo-Saxon England (Fell 1984).

People who hold this view of the world tend to include vegetarians and people who emphasise the inherent harmony of the natural world and humankind. They are likely to express their beliefs through their food choices as well as in other ways (Sims 1978; Worsley and Skrzypiec 1997; Allen and Baines 2002). They tend to be interested in nature, and many of their food choices will be based on moral decisions since eating some foods (e.g. 'organic' foods) and refusing others (e.g. meats) will be inconsistent with this ideology. About 20 per cent of the general population hold these views (Lea and Worsley 2003).

Recognition of this social discourse provides nutrition promoters and communicators with opportunities to transform basic nutrition strategies into socially desirable food consumption practices. People do not eat in isolation, they eat as part of lifestyles, and if a lifestyle is based on equity then promoted food choices must be consistent with the underlying discourse. For example, if we want to emphasise the need for iron, for this group it would be wise to emphasise ways to maintain iron status using plant foods only or mainly plant foods, or to explain how and why appropriate sources of meat might be used. An easier example might be the use of eggs as food. Within this equity discourse, the source of the eggs will be significant — can the nutrition promoter show these consumers where to access eggs that are produced by humanely treated chickens (i.e. free-range eggs)? It is not enough to propose that eggs are nutritionally useful; nutrition promoters have to ensure that the eggs supplied are acceptable to the group's belief system.

Gender equity is another manifestation of the equity discourse. In a study the author conducted in South Australia in the mid-1990s among sixteen-year-olds, strong gender equity beliefs were found in both sexes (not surprisingly more so among females). These beliefs were associated with greater intakes of fruits, vegetables and cereal grains, and lower intakes of meat products.

Materialism

Perhaps the most pervasive social discourse is materialism, defined by Belk (1983) as the pursuit of happiness through the acquisition of material objects. This discourse is particularly relevant to the pursuit of health since the self-focus of materialism tends to make people seek health benefits predominantly in terms of benefits to the self. It has long been recognised that in much of Western society it is assumed that material objects like pills and potions and technological procedures are the panacea for many prevailing illnesses. The role of social relationships in promoting health is only grudgingly accepted by many, although the work of social epidemiologists strongly suggests that people's social relationships and position in the social hierarchy have a strong influence on their health status (Marmot and Wilkinson 1999).

Convenience, value for money, fashion and efficiency have probably always been valued, but they seem to be particularly esteemed today and they certainly play important roles in the marketing of food products. People who have experienced material insecurity in childhood tend to value material objects, including food, more as adults; they appear more likely than others to eat when they are anxious (Allen and Wilson 2005). Nutrition promoters can respond to materialist world views, fostering the social and convivial aspects of food shopping, preparation and consumption.

The risk society

Closely associated with several of these discourses is a pervasive sense of uncertainty and anxiety. We live in the 'risk society' (Beck 1992); people have little trust in the food system, they are scared of chemical contamination, they are scared to handle food during cooking because they might pick up infections, they worry about their health, and they yearn to be 'reconnected' with natural systems. There are good opportunities for nutrition promoters to help allay these risk perceptions through the promotion of healthy foods and sustainable practices, helping people to understand and control their role in the food system.

Conclusions

Food behaviours are complex but understandable. Such understanding forms one of the bases of nutrition promotion. Nutrition promoters have to put themselves into the minds of food consumers so that they can appreciate the social and personal demands consumers face in their daily lives. Consumers' views and experiences are sovereign, and promotion programs must take them into account while reflecting the values of public health.

Discussion questions

3.1 Discuss the various definitions of food behaviours.

3.2 Discuss how the concrete attributes of foods/beverages can influence their consumption.

3.3 Discuss current trends in food consumption/purchasing in relation to prevailing social ideologies.

3.4 'Nutrition knowledge has little influence on food consumption.' Organise and cite evidence which refutes this common view.

3.5 'The study of food behaviours is merely an extension of "individualism". It has no place in nutrition promotion.' Discuss.

3.6 Describe the utility of Grunert's food-related lifestyle model for the understanding of consumers' food behaviours. What are its strengths and limitations?

3.7 How do shoppers purchase commonly available food products? Choose a food category and apply what you think is the most appropriate model or theory to the regular purchasing of foods in this category.

3.8 How would you promote healthy eating to each of Rudder's (1999) Australian food lifestyle groups?

4

Nutrition problems and solutions

Introduction

Nutrition promotion is a two-way street. On the one hand, members of the community have their own sets of beliefs and priorities about food and health which have to be taken into account in any nutrition promotion program. On the other hand, nutrition promoters have access to advanced nutrition and health knowledge which is derived from scientific research. The task facing promoters is to reconcile their expert knowledge with the community's interests and lifestyles so that programs are relevant while still being accurate. This is often difficult, since the food and nutrition needs of particular social groups and communities vary widely. Many areas of science appear to change rapidly, yielding divergent findings and claims. This is compounded by the rapid publication of diverse scientific findings in the mass media. The result is that many in the community are confused and demotivated by nutrition communication. Despite these popular opinions, however, the core of nutrition information is stable and quite clear.

Nutrition promoters deal with a wide variety of problems and issues, ranging from food production to consumer food purchasing. To deliver successful programs, nutrition promoters need to be confident that the scientific premises of their programs are valid and likely to remain so over fairly long time periods. Therefore in this chapter we will examine the main food and health problems facing nutrition promotion, and the food and nutrition responses currently being applied to these problems.

Food and nutrition problems

There are many food and nutrition problems that nutrition promoters have to address. A full description of them is beyond the scope of this book so only the better

97

recognised problems are outlined here. Particular communities have specific problems or combinations of problems which will require action research within these
communities in order to identify the most appropriate solutions to their food and
nutrition problems.

The prevention and amelioration of disease

At first sight, this seems to be a straightforward matter: nutrition promotion can
help to prevent major diseases, both non-communicable (like heart disease, various
cancers and type 2 diabetes) and infectious diseases such as HIV-AIDS. However,
as nutrition science advances, it is becoming clear that nutrition promotion might
play important roles in the prevention or amelioration of a variety of other conditions such as dementia, depression and sarcopenia (age-related muscle wasting).
The increasing incidence of food allergies and sensitivities among children is one
example of an under-recognised or 'new' condition. It also illustrates that nutrition
promotion is about food and *all* its constituents and their effects, not only the effects
of commonly recognised nutrients (like saturated fats).

A common challenge to nutrition promotion comes from technological
approaches such as drug therapies devised by the pharmaceutical industry. The argument runs something like this: *There is no need for individuals to do all that hard work to
change their ways, we have a very advanced drug which will simply prevent disease X. Just take a
couple of pills a day.* The fact is that there are drugs around which do just that (e.g.
statins' role in the prevention of heart disease). Unfortunately, such drugs are not
cheap, often have unpleasant side-effects and, unlike the nutrition and health promotion approach, they are truly specific—usually ameliorating or preventing only a
single disease. In contrast, appropriate small changes in dietary practices—such as
consuming a small amount of fish twice a week, or eating a couple more serves of fruit
or vegetables each day—can be more efficacious (and much cheaper) in preventing
the same disease *and* they also tend to prevent a *range* of diseases (not just one).

This example illustrates that there are agendas which can impede or oppose that
of nutrition promotion. There are vested interests in society which may oppose
public health nutrition strategies. These include some manufacturers of high-energy
snack foods and beverages (for example). They see nutrition promoters' attempts
to limit the consumption of their current products as a threat to their companies'
financial prosperity. This is a reality of life; not everyone agrees with one position.
We have to advocate for public health nutrition and find enterprising, innovative
business and community solutions. Fortunately, most manufacturers of high-energy
foods and beverages are increasingly willing to make lower energy-dense products.

It is important that every nutrition promoter should be familiar with the extent and severity of the nutrition problems which afflict humankind. There are several convenient, authoritative sources of this information. The annual World Health Reports of the World Health Organization are excellent overviews, as are reports from the World Cancer Research Fund, the International Obesity Taskforce and various national Heart and Cancer Foundations. (*Australia's Health*, published by the Australian Institute of Health and Welfare, provides an excellent overview of Australians' disease risks.)

During the twenty-first century, the world population faces a major epidemic of non-communicable metabolic diseases, including cardiovascular disease, type 2 diabetes, obesity and various cancers. Heart disease is the major cause of premature deaths and disability in most countries, far exceeding the burden of serious diseases such as HIV/AIDS. The rapid increase in the prevalence of these diseases during the past 20 years suggests that environmental factors are the precipitating influences. The situation is likely to become much worse in the next 20 years as Figure 4.1 suggests for type 2 diabetes. The present burden in Victoria, Australia, which is fairly typical of affluent societies, is shown in Table 4.1.

FIGURE 4.1 GLOBAL PROJECTION FOR THE DIABETES EPIDEMIC 2003–25

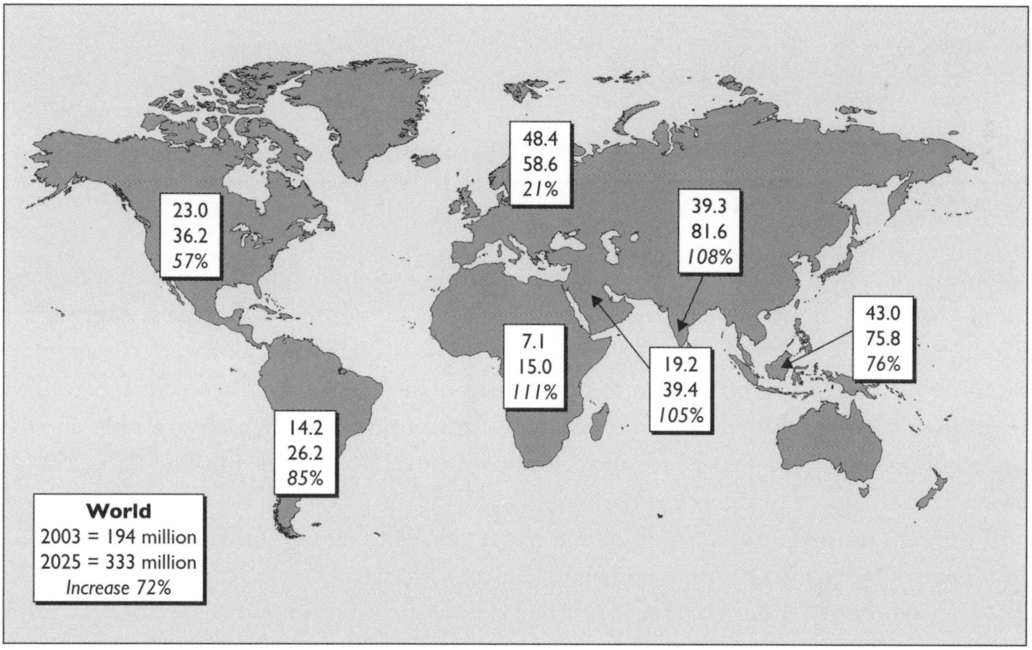

Source: International Association for the Study of Obesity, 2003.

TABLE 4.1 THE BURDEN OF DISEASE EXPRESSED AS DISABILITY
ADJUSTED LIFE YEARS (DALYs): LEADING CAUSES OF
DALYs IN PEOPLE 35–64 YEARS BY SEX, VICTORIA,
AUSTRALIA, 2001

Males	DALYs total	% of total DALYs	Females	DALYs total	% of total DALYs
1 Ischaemic heart disease	11 962	10.0	1 Breast cancer	10 163	DALYs
2 Diabetes	9 610	8.0	2 Depression	10 011	9.9
3 Depression	7 887	6.6	3 Diabetes	6 736	6.6
4 Lung cancer	5 566	4.7	4 Stroke	4 946	4.9
5 Hearing loss	5 273	4.4	5 Generalised anxiety disorder	4 605	4.5
6 Stroke	5 191	4.3	6 Lung cancer	3 324	3.3
7 Suicide	4 197	3.5	7 Ischaemic heart disease	3 145	3.1
8 Bowel cancer	4 159	3.5	8 Bowel cancer	3 108	3.1
9 Chronic obstructive pulmonary disease	3 678	3.1	9 Hearing loss	3 009	3.0
10 Prostate cancer	2 738	2.3	10 Chronic obstructive pulmonary disease	2 633	2.6

Source: Department of Human Services, Victoria.

Under-nutrition and starvation, especially among women and children, are commonly found in many 'developing' countries due to the interruption of meagre food production and/or distribution by war, civil strife or harvest failure. For example, flooding during the rainy season may make roads impassable. These circumstances may be short-lived but they often occur within populations which are chronically under-nourished. The World Health Organization, the World Bank and UNESCO, among others, have highlighted the extent of under-nutrition and various alleviation and prevention programs. Some of the main causes of under-nutrition are outlined in Figure 4.2 (see also Stephenson et al. 2000). Of particular concern is the plight of children who are orphaned as a result of HIV, other infectious diseases or war who are left to fend to themselves. There are estimated to be 11 million of them in Sub-Saharan Africa alone (Grantham-McGregor 2005).

FIGURE 4.2 THE CAUSES OF MALNUTRITION

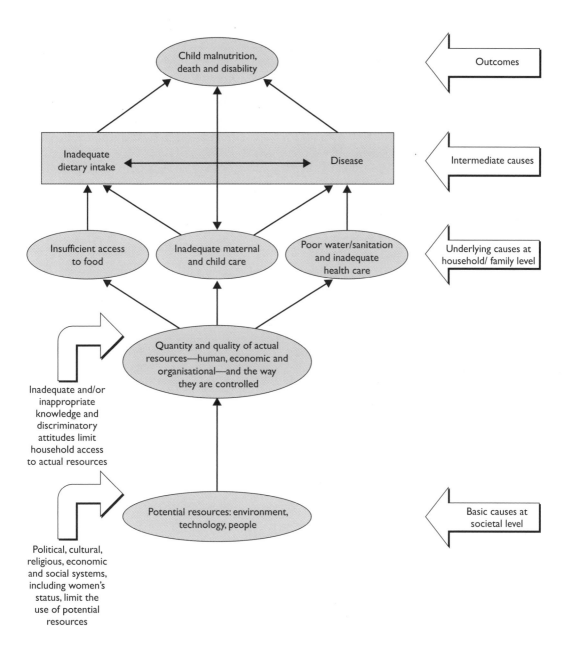

Source: UNICEF (1997).

TABLE 4.2 SELECTED MAJOR THREATS TO CHILDHOOD HEALTH, MATERNAL UNDER-NUTRITION AND DIET-RELATED FACTORS AND INACTIVITY

Risk factor	Theoretical minimum exposure	Measured adverse outcomes of exposure
Underweight	Same percentage of children under 5 years of age with <1 standard deviation weight-for-age as the international reference group, all women of childbearing age with BMI >20 kg/m^2	Mortality and acute morbidity from diarrhoea, malaria, measles, pneumonia, selected other Group 1 (infectious) diseases Perinatal conditions from maternal underweight.
Iron deficiency	Haemoglobin distributions which halve anaemia prevalence, estimated to occur if all iron deficiency were eliminated (g/dl)	Anaemia, maternal and perinatal causes of death
Vitamin A deficiency	Children and women of childbearing age consuming sufficient vitamin A to meet physiological needs	Diarrhoea, malaria, maternal mortality, vitamin A deficiency disease
Zinc deficiency	The entire population consuming sufficient dietary zinc to meet physiological needs, taking into account routine and illness-related losses and bioavailability	Diarrhoea, pneumonia, malaria
Blood pressure	115; SD 11 mmHg	Stroke, ischaemic heart disease, hypertensive disease, other cardiac disease
Cholesterol	3.8; SD 1 mmol/l (147 SD 39 mg/dl)	Stroke, ischaemic heart disease
Overweight	21; SD 1 kg/m^2	Stroke, ischaemic heart disease, diabetes, osteoarthritis, endometrial cancer, post-menopausal breast cancer
Low fruit and vegetable intake	600; SD 50 g intake per day for adults	Stroke, ischaemic heart disease, colorectal cancer, gastric cancer, lung cancer, oesophageal cancer
Physical inactivity	All taking at least 2.5 hours per week of moderate exercise or 1 hour per week of vigorous exercise	Stroke, ischaemic heart disease, breast cancer, colon cancer, diabetes

Source: World Health Organization (2002, p. 52).

Under-nutrition is common throughout the world. Protein energy malnutrition is less common than in earlier times (except under the conditions noted above) but micronutrient deficiencies are rife in developing countries, especially iodine, iron and vitamin A deficiencies. Well over half the women in poorer social strata in India, Indonesia, Southeast Asia, Africa, Latin America and elsewhere suffer from iron deficiency anaemia. Over a billion people are at risk of iodine deficiency and its consequences, and hundreds of millions suffer from vitamin A and/or iron deficiency. Other deficiencies often occur along with the 'Big 3' such as deficiencies of zinc, selenium and a variety of vitamin deficiencies. The main 'cause' of these deficiencies is the lack of a sufficient varied food supply.

Nutrient deficiencies, such as calcium, iodine, iron, zinc, vitamin A and other deficiencies, also occur within affluent countries. Here, the problem may be due to poor food distribution and production (as in Aboriginal communities in remote parts of Australia), or to demand-side problems such as financial inability to purchase available foods, or 'lifestyle' choices which result in the selection of high-energy, nutrient-poor foods. For example, people who avoid meat and do not follow sound vegetarian dietary principles may experience iron and associated deficiencies. Essentially, under-nutrition results from either lack of sufficient quantities of food and/or imbalances in food intake associated with poor food variety. Figure 4.3 clearly shows that under-nutrition is among the top 20 risk factors associated with the global burden of disease.

Many government and non-government agencies are working to ameliorate these problems, especially with regard to children's nutrition. Examples include Micronutrient Initiative Canada (see Box 4.1), the Indian government's provision of one meal per day to 100 million school children (Gopaldas 2005) and the World Bank's financial support for farmers in Mozambique to grow more food for their communities (Bundy 2005). A useful set of papers describing the issues facing schoolchildren around the world is provided by Galal et al. (2005).

BOX 4.1 GOALS OF THE MICRONUTRIENT INITIATIVE CANADA

- The virtual elimination of iodine deficiency disorders (IDD).
- The virtual elimination of vitamin A deficiency.
- The reduction of iron deficiency anemia in women by one-third.

Source: www.micronutrient.org.

FIGURE 4.3 PUTTING NUTRITION INTO PERSPECTIVE: THE GLOBAL DISTRIBUTION OF BURDEN OF DISEASE ATTRIBUTABLE TO 20 LEADING SELECTED RISK FACTORS

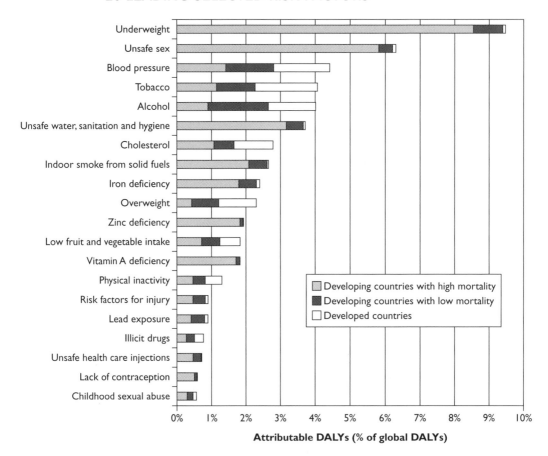

Source: World Health Organization (2002, p. 82).

Food insecurity and access issues are also related to under-nutrition (and often to 'over-nutrition'). Food insecurity refers to those circumstances in which the food supply may be interrupted (see Box 4.2). It is most apparent in developing countries. Starvation and nutrient deficiencies in poor countries result from problems in the food system which make the food supply insecure. However, food insecurity also occurs in affluent countries. For example, families may periodically run out of food when their pension payment runs out or, in the case of many refugees, they may not be able to find foods from their own culture and may rely mainly on 'fast foods'. The positioning of large supermarkets on the edges of cities which require car transport can cut poorer people off from less expensive food. So, in countries like the United Kingdom, Australia and the United States, the term 'food insecurity' covers a wealth of circumstances which result in people 'going without' sufficient food and food variety. Philip McMichael (2007) has shown how the World Trade Organization's rules have enabled large supermarkets to take over much of the world's food production during the past 20 years. While providing cheap foods for many people in developed economies, they have also contributed to the displacement of millions of farming families from their lands and livelihoods. In turn, this has led to the rise of the 'food sovereignty' and *Via Campesina* movements which encourage local food production and ownership (http://viacampesina.org/main_en/index.php).

Food sovereignty may seem a long way from nutrition promotion, but it is relevant because major health and nutritional deficits result from the displacement of agricultural workers and because nutrition promotion aims to promote health via the food and health systems.

BOX 4.2 SUMMARY OF FORMS OF FOOD INSECURITY

Definition of food insecurity

Food insecurity is 'limited or uncertain availability of nutritionally adequate and safe foods or limited or uncertain ability to acquire acceptable foods in socially acceptable ways' (Olson and Holben 2002).

Hunger is defined as: 'The uneasy or painful sensation caused by a lack of food. The recurrent and involuntary lack of access to food. Hunger over time may produce malnutrition.' (Hamilton et al. 1997)

There is a continuum of experience involving food insecurity with or without hunger according to the severity and consequences of the food scarcity situation. Food insecurity consists mainly of anxiety about having enough food to eat or running out of food and having no money to purchase more (Klein 1996). Adults who believe they are food insecure may try to avoid hunger by reducing the size of meals, skipping meals, or even going without food for one or more days. However, when food is extremely limited, the means to avoid hunger are ineffective and cause severe personal hunger that spreads to the family and children (Klein 1996).

Prevalence of food insecurity

Food insecurity can be measured by various survey tools that range in complexity from a single question to multi-item questionnaires. The standard single-item question to determine

---

Box 4.2 continued

individual and household food security is: 'In the past twelve months, have you or has anyone in your household run out of food and not had enough money to purchase more?' (Booth and Smith 2001)

In the 1995 Australian National Nutrition Survey, 5 per cent of the respondents were screened as food insecure. However, the risk was much higher in specific groups: the un-employed (23 per cent), single-parent households (23 per cent), second lowest income quintile (20 per cent), rental households (20 per cent), young people (15 per cent). In the 16–18 years bracket, 8 per cent of males and 5 per cent of females were food insecure, while in the 19–24 years bracket the prevalence of food insecurity was 9 per cent males and 12 per cent females (Moon et al. 1999).

Source: Burns (2004).

There is some debate about the distribution of food insecurity in prosperous countries. Certainly, many people from low socioeconomic backgrounds are food insecure. However, food insecurity can be found in all socioeconomic strata (ABS 2001). Over a decade ago, Crotty and her colleagues (1992) showed that poverty and food insecurity transcended socioeconomic strata: people without jobs ran out of food, but so did employed people with large mortgages. When incomes decrease or the cost of living rises, people tend to cut their food expenditure in order to pay for their rent, mortgage, transport and telephone.

Over-nutrition and excess energy intakes

Many recent nutrition promotion efforts have focused on the problem of over-nutrition associated with high fat intakes (especially of saturated fats), high sugar and salt intakes, and low fibre and micronutrient intakes. The reason for this interest is that 'over-nutrition' appears to play causal roles in the development of degenerative diseases such as coronary heart disease, stroke, diabetes (type 2) and various cancers which rose to become major sources of premature mortality in rich countries in the mid-twentieth century. These are now the leading causes of premature death in most countries of the world.

When degenerative diseases first became prevalent, researchers were particularly focused on heart disease and on the roles of saturated fatty acids in its causation. This led them to examine its dietary sources, which were mainly animal fats derived from dairy foods and meats. This led to strong attacks on these food groups. People were encouraged to minimise their consumption of them, despite their major nutritional benefits (e.g. dairy products are major sources of calcium among other nutrients; cheese is the key source of vitamin K; and red meat is the principal source of iron in many people's diets). The present supermarket era of fat-reduced produce had arrived!

More recent research suggests that the negative roles of 'fats' have been over-emphasised. It is apparent that excessive energy intakes associated with high consumption of 'fast foods' containing large amounts of saturated fats, sugars and little else, together with low intakes of fruits, vegetables, legumes and cereal grains are partly responsible for degenerative disease. Today, the emphasis in nutrition pro-motion is on the promotion of whole foods like fruit, vegetables and whole grains, and reductions in the consumption of products containing high concentrations of energy, saturated fats and salt.

Obesity is a risk factor for several degenerative diseases and mortality (Flegal et al. 2005), particularly type 2 diabetes, and is itself a major source of morbidity. For example, among American women aged between 30 and 55 years, their relative risk (compared to 18-year-olds) of type 2 diabetes rises sixfold as their BMI rises to 24, and their relative risk of coronary heart disease and hypertension rises twofold as their BMI approaches 30 (Willet et al. 1999). The recent rapid rise in obesity prevalence, especially among children (Figure 4.4) has created strong interest among nutrition promoters who are building cooperative links with physical edu-cators, educators, medical practitioners, town planners, community development workers and others.

The increase in the prevalence of obesity during the past 20 years strongly sug-gests that environmental forces are at work, hence the interest in ways of reducing energy intakes and increasing energy expenditures based on ecological models of obesity (Egger and Swinburn 1997). The ecological model of obesity causation iden-tifies environments which are either *obesogenic* or anti-obesogenic (*leptogenic*). These environments can be to reduce prevalence of obesity. Much of the emphasis is on the supply side, such as constraining the marketing of high-energy food products and the promotion of lower energy, more nutritious alternatives like fruits and veg-etables. The ANGELO framework provides a checklist of environments which may influence obesity prevalence (Table 4.3a and 4.3b). These can be further broken down into micro-environmental and macro-environmental settings (Table 4.4). Glass and McAtee's (2006) society–behaviour–biological model is also a useful guide to the likely precursors of obesity (see Chapter 6).

FIGURE 4.4 THE RISE IN EXCESS BODY WEIGHT IN EUROPEAN CHILDREN

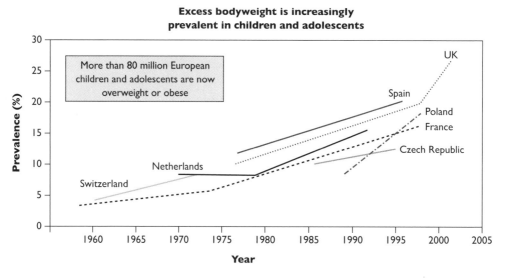

Source: International Obesity Taskforce (2002).

The International Obesity Taskforce, among other organisations, implicates the 'hostile food environment' with its provision of cheap high-energy foods and associated over-consumption. Philip McMichael (2007) concurs with this diagnosis but notes that the modern food system acts in paradoxical ways, producing different forms of malnutrition in poor and rich countries—under-nutrition in the poorer South and over-nutrition in the rich North.

TABLE 4.3a THE ANGELO FRAMEWORK

Environment type \ Environment size	Micro-environment (settings)		Macro-environment (sectors)	
	Food	PA	Food	PA
Physical	What is or isn't available?			
Economic	What are the financial factors?			
Policy	What are the rules?			
Socio-cultural	What are the attitudes, beliefs, perceptions and values?			

Note PA = physical activity

TABLE 4.3b SCHEMATIC USE OF THE ANGELO FRAMEWORK

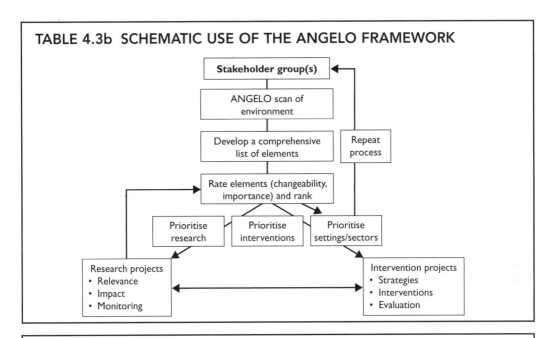

TABLE 4.4 EXAMPLES OF MICRO-ENVIRONMENTAL AND MACRO-ENVIRONMENTAL SETTINGS

Micro-environmental settings	Macro-environmental settings
Homes	Technology/design (e.g. labour-saving devices, architecture)
Workplaces	
Schools	Media (e.g. women's magazines)
Universities/tertiary institutions	Food production/importing
Community groups (e.g. clubs, churches)	Food manufacturing
Community places (e.g. parks, shopping malls)	Food marketing (e.g. fast food advertising)
Institutions (e.g. hospitals, boarding schools)	Food distribution (e.g. wholesalers)
Food retailers (e.g. supermarkets)	Food catering services
Food service outlets (e.g. lunch bars, restaurants)	Sports/leisure industry (e.g. instructor training programs)
Recreation facilities (e.g. pools, gyms)	Urban/rural development (e.g. town planning, local councils)
Neighbourhoods (e.g. cycle paths, street safety)	Transport system (e.g. public transportation systems)
Transport service centres (e.g. airports, bus stations)	
Local health care (e.g. GP, hospital)	Health system (e.g. Ministry of Health, medical schools, professional associations)

Source: Swinburn et al. (1999).

Low consumption of plant foods Closely associated with excessive energy intakes and metabolic disease prevalence are low levels of consumption of vegetables, fruit and whole grains. Most affluent populations do not consume enough plant foods. For example, in Victoria in 2006 only 47% of adults consumed the recommended two servings of fruit per day and less the 10% consumed five serves of vegetables per day (Victorian Population Health Survey 2006).

Higher intakes of fruit, vegetables and other plant foods would have at least two benefits for population health. First the risk of obesity and metabolic diseases such as bowel cancer (World Cancer Research Fund 2007) would be substantially reduced. Second, since most plant foods are locally produced and tend to produce less greenhouse gases than foods of animal origin, the long-term environmental sustainability of the food system would be increased.

The nutrition transition

Obesity is not confined to Western countries: its prevalence seems to be increasing in all affluent market societies—for example, the prevalence of obesity in children is almost as high in urban China and India as it is in the United States. Almost all countries have been undergoing economic change; some Western countries have been through a process of industrialisation (and more recently a post-industrial phase) for almost two centuries. Others, like those in East and South Asia, have undergone rapid economic and industrial change for less than 30 years. These economic changes are associated with social and epidemiological changes such as the rise of degenerative diseases. Changes in food consumption and other food behaviours, along with increasingly sedentary lifestyles and substance abuse (e.g. smoking), have been partly responsible. It is generally assumed that economic changes lead to changes in lifestyles and food selection, which in turn lead to changes in disease prevalence. This is known as the *Nutrition transition*. At some stages of the transition, it is possible for subpopulations (and individuals) to suffer from profound nutrient deficiencies while at the same time suffering from energy excess in the form of obesity. The transition presents nutrition promotion with major challenges: how do societies in the midst of rapid change ensure that their new dietary patterns are healthy? Frameworks like the ANGELO model (see Tables 4.3a, 4.3b and 4.4) help us think about ways to tackle these problems by focusing on macro- and micro-environmental settings in which food is made available for consumption.

Food safety

There used to be a time when nutritionists ignored the fact that foods are covered in microbiological organisms, some of them harmful. However, the rise of Hazard Analysis at Critical Control Points (HACCP, pronounced 'ha-sip') in food manufacturing and catering, implemented in response to outbreaks of food-borne illness, has brought home the realisation that foods can have other effects on health in addition to their nutritional effects. In developed societies, the effects of food-borne microbiological illness have less impact on population health than the various forms of malnutrition. However, in developing countries food adulteration and contamination by microbiological organisms present in water are major sources of parasitic and infectious diseases and indirect causes of under-nutrition.

In many countries there has been great investment in the promotion of food safety, which has resulted in quite stringent food laws and extensive food safety promotion processes (such as HACCP). These are designed to reduce the incidence and prevalence of serious food-borne illnesses such as salmonella contamination, listeriosis, bovine spongiform encephalitis (BSE) and other diseases. However, too great an emphasis on food safety can impede nutrition promotion, as was found in Victoria, Australia when pre-school centres were (inadvertently) encouraged not to serve fresh foods like fruit pieces for fear of contravening the *Food Safety Acts*. All legislation should be based on empirical evidence—for example, that serving fresh fruit to children *actually* puts them at *significant* risk of food-borne infection.

Environmental degradation is a massive problem which affects all countries. It includes land degradation (such as salination), climate change (associated with CO_2 and greenhouse gas emissions), water pollution and over-use, and other forms of ecological damage which threaten the survival of humans and other species. The causes of these problems are summed up in an old demographic equation:

Impact of the human population (environmental degradation) = numbers of people + damaging technology + standard of living (way of life associated with levels of economic development). (from Ehrlich 1971)

The world population is expected to level out at around 9 billion by the middle of the twenty-first century, so that leaves two of the factors in the above equation which might be changed to reduce human impact on the environment: the creation of technologies which do not degrade the environment as much as older technologies (e.g. solar, wind and water power); and changes in our ways of life. An example of a proposal to reduce CO_2 emissions is the UK Tyndale Institute's suggestion for the introduction of carbon rationing under which individuals would only be allowed to use a certain amount of fossil fuel each year in the form of petrol, air trips, heating and so on. The aim is to reduce UK carbon dioxide emissions by 60 per cent by 2040.

This planet-threatening problem is rather stark and may seem to be unrelated to nutrition promotion. However, it is intimately tied up with the area because consumers' food purchases represent a demand for food products which are produced and distributed through use of the environment. For example, 'healthy' fruit and vegetables are often produced in ways which waste water, damage arable land and use fossil fuels in the form of herbicides and weedicides, and fuel for tractors and other equipment. The increasing production of ruminant animals (cattle and sheep) to meet the worldwide demand for red meat is a major source of methane, which is a potent greenhouse gas (McMichael et al. 2007). The distribution of foods may involve packaging (and thus fossil fuel use), inefficient long-distance transport (e.g. air or truck transport) from one region of a country to another or from one country to another, as well as cooling and refrigeration in storage and lighting in display units. In summary, the production, distribution, marketing and consumption of foods and beverages have major implications for environmental degradation and climate change. If climate change is to be mitigated, there will have to be major changes in the ways food is produced and distributed. Generally, there will have to be greater consumption of locally produced foods and less consumption of ruminant meats.

Various approaches have been put forward to achieve environmental sustainability, such as the encouragement of local food production through community gardening, farmers' markets and community-supported agriculture; clear labelling of the country and region of origin of food products; labelling of the energy costs of production, distribution and packaging of products; the use of less packaging and less polluting packaging and recycling of packaging materials; and taxes and bans on the use of plastic shopping bags.

More and more research is being conducted on the fossil fuel costs of food production and transport ('product lifecycle analysis'). In particular, the concept of 'food miles' has been researched in the United States, the United Kingdom and elsewhere. 'Food miles' refer to the fossil fuel use in transport. Food now accounts for about 30 per cent of goods transported by road. Calculation of the fossil fuel use and greenhouse emissions is quite complex—for example, sea transport is among low CO_2 emitters while air transport is the highest, so simple distance travelled is only one aspect of the impact of transport on the environment (DEFRA 2005; see Figure 4.5).

FIGURE 4.5 CO$_2$ EMISSIONS ASSOCIATED WITH UK FOOD TRANSPORT, 2002

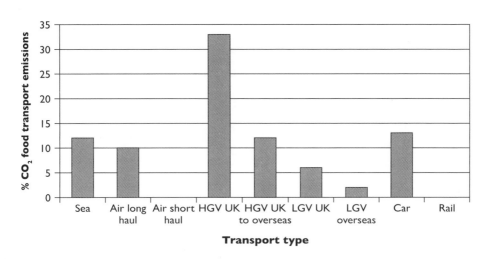

Notes: HGV = heavy goods vehicle, LGV = light goods vehicle

Source: DEFRA (2005).

Researchers at the Leopold Center at Iowa State University in the United States have compared the environmental impacts of vegetables purchased in Des Moines supermarkets with locally produced vegetables (see Figure 4.6). They have proposed eco-labels for food products which would indicate both the type of transport used and their fuel efficiency (see Figure 4.7). These are good ideas but they need more research before they can be widely implemented. The efficiency of transport needs further consideration. For example, long-distance sea transport may be more efficient and less polluting than air or road transport. However, sea transport accounts for about 4.5 per cent of global greenhouse emissions (www.guardian.co.uk/environment/2008/feb13/climatechange.pollution).

Sociological and market research evidence suggests that the number of environmentally conscious consumers is growing rapidly in most affluent countries. Australian and British consumers (among others) actually pay more for organically produced products. A substantial minority want to consume plant-based diets (Lea et al. 2006) and the younger generation, despite some of its profligate uses of packaged products, is environmentally aware (Worsley and Skrzypiec 1997). The general picture, however, is that the consumer food market is fragmented: many people are not interested in environmental sustainability, some are mildly interested but around a fifth are so interested that they consume products like meat in small quantities,

FIGURE 4.6 IMPACTS OF VEGETABLES SOLD IN DES MOINES SUPERMARKETS RELATIVE TO LOCALLY GROWN VEGETABLES

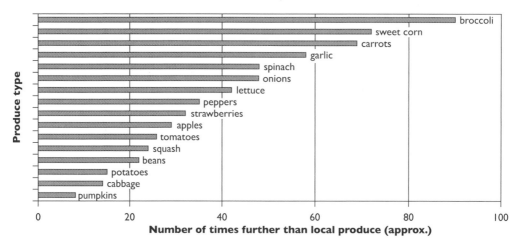

Source: Pirog and Schuh (2002).

FIGURE 4.7 PROPOSED ECO-LABELS TO INDICATE TO CONSUMERS THE ORIGIN AND TRANSPORT ENVIRONMENTAL IMPACTS OF FOODS

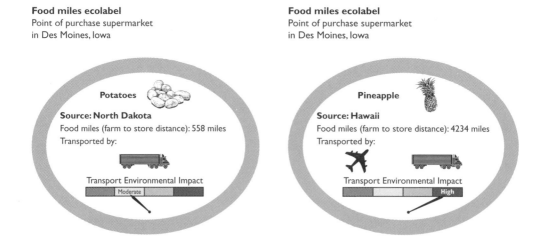

Source: Pirog and Schuh (2002).

partly because they consider it to be an energy expensive way to provide protein-rich foods ('cognitive vegetarians') (Lea and Worsley 2004). This trend is likely to grow in future and to become a more important consideration in nutrition promotion. Perhaps the new food ethics proposed in Box 4.3 may be the orthodoxy of the future?

BOX 4.3 THE NEW FOOD ETHICS

Food should be safe to eat and:
- close to natural;
- seasonal;
- locally grown;
- use as little energy as possible in production and distribution;
- low on the food chain;
- not damage ecosystems;
- avoid animal cruelty;
- avoid agrichemicals and artificial chemicals;
- have only one layer of packaging;
- use biodegradable or reusable, locally produced packaging; and
- require minimal energy in home preparation.

Source: www.bml.csiro.au/susnetnl/netwl52e/pdf.

Two decades ago, Gussow and Contento (1984) observed that it is not sufficient to provide only for population nutrition needs in the present but it is also necessary to guarantee future generations an adequate food supply. This means that the ecological and economic sustainability of the food system, especially the local food system, is an integral part of nutrition promotion. Nutrition promoters have a 'vested' interest in ensuring that pollution of the biosphere is reduced, if only to protect the viability of the ecosystems which underlie the food system. Similarly, the economic sustainability of the food system is of relevance since it maintains the population's access to a healthy food supply.

The ecological agenda can be regarded as the future tense of nutrition promotion. It is all very well to provide suitable foods for today's generation, but if we don't look after the ecosystems on which we depend then there will be no food and no life for future generations. Nutrition promotion must include the ecological dimension in its agenda—what Wahlqvist (1995) terms 'eco-nutrition', and the 'Giessen Declaration' calls 'the New Nutrition project' Box 4.4).

BOX 4.4 EXCERPTS FROM THE GIESSEN DECLARATION: THE 'NEW NUTRITION PROJECT'

This is a call for redirection of nutrition science, as follows:

1. Biological, social and environmental dimensions

Now is the time for the science of nutrition, with its application in food and nutrition policy, to be given a broader definition, additional dimensions and relevant principles to meet the challenges and opportunities faced by humankind in the twenty-first century.

2. Personal, population and planetary health

Now all relevant sciences, including that of nutrition, should and will principally be concerned with the cultivation, conservation and sustenance of human, living and physical resources all together, and so with the health of the biosphere.

3. Food systems and nutrition science

Nutrition science needs to incorporate comprehensive understanding of food systems. These shape and are shaped by biological, social and environmental relationships and interactions. How food is grown, processed, distributed, sold, prepared, cooked and consumed is crucial to its quality and nature, and to its effect on well-being and health, society and the environment.

4. The general challenges of this century

... These and other changes collectively constitute an imminent global environmental crisis on a scale not previously encountered. Great pressures on various components of the life-support system of our planet are already evident. The resultant environmental and ecosystem changes pose many threats to food systems. To understand and remedy this situation will require extending the scope and collaborative engagement of many scientific disciplines, including nutrition science.

5. The nutritional challenges of this century

Global food and nutrition insecurity and inadequacy, and even chronic hunger, have not significantly changed in the last 20 years ... General and specific nutritional deficiencies increase vulnerability to infectious diseases, especially in women, infants and children ... New epidemics of obesity, diabetes and other chronic diseases, including cardiovascular and cerebrovascular diseases, bone disease and cancers of various sites, are also now afflicting middle- and low-income countries, populations and communities ... Nutrition science can

Box 4.4 continued

address these challenges, but can do so successfully only by means of integrated biological, social and environmental approaches.

6. General principles

The overall principles that should guide nutrition science are ethical in nature. All principles should also be guided by the philosophies of co-responsibility and sustainability, by the life-course and human rights approaches, and by understanding of evolution, history and ecology.

7. Definition and purpose

Nutrition science is defined as the study of food systems, foods and drinks, and their nutrients and other constituents, and of their interactions within and between all relevant biological, social and environmental systems. The purpose of nutrition science is to contribute to a world in which present and future generations fulfil their human potential, live in the best of health, and develop, sustain and enjoy an increasingly diverse human, living and physical environment. Nutrition science should be the basis for food and nutrition policies. These should be designed to identify, create, conserve and protect rational, sustainable and equitable communal, national and global food systems, in order to sustain the health, well-being and integrity of humankind and also that of the living and physical worlds.

Conclusion

There remains much work to be done in the biological dimension of nutrition science. Much other important work now has to be carried out also in the social and environmental dimensions: this will require a broad, integrated approach.

This Declaration emphasises that the most relevant and urgent work to be done by professionals working in nutrition science and in food and nutrition policy is in its three biological, social and environmental dimensions all together.

Source: International Union of Nutrition Sciences (2005).

Food for social health

For aeons, humans have gathered together to prepare and consume food. It is during these times that social relations are strengthened, gossip is exchanged and bonding between friends and families is often facilitated. There is strong evidence that such social relationships are necessary for the maintenance of people's health (Kawachi and Berkman 2000). In today's fast food society, the conviviality aspects of foods are under-valued: people often think of foods merely as fuel or as sources of nutrients with specific health functions (e.g. phytosterols in margarines which can help prevent heart disease) rather than providing occasions for sociability, fun and relaxation. Early childhood educators and community developers tend to be more aware of the social uses of food than many nutritionists, but we can perhaps borrow from the French who, despite being immersed in 'high tech' and business, still manage to have long lunches and considerable merriment! It has been claimed that the stress-reducing effects of shared meals are health promoting—perhaps as much as any combination of nutrients the meals contain.

Problems of special population groups

In most populations, there are special groups which have particular food and nutrition needs. These groups vary somewhat according to local conditions and histories. However, there are three groupings which occur throughout the world.

Babies, infants, children and adolescents

These groups are of particular concern for nutrition promoters because the growth processes occurring during these life stages are vulnerable to nutritional insults. Recent research has shown that during pregnancy the foetus is subject to nutritional programming which can have long-term effects on disease risk (Moore and Davies 2001). It is also well established that nutritional deficiencies during early childhood which can substantially reduce growth (stunting) may result in long-term effects on stature, and possibly long-term cognitive and emotional deficits (Grantham-McGregor 2005; Pollitt 1993, 2001). Pregnant and lactating mothers may also be at risk from nutrition deficiencies which can affect both their long-term health and that of their babies. The issue of low prevalence of breastfeeding in some countries and early cessation of breastfeeding are serious problems, since breastfeeding provides the best start to a child's life.

Similar concerns apply to adolescents who may be at risk of nutrition deficiencies due to access problems (shortages of adequate foods) or to demand-side issues such as non-nutritious food choices. In developing countries, food shortages are associated with female iron deficiency anaemia, and in developed societies the use of slimming diets or ill-informed vegetarian diets appears to be associated with iron and calcium deficiencies.

While these are rather narrow nutrition concerns, nutrition promoters are particularly interested in broader food issues. These relate to the development of healthy food preferences and the acquisition of food skills such as shopping and cooking skills, as well as the ability to counter misleading food advertising and marketing. Recent nutrition promotion has begun to emphasise the establishment of preferences for less-processed foods like fruits and vegetables during childhood on the assumption that preferences established in childhood will influence the person's food consumption during their lifetime. So today there is less focus on nutrients and a greater emphasis on the enjoyment of healthy foods. This will be discussed further in Chapters 8 and 9.

Older persons, disabled people and hospital patients

Increasing population longevity and better medical technologies have been responsible for greater numbers of people living to old age and greater numbers living with disabilities. This has led to more people being treated in hospitals and health services or living in various forms of residential care.

Two things are becoming clear. First, elderly people—even when they are apparently healthy—have special nutritional issues. Frequently they suffer from the effects of degenerative diseases (such as heart disease and diabetes) and the treatments for these diseases (e.g. the side-effects of drug therapies). This appears to be related to loss of appetite, lower absorption of many nutrients and to reduced metabolism of other nutrients (e.g. the production of vitamin D in older people's skin when exposed to sunshine is about six times slower than among young adults (Nowson and Margerison 2002).

Second, many people (over 40 per cent) who are in hospital or in residential care suffer from nutritional deficiencies of various sorts (Beck et al. 2001). The nutrition status of people with severe physical or intellectual disabilities, for example, is under-researched. Anecdotal reports suggest that it is poor; however, more action research and associated nutrition and health promotion are required.

Minority, indigenous and socially disadvantaged groups

Minority groups, often of indigenous people, exist in most countries. For a variety of reasons, their access to food and their nutritional health may not be as good as for people in the 'mainstream' groups. Many governments have developed health policies to provide more equitable conditions. Nutrition promoters working in these disadvantaged groups face special problems associated with their cultural practices and lack of adequate resources. In Australia, the nutrition and health of Aboriginal people in both remote areas and in urban locations remains much poorer than that of other Australians, though a great deal of work is now being done to improve their health status (see special issue of *FoodChain*, www.nphp.gov.au/workprog/signal/foodchain.htm).

Newly arrived refugees and migrants from non-English speaking countries face unique problems. Many refugees suffer the effects of trauma in their countries of origin and they are unfamiliar with life in the host country, including its food ways. These people also face considerable linguistic transitions, often relying on their children to translate important information such as personally embarrassing health information. Food insecurity is a big problem in these groups (Burns 2004).

Other population groups which frequently require special consideration include low socioeconomic status groups such as pensioners with young children and people living in rural and remote locations or in outer suburbs, who are a long way from health services and food shops. Over the past few years, increasing interest has emerged in the concept of 'fresh food deserts'. This refers to the observation that in some cities some groups of people have difficulty in accessing fresh, healthy foods. They may live a long way from stores which sell these foods cheaply and they may not have a car, or the local public transport may be so limited that it is not useful for food shopping. Alternatively, people may live near retail food stores but these outlets may sell fresh foods at inflated prices, in effect denying poorer people access to these foods. The initial work in this area was done in the United Kingdom and the United States, and the existence of 'food deserts' remains under-examined in Australian cities. Opinions are divided over the practical importance of 'food deserts'.

The promotion of health equity

The special groups above are examples of groups of people who do not have access to health resources such as food. There is considerable debate in health promotion about the best ways to lessen health inequities in society. They include:

- *The treatment and assistance of people in these groups* — This can provide speedy relief but has the disadvantages that the group may be stigmatised (as needy) and nothing is done to prevent further occurrence of the problem since only the immediate problems (signs of risk factors) are dealt with.
- *The whole-of-population approach* — Promotion is conducted throughout the whole population (e.g. a social marketing campaign may be run to promote fruit and vegetable consumption). This approach can be effective — for example, Go for 2 and 5 in Australia — but it may actually increase inequity because advantaged groups may be able to gain greater benefits than more marginalised groups. This can give rise to sub-population approaches in which certain groups are targeted more than others.

- *A social determinants approach*—This approach looks behind the risk factors (like low levels of fruit and vegetable consumption) to examine the social causes of health inequities such as the poorer educational opportunities for marginalised groups, and proposes that education is available for everyone regardless of their background. Similarly, schemes may be enacted to ensure that all children receive good health care, especially during their early years (Wilkinson and Marmot 2003). This approach has not been extensively tried.
- *The lifecourse approach*—This complementary approach identifies health inequities present at each life stage and uses a variety of program logics to reduce them (Boyd 2008).

Nutritional responses to food and health problems: Nutrition taxonomies and guides

Historically in Western countries, especially Anglo-American countries, much of the response to nutrition problems has been in the form of nutrition education. It certainly has not been the sole response, but it has tended to dominate. In developing countries, there has been a greater emphasis on the food supply—either growing food or supplying foods or supplements to needy people.

The context of Anglo-American nutrition promotion during the twentieth century was gendered and individualist. As sociologist Tom Laws has remarked: 'During the Industrial Revolution, men were sucked out of the family'—that is, out of domestic life (Laws, personal communication 2004). Men were breadwinners and women stayed at home and looked after the children. The result was that family health and child care were seen as the responsibility of women. Observations of starvation among children and workers, particularly after the introduction of mass primary education in the United States and Western Europe in the last third of the nineteenth century, led to the great American home economics movement which taught girls and women how to manage households. It was the first life skills education movement, but because of the prevailing gendered view of humanity it was for girls and women only. It exerted huge influence for almost a century. Many of us still live with its effects—for example, adhering to the 'five food groups'. However, several factors combined to undermine it from the 1960s onwards. These were the renaissance of the women's movement in the 1960s and 1970s, which challenged the ascription of roles on the basis of gender, the post-World War II economic expansion which opened up a wider job market for women, and the development of the welfare state which enhanced the economic independence of women through the provision of pensions to single mothers, divorce law reform and education opportunities for girls.

The other theme which encouraged nutrition education (rather than promotion) was the overriding individualist philosophy of Anglo-American countries. Individual effort and competition is highly valued in these societies. This is associated with the view of education as the great equaliser: people need to be prepared for life through access to education so that they can make free, well-informed choices. There is nothing particularly wrong with this view, except to say that it is not the only way we can view life. Cooperatives, governments, associations, companies and corporations, and other institutions can also affect how we live. The individualist view tends to ignore the power of these combinations, pitting the individual against powerful economic forces such as those who control the food supply. Nevertheless, much nutrition promotion focused on education during most of the twentieth century, tending to ignore the settings and environments which influence individuals. Within this tradition, several nutrition frameworks have been developed to provide guidance for policy-makers and others. Here are some examples.

Dietary guidelines

A useful but conservative way to understand the range of nutrition problems within any country is to consult the dietary guidelines for that country. These sets of guidelines provide rapid introductions to the nutrition consensus since each guideline summarises nutritionists' knowledge of a particular topic area. Note that there are often separate sets of guidelines for specific age groups, such as older people and children.

Several countries have devised dietary guidelines which were originally intended to guide public health workers' policy and promotion efforts, though nowadays they have a bigger audience. The Australian Dietary Guidelines are a typical example. They were first issued in 1979 and published in their latest form in 2003 (Box 4.5). Although several studies have shown that the majority of the public have little awareness of them, many people appear to have attempted to alter their food intakes in ways which are consistent with them.

BOX 4.5 DIETARY GUIDELINES FOR AUSTRALIAN ADULTS

Enjoy a wide variety of nutritious foods

- Eat plenty of vegetables, legumes and fruits.
- Eat plenty of cereals (including breads, rice, pasta and noodles), preferably wholegrain.
- Include lean meat, fish, poultry and/or alternatives.
- Include milks, yoghurts, cheeses and/or alternatives. Reduced-fat varieties should be chosen where possible.
- Drink plenty of water.

Take care to do the following:

- Limit saturated fat and moderate total fat intake.
- Choose foods low in salt.
- Limit your alcohol intake if you choose to drink.
- Consume only moderate amounts of sugars and foods containing added sugars.
- Prevent weight gain: be physically active and eat according to your energy needs.
- Care for your food: prepare and store it safely.
- Encourage and support breastfeeding.

These guidelines are not in order of importance. Each one deals with an issue that is key to optimal health.

Two relate to the quantity and quality of the food we eat—getting the right types of foods in the right amounts to meet the body's nutrient needs and to reduce the risk of chronic disease. Given the epidemic of obesity we are currently experiencing in Australia, one of these guidelines specifically relates to the need to be active and to avoid over-eating.

Another guideline stresses the need to be vigilant about food safety and, in view of the increasing awareness of the importance of early nutrition, there is a further guideline that encourages everyone to support and promote breastfeeding.

Source: NHMRC (2003a).

The guidelines are intended to act as guides for the general population—for example, that the population at large should consume more fruit and vegetables. Obviously, this advice might not apply to particular individuals, such as those who already consume lots of fruit and vegetables. This is why the guidelines need to be used carefully.

While they contain much useful scientific information, dietary guidelines are political documents. They are often influenced by the activities of agriculturists, manufacturers, government bureaucracies, researchers' viewpoints and various health service lobbies. Often their content is the end result of compromises between the influences of various lobby groups. A related set of recommendations is the nutrient reference values (Box 4.6), which are used to examine the nutritional adequacy of individuals' and populations' food consumption.

BOX 4.6 NUTRIENT REFERENCE VALUES IN PLAIN ENGLISH

EAR Estimated Average Requirement

The EAR is the median usual intake estimated to meet the requirement of 50 per cent of healthy individuals. If the average healthy individual is not meeting the EAR for a nutrient then they are not likely to be consuming adequate amounts.

RDI Recommended Dietary Intake

The RDI is the average daily dietary intake level sufficient to meet the nutrient requirements of nearly all healthy individuals in the population. If an individual is consuming nutrients equivalent to the RDI then there is a good chance the individual has an adequate intake. However, if an individual is consuming a nutrient above the RDI there is a chance they are consuming too much.

AI Adequate Intake

When there is not enough information about a nutrient to determine an EAR (and therefore an RDI) an AI is used. If an individual consumes at or above the AI then the intake is probably adequate.

EER Estimated Energy Requirement

The EER is the average dietary energy intake required to maintain a balance between food intake and physical activity. The EER is specific to age, gender, weight, height and level of physical activity.

UIL Upper Intake Limit

Usual intakes above this level may place an individual at risk of adverse effects from excessive intakes. It is the highest amount recommended for daily consumption to prevent potential toxicity.

Source: Ward (2006).

Critique of dietary guidelines

The guidelines are a useful set of statements because they are related to a large number of issues which are of interest to a range of population groups. Therein lies a major problem. Members of different groups tend to have different concerns, which may lead them to place a different emphasis on particular guidelines. For example, mothers of babies are likely to be much more concerned with the breast-feeding guideline than middle-aged men.

A related problem is that they are very general prescriptions which assume a sound basis of nutritional knowledge. For example, in the first guideline the terms 'wide variety' and 'reduced fat' are used. The variety refers to the need to have a range of foods from several groups of foods. The consumer may not have this knowledge. Similarly, the term 'reduced fat' is likely to be quite meaningless to many people.

The dietary guidelines may be too bland to meet many people's needs as they cannot take into account the peculiarities of individuals' lives (e.g. 'I don't know what "eat a variety of foods" means or how to go about it'). Do people actually use the guidelines in their daily lives? If they do, which ones do they use? Little research has been conducted to determine their effectiveness in guiding people's food consumption behaviours. Perhaps the injunctions to reduce fat intake have been acted upon but this may have more to do with the pursuit of beauty and associated social acceptance!

Additional criticism of the guidelines includes the claims that they are too prescriptive and that they may discriminate against women (Crotty et al. 1992). Certainly they are a set of general prescriptions, but whether they are too prescriptive probably depends upon the ways in which they are promoted. The guidelines, after all, are very general, allowing wide latitude in their interpretation.

Similarly, the accusation of gender bias depends on the ways in which the guidelines are implemented and interpreted. If they are targeted only at women in the belief that they are the sole 'gatekeepers' of household food intake then this could be unfair and unrealistic. Such a stance places too much responsibility on the woman for the 'correct' dietary behaviours of her male partner in particular, and those of her children. She may not have any power to influence their behaviour and may be obliged to prepare foods that she believes to be 'unhealthy'. This is likely to increase the levels of guilt women feel over food shopping. A recent Australian report suggests that as many as one in two parents feels some guilt over the foods they give children at school (Worsley 2007).

Similar claims could be made about the promotion of the guidelines to disadvantaged groups such as financially deprived families, who may find it difficult to afford a wide variety of food, or members of ethnic minorities who may find access to their own culture's foods difficult.

Again, these are not reasons for the abandonment of the guidelines but they underline the importance of their sensitive implementation. The challenge is to make it easier for people to buy and consume foods which conform to the guidelines. Perhaps the best way of doing this is to ask: 'Who is this program for?' and then to ensure that the program is framed to include the needs of the particular population group.

The danger is that the guidelines will be focused only, or mainly, on the food buyer or consumer. They could be seen to be the only ones who have to decide between healthy and less healthy alternatives in their buying and consumption. However, such a singular emphasis would be unfair since the food consumer may have little power or ability to easily choose a healthy diet. It is unreasonable to expect consumers to invest very much time and effort into nutrition issues because they have so many other immediate and pressing issues to attend to, such as their relations with their friends and families, their work, their carer roles, other health concerns and much more.

It is more reasonable to expect that the environments in which consumers choose and consume food should be more compatible with the dietary guidelines. This means that environments such as supermarkets and catering premises should provide a wide range of foods which enable people to easily follow the guidelines (e.g. that the fruit and vegetables are fresh, attractive and reasonably cheap) and that these foods should be highlighted in ways which enable consumers to make healthy choices easily (e.g. low-fat products could be promoted more than higher fat alternatives). That is, the food environment should support the guidelines.

How relevant are dietary guidelines to consumers' needs?

There is a large body of knowledge in nutrition and food science which might be of relevance to the general population. Depending on how they are defined, there are well over 40 nutrients which are important for human health and there are thousands of compounds in foods which may affect physiological and psychological functioning. In addition, there are many factors and processes which influence the production and distribution of foods and their nutrients. So nutrition promoters have many kinds of information to offer the general public.

Food consumers have broad interests in food and health issues, covering food safety, environmental issues, equity and fairness as well as conventional nutrition-related problems such as heart disease (Worsley and Scott 2000). These appear to vary in their salience according to the consumer's life stage. Several aspects of nutrition and related food sciences appear to be useful for various population groups, including:

- the functions of basic nutrients in the body;
- the main sources of these nutrients in commonly available foods;

- the relationships between nutrients and food constituents and common diseases;
- the likely needs of people in different life stages for nutrients and other food constituents;
- guidelines about daily eating and food preparation activities and nutrient intakes, such as how much iron one needs each day, and notions of 'balance' and 'variety';
- the problems which may arise through following misleading nutritional concepts; and
- the systems which produce, distribute and handle food—so as to allow consumers more control and trust in the food system. This includes ecological factors related to the sustainability of food production.

These broad areas of food and nutrition science are interrelated, and are probably necessary in varying degrees in today's societies for people to perform daily tasks in a healthy manner without damaging the ecosphere or other people's well-being. They are related to people's needs and concerns during particular life stages. For example, parents of young children tend to be concerned about children's vulnerability to childhood diseases and their healthy development. They are likely to be more interested in nutrients which aid their children's growth and immunity than in the factors which help prevent heart disease. Thus only certain sets of knowledge are likely to appeal to them.

Nutritional taxonomies

Several taxonomies of healthy foods groups have been produced by nutritionists over the years to help consumers select healthy foods. Here are three prominent historical examples.

The Basic 4 scheme

This scheme was developed in the United States during a time of protein energy malnutrition in the mid-twentieth century. Before that period, various taxonomies had been used in the United States which included as many as twelve food groups. The development of food guides had been influenced by prevailing malnutrition during the first half of the century. For example, during World War I many recruits had to be turned away because of their poor nutrition status (e.g. they suffered from diseases such as rickets).

The five food groups (see Box 4.7) consisted of milk and dairy products, meat, fruit and vegetables, and grains (bread and cereals). A more recent example is the British Eat Well Plate (Figure 4.8). Most countries have similar guides.

BOX 4.7 THE FIVE FOOD GROUPS

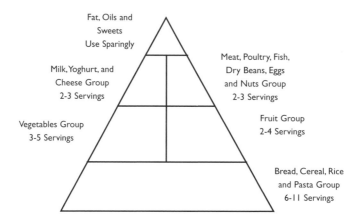

Fat, Oils and Sweets
Use Sparingly

Milk, Yoghurt, and Cheese Group
2-3 Servings

Meat, Poultry, Fish, Dry Beans, Eggs and Nuts Group
2-3 Servings

Vegetables Group
3-5 Servings

Fruit Group
2-4 Servings

Bread, Cereal, Rice and Pasta Group
6-11 Servings

Grain

Grains, breads, cereals, pastas and other grain-based foods are the foundation of a healthy diet. Grains are excellent sources of carbohydrates, vitamins, minerals and fibre. The most food servings should be chosen from this part of the food pyramid every day. The carbohydrates in grains help fuel muscles and protect against fatigue.

Fruits

Fruits are rich in vitamins, fibre and carbohydrates. The carbohydrates found in fruits aid in recovery after exercise.

Vegetables

Vegetables are an excellent source of vitamins, minerals and fibre.

Meats and other protein sources

Protein-rich foods help repair muscles, ensure proper muscle development and reduce the risk of iron deficiency.

Dairy

Dairy products are also quick and easy sources of protein. They are rich in calcium, which is a mineral that helps maintain strong bones and reduce the risk of osteoporosis.

Source: (1992) USDA; www.winforum.org/food-groups.html; www.mypyramid.gov.

FIGURE 4.8 THE BRITISH EAT WELL PLATE

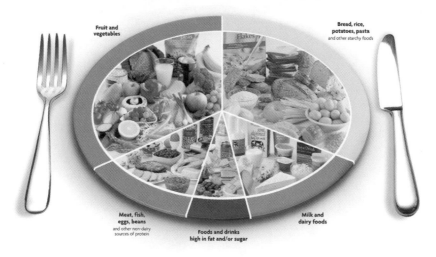

Source: http://www.food.gov.uk.

The Australian Guide to Healthy Eating (NHMRC 1998), promoted by Nutrition Australia, is the official Australian government guide to healthy eating (see Figure 4.9). Extensively tested among several Australian population groups (NHMRC 1998), it provides a useful basis for education and counselling activities. It is typical of guides which governments have issued to enable consumers to consume healthier foods. It provides a useful basis for education and counselling activities. Similar consumer guides have been produced in other countries (e.g. the Chinese 'pagoda' of healthy eating, and the American and UK guides). Their introduction has often met with opposition from producer organisations who have felt their products were maligned in some way—for instance, American beef producers objected to the placing of meat products close to the top ('less desirable part') of the US food pyramid.

FIGURE 4.9 THE AUSTRALIAN GUIDE TO HEALTHY EATING

Logo schemes

From time to time, people have suggested that these nutritional taxonomies are impractical for daily tasks like shopping. So they have invented simpler schemes which have been used with food product labels. One example is the Tick logo used by the Australian and New Zealand National Heart Foundations (see Box 4.8). They award a Tick logo to manufacturers who try to reduce the amounts of fats and salt in their products.

BOX 4.8 THE TICK LOGO USED BY THE AUSTRALIAN AND NEW ZEALAND HEART FOUNDATIONS

CERT TM

The Heart Foundation Tick is used to highlight healthier foods in the supermarket and eating out.

To earn the Tick, food companies must meet strict nutrition and labelling standards set by the Heart Foundation and foods eaten out must also meet the strict processing standards that ensure meals can meet exacting standards time after time.

Foods with the Tick are subject to the most rigorous independent auditing in Australia. Independent, random tests for saturated and trans fat, sodium and energy (kilojoules) and, where appropriate, fibre, calcium and protein ensure approved products maintain the standards required to retain their licence.

The Heart Foundation Tick has about 55 different categories, each with its own unique nutrition criteria (the criteria for breakfast cereals are different from the criteria for vegetable oils). They are reviewed systematically to ensure they continue to challenge the food industry and to keep pace with Australian eating patterns.

Broadly, Tick-approved products fit into three main categories:

1. *Fresh foods*, like meat, fruit and vegetables, eggs and nuts and seeds—the Tick reminds people to include these foods as part of a healthy eating plan.
2. *Everyday foods*, like yoghurt, bread, breakfast cereals and margarine, where often there is so much choice that making the healthier choice can be overwhelming.
3. *Occasional foods* like oven fries, takeaways and ice cream, so that there are better choices when we want to indulge too.

Source: S. Anderson, Heart Foundation of Australia (personal communication 2007).

Survey evidence suggests that the Tick logo has been well received by consumers, many of whom use the logo to choose food products during their shopping. Young and Swinburn (2002) have shown that, in New Zealand, many products have been altered because of the Tick scheme in the past decade; many tonnes of saturated fat are being removed from the food supply.

Critiques of food taxonomies, food guides and logo schemes

The main criticism of these approaches is that they focus on individuals' knowledge of food and nutrition — that is, they try to inform people about food and nutrition. In itself this is no bad thing, but sole reliance on this approach ignores other factors which influence food consumption. These include motivation and the influence of local settings.

Many people may know which foods they should consume, but they are not motivated to do so — healthy eating is not on their agenda. A number of motivational techniques and programs can be used to attempt to influence people to eat more healthily. These approaches have more to do with social influence and emotional appeals that to the transfer of nutrition knowledge. Some of them, like motivational interviewing, peer-led education, mass media persuasion campaigns and financial incentives, are discussed in Chapters 8 to 13.

The influence of the local environment, including purchasing and usage situations, on people's food consumption is central to nutrition promotion. Thus the price of foods, the availability of foods in stores and in the home, people's geographical and financial access to foods and their familiarity with them are all aspects of the 'micro-environment' which can be altered to increase the population's consumption of healthy foods. This doesn't make the dietary guidelines and food taxonomies redundant, but they can be applied in novel ways. For example, in addition to teaching consumers about the virtues of various food groups and dietary guidelines, local food stores (and other food access sites such as school canteens) can be assessed according to the degree that they support national guidelines and food group schema. A food store can be assessed according to whether it makes available a range of fresh fruit and vegetables as recommended by the dietary guidelines or the national food pyramid. How much are confectionery and other high-energy foods over-represented compared with fruits and vegetables? Do the relative prices of these two categories further impede compliance with the dietary guidelines?

Food provision schemes

Not all nutrition promotion programs have relied on the provision of guides, knowledge and skills only. There are many prominent examples of food provision schemes (which are usually combined with nutrition education). Examples include the North Carolina Food Bank, the US Women, Infants and Children (WIC) program, the British food rationing program of World War II and the Gujerat School meals program in India (Gopaldas 2005).

Conclusions

The problems facing nutrition promoters are to be found in both the food supply and in the demand for foods. They range from nutritional deficiencies and excesses to the types of foods that are available for people to choose. The responses of nutritionists have largely been on the demand side to equip individuals with the knowledge and skills to make healthy food choices. While these are essential, sole reliance on nutrition education for the general population ignores the power and influence of governments, companies and market forces. It is clear that, left to themselves, market forces do not deliver adequate nutrition for all; government intervention is required to provide the framework for equitable distribution of foods within the population. Food and nutrition knowledge and life skills education are required so that consumers can choose healthy foods, but so are food and nutrition policies which provide nutrition-promoting settings in which individuals can exercise their freedom of choice.

Dietary guidelines and nutritional taxonomies provide essential policy directions. However, the key challenge is in their implementation. To help members of the public consume foods in the ways recommended by nutrition scientists, nutrition promoters need to understand the influences on human and institutional behaviour and methods to bring about behavioural change. These are outlined in Chapters 5, 6 and 7. They also need to appreciate the problems and opportunities available in the variety of settings in which humans live. These are described in Chapters 8 to 13.

Discussion questions

4.1 What are the key elements of the traditional cuisines which have been developed in different parts of the world during the past 5000 years? What do they tell us about the ways in which humans have optimised their survival? How are they different from so-called fast food diets?

4.2 Describe the main supply-side and demand-side influences on human food selection. What opportunities do they present for nutrition promotion?

4.3 Describe the special nutritional and food needs of special groups in the general population.

4.4 What is the nutrition transition? What challenges does it present for nutrition promoters?

4.5 What are relationships between ecological systems and nutrition promotion?

4.6 Describe the Australian dietary guidelines and how they may be used by nutrition promoters. What are the advantages and disadvantages of such guidelines?

5

Theories and dietary change:
INDIVIDUAL-LEVEL THEORIES

Introduction

Why does nutrition promotion need behavioural and social theories?

Nutrition promotion is a set of activities which endeavours to bring about changes in human food consumption to maximise the population's health over the short and long term. This means that it deals with individual humans and groups of humans in social institutions, usually to bring about change. It follows, then, that nutrition promoters require a profound understanding of behavioural and social sciences so they are in the best position to influence food and health-related activities. Much of the daily subject-matter of nutrition promotion (such as needs assessment, working with groups, education and persuasion, communication, evaluation and policy implementation) has been examined for at least a century by a range of social and behavioural scientists. The danger of ignoring them is to waste effort in reinventing the wheel. The opportunity for nutrition promoters is to apply behavioural and social theories in the rapidly changing food and nutrition domain for the public good.

The focus of nutrition promotion varies. Sometimes it is on the individual, but at other times it is in the community or in broad organisational settings like the workplace. Or it may be at the level of national government where national nutrition policies are designed and implemented. There are social-behavioural theories which are relevant at each level. Individual-level theories are described in this chapter, and environmental and broader social theories are discussed in the following chapter.

Theories and paradigms

There are many theories and models which have been applied to the understanding of dietary change, though few of them actually deal with the *process* of dietary *change*. Theory-making is part of human (and animal) nature because we like some pre- dictability in our lives. Theories summarise our knowledge of the world. For example, when we get up in the morning we expect the sun to rise and that there will be daylight. If there isn't, then we have explanations ready to test, such as 'there is very heavy cloud', 'I've woken up in the middle of a solar eclipse', 'those are heavy curtains' or 'I'm not awake yet, go back to sleep!' A theory is a set of generalisations based upon observations of previous events which enable us to explain and predict future events. In the physical and chemical sciences in which events can often be tightly controlled ('closed' systems), theories can be tested rigorously, but human social affairs—including food behaviours—can be influenced by many factors (hence they are sometimes called 'open' systems), so theory formation and testing may be more problematic.

Nevertheless, it is useful to make a checklist of factors which are likely to affect various processes and outcomes involved in dietary behaviours. Baranowski et al. (1999) examined individualist attitude-behaviour models of dietary behav- iours and came to two conclusions. The first was that studies which employed any kind of theory explained a higher proportion of dietary behaviour than those which had no explicit theory. Second, they concluded that no current behavioural theory explained more than about 30 per cent of the variability in dietary behav- iours. The conclusion is that, although an organised set of expectations is useful, no current individual-level theory adequately explains dietary behaviours. Nevertheless, several behavioural concepts are useful for health and nutrition promotion.

The usefulness of theories for nutrition promotion depends upon several con- siderations. The first is our basic assumptions, or *paradigms*. Do we believe, like individualist philosophers, that food behaviours are fully under individual control— that we are free to choose the foods we want so long as we use sufficient personal resources, like willpower, skills and knowledge, to achieve them? That is, do we attribute *agency*, the ability to choose actions, to humans? Or do we see humans as passive beings who are acted upon by outside forces beyond their conscious control? This alternative view suggests that it is the influence of environments that counts. If we want to change dietary behaviours, we have to alter those environ- ments. Public health leans more towards this view, with its concepts of *vector* (that which carries diseases), *exposure* to vectors and *environment*, though the concept of *host* (the individual body and its processes) is also implicit in public health (Egger and Swinburn 1997; see Figure 5.1).

These are not minor distinctions. For example, if we want to help people to lose weight and if we assume that individuals are powerful agents for change, then we will base our interventions on programs which try to influence their attitudes, beliefs and motivations—apparently with little success to date (Jeffery 2004). However, if we see people as being influenced by environmental forces like billiard balls being struck by cues, then we will look for environmental influences over weight loss such as the support of social groups, or the influence of marketers of high-energy food and beverage products or even the satiating properties of foods. This environmental approach, seen in the renaissance of social epidemiology (Berkman and Kawachi 2002), has come back into favour in the early twenty-first century.

FIGURE 5.1 THE EPIDEMIOLOGICAL TRIAD AND ITS IMPLICATIONS FOR THE PREVENTION OF OBESITY

Hosts—biology, behaviours, medical interventions
> *ways to intervene: education, behavioural and medical interventions*

Vetors—energy density, portionsize, time-saving (eg washing machines, cars) and time-using machines (eg TV)
> *ways to intervene: technology, engineering*

Environments—physical, economic, policy, social-cultural
> *ways to intervene: policy, legislation, social change*

No particular paradigm is inherently better than another. However, one paradigm may be more suited to the explanation of one level of behaviour than another. For example, the individualist paradigm may provide a better explanation of purchasing behaviours in supermarkets than purely environmental explanations. Conversely, the environmental paradigm may provide better explanations of the differing prevalence of obesity between countries since differences in transport systems and town planning may facilitate physical activity more in one country (e.g. the Netherlands) than another (e.g. the United States). In reality, dietary change is likely to be promoted through a mix of environmental and individual factors, such as changes in knowledge held by members of the population and changes in the relative availability and promotion of various types of food products (such as low-energy density and high-energy density foods). In health promotion, the prevailing view is that individual influences over behaviour are constrained or limited by the social and physical environments ('settings') in which the behaviours occur.

In this and the following chapter, several examples of theories and models which can help us understand the ways in which food behaviours may be influenced will be described. This account is not intended to be exhaustive; several theories are not included here either because their focus is not on behaviour change processes or because they do not provide accounts of behaviour which are relevant to nutrition promotion.

Again, it is important to remember the paradigmatic assumptions which underlie the models described below. Accordingly, the various theories are examined in two separate chapters: individual-level theories in this chapter and broad environmental theories in the following chapter. The aim is to briefly describe each theory and to assess its utility for nutrition promotion. A useful, short guide to several prominent health promotion models is *Theory in a nutshell* (Nutbeam and Harris 2003).

Rothschild's model

A very simple framework which integrates the key features of both individual and social ecological approaches has been provided by Rothschild (1999), who proposes that three interacting sets of determinants influence eating behaviours. Box 5.1 illustrates this framework in relation to children and adolescents' eating, but the principles apply to adults as well. This simple model is worth keeping in mind when the plethora of models and theories is considered.

BOX 5.1 THE APPLICATION OF ROTHSCHILD'S FRAMEWORK TO CHILDREN'S NUTRITION PROMOTION

Why do we eat what we eat? Rothschild proposes three sets of key interacting determinants: motivation, ability and opportunity.

Motivation: Why would children and adolescents want to eat a healthful diet?

- **Attitudes**—weighing up the consequences of eating short term: taste, satiety, pleasure; children learn to like bitter and high-energy foods; lots of learning opportunities from observational learning to classical conditioning. Health-related expectations are important in adults, much less so in children and adolescents. Cost is not important for children but is important to adolescents and adults.

Box 5.1 continued

- **Perceived social influences**—subjective norms are expectations about what 'important others' want us to do. Parents' eating strongly influences children's dietary habits. Children and adolescents take their cues from their families and peers and their school environment. In the absence of knowledge, people tend to judge the adequacy of their behaviours by comparisons with others (optimism bias).
- **Self-efficacy or perceived behavioural control**—perception of confidence in one's abilities and skills to engage in certain behaviours. Self-efficacy for food behaviours is quite specific (e.g. confidence to cut back on fatty foods is quite different to confidence that one can lose weight).

    Many food behaviours are not intentional, they are subconscious habits. Environmental factors (e.g. the smell of a hot bread shop) may trigger them. Habits have differing strengths and are modifiable.

Ability: What enables children and adolescents to eat a healthy diet?

- **Skills** such as cooking and shopping skills as well as the skills relating to asking for help. Skills can be learned. They depend on practical or procedural knowledge such as knowing how much salt in a food is too much, or what are appropriate daily intakes of fruits and vegetables. In the absence of such knowledge, people judge their behaviours through comparisons with others.

Opportunity: Availability and accessibility of healthy choices

Environments present children with both opportunities for healthy eating and barriers. These include:

- opportunities for objective self-assessment (e.g. of energy intake);
- access to and opportunity to purchase healthy foods in one's environment;
- strong parental influences on children—parents can provide healthy foods and endorse clear rules of eating behaviour;
- schools, which may have healthy food policies that prevent marketing and availability of unhealthy food products like soft drinks in vending machines;
- neighbourhoods, which can also have a strong influence, particularly if they are 'food deserts' in which access to healthy foods may be limited;
- macro-level environmental influences, which often have negative effects such as advertising of fast foods to children and the marketing of these foods in schools. In turn, these are contained to various degrees by government policies.

Sources: Based on Brug and Klepp (2007); Rothschild (1999).

Individual-level models

Most individual-level theories and models deal with the behaviours in individuals' daily lives, such as their preferences for foods, their shopping choices, their daily food consumption and so on. Most of these theories have *not* been derived from observations of food behaviours. Instead, various investigators have *applied* them, with varying degrees of success, to dietary behaviours.

Learning theories[2]

A central aim of nutrition educators is the acquisition by children of preferences for 'healthy' foods. Learning theory, developed mainly by American psychologists in the first half of the twentieth century, helps us understand these processes. Psychologists like Paul Rozin and Leanne Birch applied classical and operant conditioning principles derived from animal studies to the acquisition of food preferences by young children. Rozin, in particular, looked at the development of food preferences and at the association of disgust reactions to novel foods (subsequently called 'neophobia') which reduce the variety of objects accepted by the young child as foods (Rozin and Vollmecke 1986). As part of his 'disgust' studies, Rozin developed the concept of sensory-sensory conditioning, which is a common form of learning. Subjects' food preferences can be altered simply by exposing them to disgusting or unpleasant objects in the presence of the desired food. In his most famous study, the *roaching* experiment (see Figure 5.2), the taste of fruit juice became notably less pleasant when it was paired with the mere sight of a dead cockroach!

Following this classical experimental psychology pathway, Leanne Birch (1999) proposed that three processes are involved in the acquisition of new food preferences: exposure, modelling and reinforcement. Children learn to like new foods through *repeated exposures* to those foods in an encouraging parental environment, through *watching* their parents eating and enjoying food (*modelling*) and through the *reinforcement* that parents and other children give them when they eat new foods (Birch et al. 1995).

---

2 This section is based on Worsley and Crawford (2005).

FIGURE 5.2 AN OUTLINE OF ROZIN'S 'ROACHING' EXPERIMENT

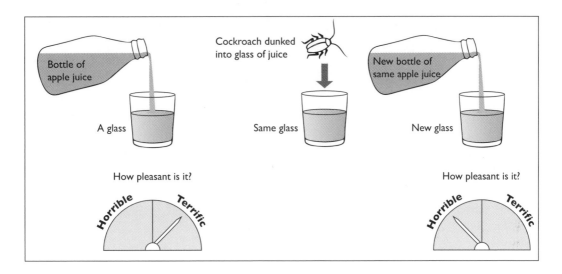

Source: Carey et al. (2003).

Exposure

In addition to genetically inherited taste preferences, infants learn taste preferences from their exposure to the flavours of the food that their mothers eat in utero (via the amniotic fluid) and to flavours carried by breastmilk. If a child is given a new food that his or her companions do not like, the child will probably develop a dislike for that food. However, when a food is presented among companions who like that food, the child will probably also come to like that food. Birch showed that pre-school children soon begin to eat foods they previously disliked (such as Brussels sprouts) when they are paired with peers who enjoy these foods. A new food may have to be presented to a child on as many as ten occasions before the child learns to accept it (Birch and Fisher 1998).

Modelling

Children and adults like to observe and emulate people they admire, so adults should be seen to eat and enjoy the foods they want their children to eat. In addition to friends, parents and teachers, the mass media also influence children, adolescents and adults. Cartoon characters, sports heroes and film stars, for example, are used to promote a narrow variety of food and beverages to children (Lewis and Hill 1998; Story and French 2004a and b). Younger children often model their behaviours on their favourite cartoon characters ('virtual persons'), while teenagers in particular tend to aspire to be like members of 'reference groups' such as football, netball, music or film stars. These virtual characters, whether present in advertising or in programs and movies, appear to have a strong influence over many children and teenagers (Soloman 1994). These may partly explain why long hours of exposure to television programs are associated with increased risk of obesity in children (Dietz 1996).

Reinforcement

People's behaviours can become entrenched or habitual through reinforcement. Usually, performance of the behaviour is followed by a reward. Reinforcement is part of everyday life. For example, parents 'shape' or encourage children to act in certain ways and not others—a process known as 'socialisation' (Parke 2004). Often, when toddlers eat a new food they are reinforced in some way by their parents—perhaps by a smile, a nod, a hug or some other sign of approval. Reinforcements shape behaviours, including eating behaviours.

Relevance to nutrition promotion

In recent years, these three learning principles have been introduced into many early childhood and school nutrition promotion programs. (see Chapter 8). In part, this is a response to parents' perceived inability to influence their children to consume 'healthy' foods and it is also part of a broader realisation that children have to be taught how to choose and enjoy foods. The learning processes involved are not wholly in our consciousness. While some may see this sort of 'arranged experience' as manipulative, such processes are an important way for parents, carers and educators to establish healthy food preferences in their children. In many 'free market' economies, complete strangers who are uninterested in the child's welfare are allowed to shape children's preferences through television advertising and other marketing techniques.

Cybernetic models of behaviour

A very old approach to behaviour change is based on the observation that many human behaviours are goal-directed. For example, a person may decide to lose some weight during a certain period. If the person hasn't lost the desired weight after a certain time, they can estimate the extent to which the goal has been missed and take corrective action. This basic feedback loop is a TOTE unit (Test–Operate–Test–Exit) (Powers 1979; Carver and Scheier 1998). The simplest analogy is that of the water heater: a temperature sensor measures the water temperature; if it is lower than the pre-set (goal) temperature, the heater is activated. A second test is then conducted: if the temperature has reached the goal temperature, the water heater is turned off (exit from system). If it isn't, the water heater stays on until another test is performed. Much conscious goal-directed human behaviour works in this way, both at the individual level (such as body weighing and subsequent dieting behaviour) and at the organisational level (e.g. organisational performance reviews). The key principle here is that feedback of the results of their behaviours enables people to change those behaviours. Feedback is essential for change.

Relevance to nutrition promotion

Feedback is often given by clinical dietitians to encourage dietary change in their clients. For example, a goal for daily fruit intake may be negotiated and then recorded on a daily basis, with this information fed back to the client on a weekly basis. The attainment or discrepancy of intake from the agreed goal can be used to maintain the client's motivation to change their food consumption. Often success will be paired with some form of reward, such as praise (see self-monitoring below). Feedback loops can be used for all manner of purposes. For example, an organis-ation's food purchasing patterns could be compared against the *Australian Guide to Healthy Eating* (to see whether the foods purchased are in the same proportions to those suggested by the *Guide*). Success or failure in meeting the guidelines can be fed back to those responsible for food purchasing, and perhaps linked to an award scheme (see Heartbeat Awards in Chapter 11).

Self-monitoring approaches

These approaches are based on cybernetic and social cognitive theories of behaviour (Bandura 1986; Carver and Scheier 1998). Essentially, they are personal change strategies. Participants are encouraged to become aware of aspects of their current diet, such as their daily intake of fruit. Then they are asked to try to change this aspect in a specific way (for example, 'I will eat an apple at morning recess instead of my usual chocolate biscuit'). They record their attempts and, if they succeed (that is, if they reach an agreed target level, such as eating an apple for recess for three days in a row), they are rewarded in some way (perhaps earning a gold star or a non-food reward). This approach is quite effective so long as it is not carried to extremes. It teaches the learners about the virtues of goal-setting, makes them think about the future and can employ powerful reinforcers (such as teacher, parent or peer praise). An early example of self-monitoring is the STAR system (**S**ee, **T**arget, **A**pply yourself, **R**eward yourself), used in the South Australian Body Owner's Program in 1980 (Worsley et al. 1987; see Figure 5.3).

FIGURE 5.3 THE STAR PLAN: APPLICATION OF SELF-MONITORING TO DIETARY CHANGE

THE STAR PLAN

See
Target
Apply   = **STAR**
Reward

Source: Coonan et al. (1984).

Award and accreditation schemes

Accreditation and award schemes are good examples of this approach at the organisational level. Child-care centres, for example, may examine the state of their staff's food and nutrition skills, enter a training scheme and, when successful, be given a healthy eating award for a period of time, as in Western Australia's Start Right, Eat Right award system.

Expectancy value theories

Several attitude-behaviour theories are based on the utilitarian proposition that people act according the likelihood (or expectancy) of various consequences and the value to them of those consequences. For example, if I eat chocolate often, I may get pimples. I have to weigh up how likely this will be and how bad or good the eruption of pimples would be for me. Social cognitive theory and the theory of planned behaviour are popular examples of this type of model.

Social cognitive theory (social learning theory)

This theory (Bandura 1986; see also Figure 5.4) has been the theoretical basis of many American healthy eating programs (for example, the CATCH program described in Chapter 8). It focuses on the interaction between the individual and the environment, particularly how the reinforcers in the immediate social environment can shape an individual's behaviour. It emphasises the influence of other people on the individual's behaviours (the social situation). For example, if the majority of children in a class like broccoli or zucchini, then each child will feel pressured to conform to this social norm. Bandura (1986) coined the term *reciprocal determinism* to highlight the continuing interaction between an individual and the social environment.

FIGURE 5.4 DIAGRAM OF SOCIAL COGNITIVE THEORY

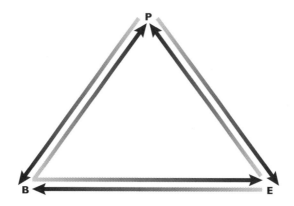

B represents behaviour, P represents personal factors in the form of cognitive, affective and biological events, and E represents the external environment.

Source: Bandura (1986); www.istheory.yorku.ca/socialcognitivetheory.htm.

Two other cognitive factors also play important roles in this theory. *Observational learning* (or learning by observing others) can change people's expectations about the importance of certain behaviours. For example, if a child sees another child being praised by their mother for eating fruit, he or she is more likely to eat fruit because doing so may earn praise. *Self-efficacy* is the belief in your own ability to successfully perform a behaviour. Observational learning and participatory learning (practising the behaviour) are likely to increase self-efficacy and thus bring about changes in behaviours. Note that the principles of *feedback* and *goal-setting*, which enable learners to plan to change their behaviours, are also important constructs in this model.

The theory of planned behaviour

This theory (Ajzen 1991; see Figure 5.5) is probably the simplest of the expectancy-value theories. It proposes that people's behaviours depend on their *intentions* to perform behaviours and in turn their intentions are largely determined by their *attitudes* towards the behaviours—that is, whether they like or dislike undertaking the behaviour. It is not so much whether a person likes fruit, but whether they like *eating* fruit that matters.

FIGURE 5.5 DIAGRAM OF THEORY OF PLANNED BEHAVIOUR

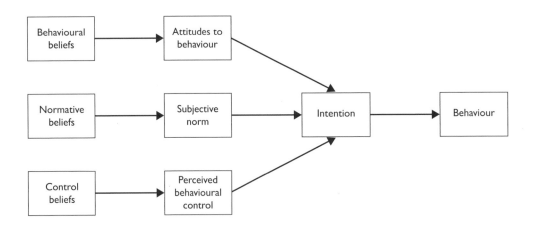

Source: Ajzen (1991).

A person's attitudes are products of their beliefs about the behaviour. For example, what are the perceived consequences of eating fruit? How likely it is that eating fruit will reduce people's risk of bowel cancer, give them smoother skin or leave a nasty taste in their mouths? These are *likelihood* or *expectancy* estimations. People usually evaluate the consequences of a behaviour in terms of how good or bad it would be for them. An adult might perceive eating fruit as likely to reduce the risk of bowel cancer, whereas a teenager probably would not consider that consequence to have any value because he or she does not think about bowel cancer.

In their minds, people process information about the likely consequences of their behaviours and the value of those consequences for themselves. They integrate these beliefs into a single attitude about the particular behaviour—for example, 'I do not like eating fruit'.

The theory categorises attitudes and beliefs into two streams: the non-social consequences (for example, the unfamiliar taste of a fruit, disease risk reduction); and social consequences (called *subjective norms*—for example, 'if I eat fruit, my friends will tease me'). It also acknowledges that people may have positive attitudes towards certain behaviours but may be unable to perform those behaviours ('I like fruit but I can't buy any to eat', or low *self-efficacy*).

Relevance to nutrition promotion

These two models are included here because they have often been applied to dietary behaviours. Whether they are useful for understanding the *process* of change is a matter of opinion. They appear to describe the state of consciousness associated with behavioural acts. In that sense, they are *actuarial* models (summarising the state of play), not dynamic *process* models. Certainly, understanding a person's views of the consequences of novel food actions is likely to be useful for promotion, but is it sufficient to bring about a change in behaviours?

The evidence regarding the application of these theories in nutrition promotion is equivocal: some interventions have been associated with dietary change, others not (Worsley and Crawford 2005a). Generally, dissatisfaction has been expressed about the utility of these models for an adequate understanding of change. They suffer from three major problems. First, they assume that the final influences on behaviour are conscious processes. In fact there is strong evidence that liking for foods (and thus consumption) is based in non-conscious processes such as sensory processes and autonomic conditioning. Second, many food behaviours—like shopping—are *habitual* processes which are *not* consciously reasoned processes. Often little thought is given to the purchase of 'everyday' foods. In contrast, these theories assume that a great deal of conscious thought is involved. Third, they are individualist models, tending to ignore the importance of the settings in which behaviours occur. More serious objections to these models have been raised by Weinstein and Rothman (2005) and Jeffery (2004), who have questioned their validity and utility. Weinstein and Rothman criticise the application of these theories, which is often based on the assumption that social cognitive factors predict behaviours, on the basis that the reverse proposition—that behaviours influence beliefs and attitudes—is rarely considered (as Bandura originally proposed in his 'reciprocity' principle).

Despite these limitations, the notions of perceived consequences of behaviours, self-efficacy, observational learning and especially self-monitoring and feedback processes are worth consideration in nutrition promotion. However, the food-related lifestyle model (Chapter 3) is a more comprehensive model which is probably more relevant for nutrition promotion.

The transtheoretical theory (stage of change model)

This theory (Prochaska and DiClemente 1984) is one of the few individualist models that has been designed to explain behavioural change. It has been widely applied to dietary behaviour change (Horwath 1999). For behaviours like eating fruit, it proposes that people can be in one of several stages of change. At each stage, the person has to ensure that the perceived advantages of doing the behaviour outweigh the disadvantages. This is called 'decisional balance'. In clinical settings, the therapist usually provides support in changing the decisional balance, often using aspects of cognitive behaviour therapy (e.g. teaching the person to think of the good effects of fruit when they ruminate on its bad aspects). The stages include:

1. *Precontemplation*—The person is not even thinking of changing their behaviour. This suggests the need for awareness-raising (for example, 'if you cut your fat intake, you may lose weight').
2. *Contemplation*—The person is considering change. Typically, people do not change because they perceive barriers to making the change. Therefore the health promoter needs to emphasise the benefits of the proposed change.
3. *Determination or preparation*—The person makes a serious decision to change. Ways to minimise the barriers to change need to be found at this stage.
4. *Action*—The person initiates the behaviour change. A program of change, setting goals and helping strategies needs to be developed.
5. *Maintenance*—The person is maintaining the change but may relapse from time to time, which may cause them to cease to perform the new behaviours. Strategies to overcome relapses are likely to be required.

Behavioural change strategies

The various cognitive and behavioural strategies derived from the transtheoretical model have been summarised by Ounpuu (1996), as shown in Table 5.1.

TABLE 5.1 BEHAVIOURAL CHANGE STRATEGIES		
Process	**Description**	**Example**
Self-monitoring*	Self-recording behaviours associated with the problem and managing the problem. It should also include information about time, places and feelings associated with eating.	Keeping a record of fruit intake and the feelings associated with eating it.
Cognitive strategies		
Consciousness-raising	A cognitive process of change that involves raising awareness about self and the problem—in this example, eating fruit daily.	Reading about eating more fruit.
Dramatic relief	An affective process that involves creating an emotion that encourages the individual to express their emotion about their problem(s).	Feeling worried about getting sick if you don't eat enough fruit.
Self-re-evaluation	A self-evaluative process that involves reappraising what is thought and felt about the problem.	Feeling good about yourself if you eat fruit.
Self-liberation	A process which involves willpower, making a commitment to change or ability to change—for example, eating fruit daily.	Talking to yourself about eating fruit, making a resolution—for example, more fruit consumption starting on Sunday.
Problem-solving	Self-correction of problem. Includes defining the problem, generating possible solutions, evaluating outcome and the re-evaluation of alternative solutions if the chosen one is not successful.	Includes a number of strategies listed above and below.

Table 5.1 continued

Behavioural strategies

Environmental re-evaluation	An evaluative process which involves considering the impact of one's problem—for example, the effect inadequate intake of fruit has on the physical and personal environment.	Consider the idea that in general, people would benefit from eating more fruit. Watch documentaries involving fruit.
Stimulus control	A behavioural process that involves using cues to take control and promote the behaviour change.	Keeping fruit and vegetables in a bowl on the kitchen bench.
Social liberation	Increasing alternatives for non-problem behaviours in society.	Advocating for fruit to be available at worksite or local coffee shops.
Helping relationships	A humanistic process of change that involves accepting or seeking help from others.	Accepting help from others in serving fruit.
Reinforcement management (contingency management)	A behavioural process that involves providing rewards for oneself or being rewarded by others.	Spending time with other people who encourage the consumption of fruit.
Counter-conditioning	A behavioural process that involves substituting the required behaviour for the problem behaviour—for example, dried fruit for crisps.	Snacking on fruit instead of high-fat foods such as crisps. Making positive statements about fruit consumption.
Interpersonal control	A behavioural process that involves seeking out people who promote eating vegetables and fruits or avoiding people who discourage that behaviour.	Spending time with other people who encourage one to eat more fruilt.
Planning ahead*	A behavioural process that involves being proactive and planning ahead.	Consider fruit as part of meal planning.

* Green and Kreuter (2005).

Source: Adapted by Sylvia Pomeroy from Ounpuu (1996).

Relevance to nutrition promotion

Potentially, the model should be useful for nutrition promotion since people may move through a series of stages of dietary change. However, several difficulties have been observed in practice. First, it is not clear how many stages of dietary change there are. Some studies suggest that most people are in the precontemplative stage and remain there. Other studies have shown that, after moving through some of the stages to action, participants may move back to precontemplation (Ounpuu 1996). It is possible that the number of stages may vary with culture or within settings, or even that the process of dietary change may not occur in stages. For example, instead of discrete stages we could think of change as occurring incrementally along a continuum.

Perhaps the main problem with the transtheoretical model is that it was developed in clinical environments, through observation of substance-abusing patients such as patients in tobacco smoking-cessation clinics. It was not developed from observational studies of dietary behaviours—which are more varied, and less addictive and stigmatised, than substance abuse behaviours. Many investigators have simply borrowed the ready-made model and applied it to dietary behaviours in the absence of thorough observation of dietary behaviours over time. It is also unlikely that the strong therapeutic support found in substance abuse clinics is available in most nutrition promotion settings.

Tailored mass communication

The use of tailoring approaches—particularly tailored communication messages (Brug, Campbell et al. 1999; Brug, Steenhuis et al. 1999; Horwath 1999)—is an interesting application of individualist models (especially the transtheoretical model) to population health and nutrition promotion. It has been made possible by electronic information systems. The theory has been used recently to tailor nutrition communication messages to individuals in the population. The messages are built from surveys of individuals' characteristics suggested by the theory, such as whether they have thought about altering aspects of their diet (e.g. fruit intake) or the difficulties they have experienced when doing so, and then tailoring advice to match their stage of change. However, tailoring does not depend solely on the transtheoretical theory—other theories, such as the theory of planned behaviour, can provide effective templates for tailored mass communication (e.g. Brug, Campbell et al. 1999; Brug, Steenhuis et al. 1999). Tailored nutrition promotion is about twice as effective as simpler mass communications in bringing about dietary change (Horwath 1999; see review by Rubinelli and Haes 2005).

Decision-making models

Decision-making models are about the ways people make decisions while they are engaged in complex behaviours in settings, for example, shopping for food in a supermarket (Underhill 2000). They can be regarded as developments of cybernetic models of behaviour but instead of only one goal being achieved there may be a series of 'mini-goals' to be achieved before the ultimate product goal. For example, the food shopper may be looking for a meat product that will make a nice evening meal, and she or he may examine and reject many products in the shop before finally selecting one for purchase

Bettman et al. (1986) have proposed a series of decisional paths that most people undertake when they go shopping in search of products with particular characteristics. Decision-making models describe the most likely paths that people take (Rice, 1993). Bettman's model illustrates the complexity of the purchasing process (Figure 5.6).

FIGURE 5.6 BETTMAN'S MODEL OF INFORMATION PROCESSING

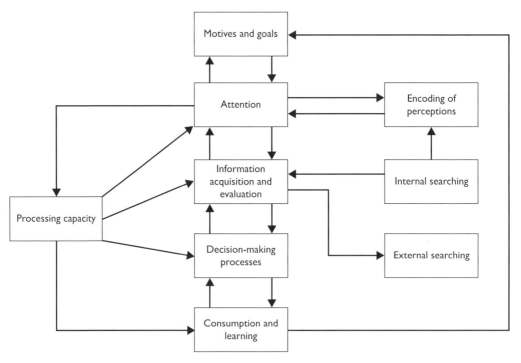

Source: Bettman (1986).

Relevance to nutrition promotion

The most obvious settings for the application of decisional models are those in which people have to make a series of choices in search of foods or beverages, such as supermarkets, cafes and restaurants. For example, in supermarkets signage or other information can be placed in positions which are most likely to be observed by shoppers and which will influence their buying decisions. For example, if a margarine is labelled as 'low in saturated fat' at the point of purchase, consumers who are interested in nutrition are likely to choose it over other similar products. Decisional models help us understand the search processes individuals use when reading nutrition information on food product labels—the presentation, layout and framing of such information is crucial to gaining shoppers' attention and thus influencing their product choices. We will discuss examples of this approach in Chapter 13, which examines nutrition communication programs in retail settings.

Cognitive consistency models

Several social psychological models first developed in the mid-twentieth century have been used by marketers and others in attempts to influence people's attitudes and behaviours (Greenwald et al. 2002). Collectively, these are called cognitive balance or cognitive consistency theories, with the most famous example being *cognitive dissonance theory* (Festinger et al. 1956; see Box 5.2). Their underlying assumption is that people find contradictions (or 'dissonance') between their beliefs and behaviours disturbing and so are motivated to change their behaviours (or beliefs) to reduce this uncomfortable internal conflict. Cognitive dissonance, then, can motivate behaviour change.

BOX 5.2 THE COGNITIVE DISSONANCE MODEL

If Belief A is consistent with Belief B then there is no conflict between them
Belief A and Belief B are consonant if A follows from B or B from A.

HOWEVER
If Belief A is inconsistent with Belief B
THEN there may be psychological discomfort > DISSONANCE.
This is motivating.

We tend to avoid information which may increase the dissonance and we may seek information which is likely to reduce it, including altering belief A or B.

For example, consider Tom, who habitually eats large amounts of fried and take-away foods despite knowing that this habit can cause heart disease. One day his obese best friend has a heart attack and dies. This forces Tom to think about his own life and health. He feels uncomfortable because his behaviours do not tally with what he knows about diet and heart disease. He could reduce his dissonance by (a) reducing his intake of fatty foods so that his behaviour is more consistent with his beliefs and knowing that he is being sensible, or (b) seeking information that is consistent with his unhealthy habits (e.g. that fats really are not all that important, and it is exercise that matters (and he does do a bit of exercise). Of course, if Tom's friend didn't die Tom wouldn't have experienced much dissonance and so he probably wouldn't have done anything. The amount of dissonance reduction is directly related to the amount of dissonance.

Dissonance theory can be used to influence people to change their behaviours or their attitudes and beliefs. Mass media communications can be used to reach 'difficult to reach' groups, such as parents of schoolchildren, with the intention of raising cognitive dissonance. A media message might propose something along the lines of:

> You say you are a good parent, but you feed unhealthy foods to your children. What are you going to do about it? Phone the number at the end of this advertisement for ways to improve the healthiness of the foods you give your children.

A development of the cognitive consistency approach is *motivational interviewing*. During interviews about their eating habits, interviewees are asked to translate the interviewer's questions into personal statements—for example, the question 'What is your favourite food?' is translated into 'What is my favourite food?' The interviewee then gives an example, such as 'Chips are my favourite food'. The interviewee is asked to explain why chips are their favourite food. The aim is to raise dissonance or discomfort about their current dietary habits so that they will feel motivated to change them in order to reduce dissonance.

Education models—group and discovery learning

Mass education has initiated some of the greatest changes in society over the past 150 years. Many accounts have been written about its effects on modern life, from self-actualisation to economic development. Educational methods have been used widely in nutrition promotion, mainly among children. Several reviews during the past 30 years have shown that nutrition education has been quite successful in bringing about changes in knowledge, beliefs, attitudes, motivations and dietary behaviours. Unfortunately, little of the vast effort put into nutrition education throughout the world has been adequately evaluated or reported, so its true impact is unknown.

Various models have been put forward to account for the education process. Skemp (1979), for example, distinguishes schooling, education and self-actualis-ation as key outcomes. Schooling is about the imparting of information which is necessary for the smooth running of society, such as knowing that drivers should stop at red lights and drive on the correct side of the road—if they don't know this they are a danger to other road users. Education is learning which mainly benefits the learner (e.g. studying nutrition in order to make a career). Self-actualisation is probably achieved by relatively few people. It is about taking control of our learning so our actions are no longer *unconsciously* influenced by environmental influences. For example, the control of our dietary behaviours would be based on our personal history of conscious decisions to choose healthy foods rather than on the influence of advertising and other external forces.

Education is the facilitation of the development of *cognitive frameworks* (or *schemas*) which enable individuals to assimilate and organise information. An example of a nutritional schema is the conservation of energy principle—that energy taken in by the body must be used in the form of activity and metabolic processes. This enables people to understand that excess energy will be converted to fat, leading to obesity. It also enables them to understand information about the energy content of food products (displayed in nutrition information panels on product labels).

Anderson (1983) proposed the 'spreading activation' model of human memory, according to which long-term memory can be compared to a fishing net spread out over a beached boat on a shore. If we pick up (or activate) part of the net, several knots will be pulled up together. Similarly, concepts of food are linked together in memory and are likely to be 'activated' when one of the concepts is mentioned—for example, when 'calorie' is mentioned in conversation we may think of concepts like 'waistline', 'too much chocolate' and so on. Education introduces new concepts and links them, hopefully in an organised manner, into a 'net' of concepts—or knowl-edge. Once a comprehensive schema or framework is established, we can carry on learning from new events because we can store (or 'assimilate') the new information in long-term memory. Unfortunately, many people's nutritional schemas are so limited that they cannot assimilate new information and so they become confused.

Many methods have been developed to achieve these educational aims. Bloom (1956) suggested that the education process has three aspects. The first is learning the language—becoming aware of, and able to name, the basic concepts. In nutrition education, this has often consisted of learning a list of nutrient names along with brief descriptions of their properties (e.g. 'vitamins A, D, E and K are fat-soluble vitamins'). This sort of approach is often associated with rote learning. It has merit, because the nutrition language does have its own words which have specific meanings. However, we have to ask whether the aim of nutrition education is to make everyone into a 'mini-nutritionist'. For lay people, the key issue is how to select healthy foods, so perhaps the most basic 'facts' they ought to learn is the food group taxonomy and the basic health properties of these groups of foods. Nutrient terminology is important because it is included in everyday parlance, but is it as important as basic knowledge of the food groups?

A second, deeper aspect of education is the ability to *analyse* problems. Again, nutrition education involves many dilemmas which people have to deal with every day, such as wanting to eat high-energy foods which will make them overweight. Analysis of these quandaries helps people learn basic principles of nutrition such as energy balance (conservation) and appetite control processes.

A third aspect of education, a form of deep-level learning, is *synthesis*. Again, nutrition promotion is full of possible solutions to problems which affect individuals, local communities and the broader world. For example, learners could shift their focus on individual weight management to the population problem of obesity prevention. Similar nutrition principles are involved but new schemas about public good, food and nutrition policy need to be brought to bear if solutions are to be created.

It is logical that the sequence of learning will be from naming and definition of concepts through analysis of problems to synthesis of solutions. But learning may not occur in this linear way. For example, adult learners are often best motivated if they are presented with a problem and asked to solve it. After much trial and error, they often have to resort to examining the meaning of basic concepts (like 'kilojoule') before they can analyse the key aspects of a problem (e.g. being overweight) and find a solution (e.g. reduce their intake of high-energy products).

Studies of nutrition education among school students strongly support the use of *discovery* and *cooperative learning* in nutrition education. Rather than being told in a didactic fashion what the 'facts' are, if children are given problems to solve (under guidance) they learn better, particularly if they work as a cooperative learning group in which inputs from all members are encouraged and valued. This form of learning is effective but more demanding of the teacher's time and effort than didactic methods. Combinations of didactic (and rote) learning approaches and problem-based cooperative learning may be the most cost effective. One danger of problem-based learning is that if students' efforts are unguided by experts in the area, they will waste a lot of time and are likely to make incorrect conclusions. That is, they may link concepts in ways which are demonstrably incorrect (e.g. that green vegetables are just as effective sources of iron as red meat).

Relevance to nutrition promotion

An appreciation of the issues involved in nutrition education implies that nutrition promoters need to carefully examine the learning needs of their clients and to consider the frameworks (schemas) of food and nutrition information that will be most acceptable and useful for their clients. Clients may not always view food and health in the same ways as nutrition promoters, and they may ascribe less value to health and nutrition promotion. A good example of such differences is to be found in the FoodCents project described in Chapter 11.

Adult learning issues

While a lot of attention is placed on nutrition education for children, adults are also capable of learning, either as parents or people at risk of degenerative disease. Adults as learners face particular problems. As a result, an area of education is now devoted to them: adult learning, or *andragogy* (Knowles et al. 1984; Brookfield 1994). Some of the key characteristics of adult learners are summarised in Box 5.3.

BOX 5.3 CHARACTERISTICS OF ADULT LEARNERS

Adult learning was pioneered by Malcom Knowles, who proposed the following characteristics of adult learners:

- Adults are *autonomous* and *self-directed*. They need to be free to direct themselves. Their teachers must actively involve adult participants in the learning process and serve as facilitators for them. Specifically, they must get participants' perspectives about what topics to cover and let them work on projects that reflect their interests. They should allow the participants to assume responsibility for presentations and group leadership. They have to be sure to act as facilitators, guiding participants to their own knowledge rather than supplying them with facts. Finally, they must show participants how the class will help them reach their goals (e.g. via a personal goals sheet).

Box 5.3 continued

- Adults have accumulated a foundation of *life experiences* and *knowledge* that may include work-related activities, family responsibilities and previous education. They need to connect learning to this knowledge/experience base. To help them do so, they should draw out participants' experience and knowledge which is relevant to the topic. They must relate theories and concepts to the participants and recognise the value of experience in learning.
- Adults are *goal-oriented*. When enrolling in a course, they usually know what goal they want to attain. They therefore appreciate an educational program that is organised and has clearly defined elements. Instructors must show participants how this class will help them attain their goals. This classification of gaols and course objectives must be done early in the course.
- Adults are *relevancy-oriented*. They must see a reason for learning something. Learning has to be applicable to their work or other responsibilities to be of value to them. Therefore, instructors must identify objectives for adult participants before the course begins. This also means that theories and concepts must be related to a setting familiar to participants. This need can be fulfilled by letting participants choose projects that reflect their own interests.
- Adults are *practical*, focusing on the aspects of a lesson most useful to them in their work. They may not be interested in knowledge for its own sake. Instructors must tell participants explicitly how the lesson will be useful to them on the job.
- Like all learners, adults need to be shown *respect*. Instructors must acknowledge the wealth of experiences that adult participants bring to the classroom. These adults should be treated as equals in experience and knowledge and allowed to voice their opinions freely in class.

Source: Lieb (1991).

Relevance to nutrition promotion

Adults often take important dietary decisions, both for themselves and for others. They often require help, advice and new learning in order to make these decisions (e.g. 'My ageing mother is losing weight, what can I do? Has she a nutrition problem? Where do I go for help? What should I feed her? Should she take vitamin pills?'). Because of their longer personal histories, they appear to learn in different ways from children. If nutrition promoters are to reach them, they must adapt to their learning styles.

Many adults suffer from at least one of three disadvantages. First, they often have poor self-confidence in their abilities, perhaps because of a series of unfortunate educational experiences. As a result, new concepts often frighten them. Second, they have an unfortunate inability to forget things. Older people in particular may have been schooled in the basic four or five food groups, and may hold beliefs such as 'sugar causes diabetes', 'old age causes decline in health' (as opposed to disease processes), 'heart disease is caused by consumption of too much fat and cholesterol', 'artificial chemicals cause most diseases'. These sorts of beliefs are not unique to the older generation, but the longer people live the more chance they have of acquiring such notions. More recent food taxonomies, and findings which discredit older orthodoxies, are often difficult to accept because the adult may have deeply held contrary views. Third, adults may present relational difficulties for teachers who may be used to dealing with children. For example, adults may be more questioning than children, and may expect more explanation than children. A related difficulty is that many adults are not familiar with the newer educational technologies (e.g. computers, the internet).

Conclusions

Many individual and small-group models and theories have been developed to explain health and food behaviours. Some of them can be quite powerful predictors of food selection behaviours in particular settings. In recent years, however, the limitations of too strong an emphasis on processes within individuals have become apparent and more emphasis has been placed on the understanding of social and environmental influences. These are discussed in Chapter 6.

Discussion questions

5.1   Discuss the advantages and disadvantages of expectancy value models of behaviour for nutrition promotion.

5.2   'Nutrition education is largely a waste of time.' Discuss.

5.3   Discuss the roles of learning principles for nutrition promotion among children.

5.4   Outline the transtheoretical model in relation to nutrition promotion.

5.5   What are the key principles of adult learning as they might relate to nutrition promotion?

5.6   Describe, with relevant examples, Bloom's education taxonomy and Skemp's views of the educational process.

6

Theories and dietary change:
ENVIRONMENTAL MODELS

Some models and theories are 'beyond the individual'. They deal mainly with the environmental settings and systems which affect the population's behaviours and health status. They suggest ways that nutrition promoters can use environmental influences or society-wide processes like mass communication to bring about changes in dietary behaviours. In Chapter 2 we examined concepts of the environment and society which are useful contexts for understanding the theories described here.

In everyday usage and in health sciences, the term 'environment' refers to things which are outside the individual. The environment is composed of the physical and social contexts in which an individual (or an organisation) behaves or operates. However, the food and eating environments are also influenced by another set of processes which, in some respects, are quite outside the normal awareness of individuals. These are the complex sensory and physiological processes that regulate appetite and eating. So this chapter opens by describing this internal milieau which forms our internal environment, before turning to an examination of external physical and social influences.

Internal environments: influences from within the body

A great deal is known about the biological regulation of food intake. It is not appropriate to describe these processes in detail here, but it is worth pointing out that several of these processes can be utilised in nutrition promotion. Two sorts of processes are relevant: sensory and satiety processes.

Sensory processes

Our liking for foods, and our subsequent consumption of them, are regulated by complex sensory and central psychophysiological processes. Information about the appearance, taste, smell and texture of foods—along with their price and perceived quality—is transmitted to our brains, where it is analysed and synthesised, often coming into consciousness as a liking or a dislike for that food. Usually we have prior expectations of what a food will taste, smell or look like (e.g. that liquid in a brown bottle might taste 'beery'); if the sensation does not match the expectation, we usually reject the food. This sort of testing is going on when we are shopping for, preparing or consuming food (Box 6.1).

BOX 6.1 SENSORIMETRICS—NOT ALL IN THE MIND!

What makes a food taste pleasant or acceptable, and how much of a difference between the experience of a food and our expectation is detectable or noticeable? The answers are found in *psychophysics* (or *sensorimetrics*). The pleasantness of a food depends on which compounds are in the food—for example, the amounts of herbs, spices, fats, sugars, salt and so on. Sensory psychologists, chefs and household cooks often try to determine the most pleasant concentrations of these compounds in foods. Tasting a food is a little like having a shower. When you get into the shower and adjust the water temperature, it may be too cold. Then you turn the hot tap on more and it may get too hot. Little by little, you adjust the temperature until it is just right. This is what psychologists call the 'bliss point'. So it is with food. A beverage manufacturer who wants to market a new beverage may organise a group of trained consumers (a taste panel) to taste several versions of the product. Some will have small amounts of sugar, others will have more. If the sugar level is too high or too low, the beverage won't taste good. There will be one point (the 'bliss point') when the sugar concentration will deliver a very pleasant sensation for most of the taste panel (see Figure 6.1). The manufacturer can then use the recipe with this level of sugar.

In reality, there may be other sensory qualities to consider, such as the mineral content and the gaseousness of the beverage. Even the appearance of the packaging can influence the pleasantness of the taste of the product. For example, if wine is put into a beer bottle, people will report that it doesn't taste good (it tastes 'beery'). Similarly, a strawberry milk drink might best be packaged in a pink-red package as this will reinforce the 'strawberriness' of the taste better than if the package were, say, blue or green. The key point to note is that, although sensory properties have major influence over food acceptance, they are largely outside consciousness. A good example of this lack of awareness is the fact that many breads contain a lot of salt but they do not taste salty. Potato crisps taste much saltier because the salt is on the surface of the crisps.

Box 6.1 continued

FIGURE 6.1 THE 'BLISS POINT'

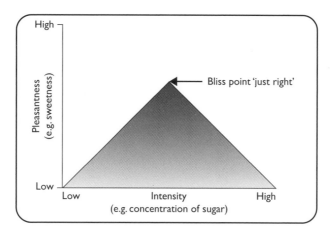

Source: Carey et al. (2003, Ch. 5).

Sensory psychologists establish the amounts of compounds which make a food product pleasant for most people, and they can also assess how much the concentration of a compound can be varied before the average person notices the difference. This is called the 'just noticeable difference', or 'jnd'. Particular products have specific jnd curves. The amount of salt in white bread can be reduced by as much as 25 per cent before most people notice the change, because the jnd is large. The jnd concept can be applied to other food properties as well as taste, such as the size of the product—its perceived 'value for money'. If the cost of chocolate rises, does the manufacturer increase the retail price, or reduce the amount of chocolate and increase the size of the packaging? In part, the answer to this quandary depends upon the relative jnds for the price, weight and perceived volume of the product. The ethics of such practices are debatable, however.

Source: Carey et al. (2003).

Relevance to nutrition promotion

A key point is that people's liking for food is based on processes that are largely out of their awareness. Second, changes can be made to the composition of foods without most people being aware of them or having to do anything about them. One of the major problems with the current food supply is that many products contain excessive amounts of fats, salt and sugars. For example, we know that, along with other processed foods, bread is a major source of salt for most people (in bread-consuming countries) and that excessive salt consumption leads to high blood pressure, stroke and heart disease.

Recent research clearly shows that small dietary changes, such as increased consumption of fruit and vegetables and small reductions in salt intakes derived from bread and other processed foods, can bring about significant reductions in systolic blood pressure in most people, particularly among people using anti-hypertensive medications (Nowson et al. 2005). Similar changes in the concentrations of other nutrients such as saturated fatty acids may yield similar health gains (Bao et al. 1998).

Given the popularity of processed foods, there are opportunities to reduce the amounts of 'harmful' nutrients in the population in a staged manner over time, since this is unlikely to harm manufacturers or consumers. Currently in the United Kingdom there is an industry–government campaign which aims to halve the amount of salt in processed food over the coming years (www.salt.gov.uk). The aim is to reduce population blood pressure and so reduce morbidity and mortality from chronic diseases (Scientific Advisory Committee on Nutrition 2003).

The jnd curve will play a large part in this process. Several manufacturers have already reduced the amounts of salt and saturated fats in their products using these psychometric principles. The World Action on Salt and Health (WASH) program aims to bring about reductions in the salt content of manufactured foods in a number of countries (see Chapter 12). Major dietary changes can be brought about in the population without individuals having to take voluntary action.

Satiety processes

The term 'satiety' can refer to the time taken to stop eating after commencement of a meal (intra-meal satiety) as well as to the time between meals (inter-meal satiety). Blundell and colleagues (Blundell et al. 1996) have described a series of psycho-physiological processes which affect human satiety, termed the *satiety cascade*. These include the tasting and chewing of food, swallowing, gastric distension and release of endocrines during digestion. Most satiety processes are outside human awareness, though the final end-point — 'feeling satisfied' — is in consciousness.

Research during the past two decades has shown that the ways in which we eat food, and the composition of food, can affect satiety. The rate of chewing of food and the size of mouthfuls of food can influence the signals sent to the brain to shut down eating (Smith and Thorwart 1998). Water and certain macronutrients are important food components. Barbara Rolls (Rolls et al. 1998) has shown that humans 'eat to gastric volume'—as the stomach distends, more signals are sent to the brain to shut down eating. Therefore, the addition of bulking agents such as water and fibrous material is likely to increase intra-meal satiety. The buffet meal is in part based on this principle since water-based courses (e.g. soup) and bulky foods (e.g. bread) tend to be served before the more energy-dense main courses. The more soup and bread people eat, the less high-energy foods they eat later in the meal. These processes can have important long-term effects—for example, habitual consumption of satiating foods like those contained in a substantial breakfast may be associated with reduced body mass indices (Figure 6.2) and thus reduced risk of metabolic disease. The respondents in Figure 6.2 were asked whether they had more or less than a standard breakfast of cereal, orange juice, toast and jam, and tea or coffee.

FIGURE 6.2 DOES BREAKFAST MATTER? RELATIONSHIP BETWEEN BREAKFAST AND BMI FOR MEN IN 1976

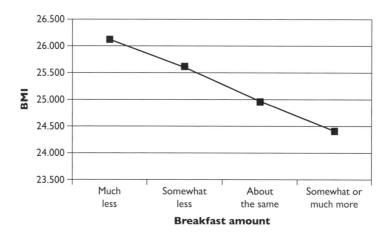

Source: Kent and Worsley (2006).

Relevance to nutrition promotion

Potentially, the satiating effects of food consumption styles, and especially of the components of foods, have major implications for nutrition promotion and the prevention of obesity. However, they have rarely been utilised in nutrition promotion programs. One of the main problems today is that people consume more energy than they require. The great abhorrence of dietary fats, which was encouraged by nutritionists in the 1970s and 1980s, probably led to the widespread consumption of fat-reduced products—the missing fat often being replaced by simple sugars, which are poor satiating agents. However, if meals such as breakfast contain satiating components such as dietary fibre or proteins, people are less likely to ingest as much energy during the day than if they do not include these components because of their satiating effects. There is some irony in the fact that the traditional English breakfast of bacon and eggs (full of proteins) may be less obesogenic than a breakfast composed of fashionable 'lite' foods like bagels or no breakfast. The inclusion of water-rich foods and water, and low-energy beverages like fruits and soups, tea and reduced fat milk drinks, is also likely to reduce the amount of energy consumed during each meal. The satiety indices of some common foods are shown in Box 6.2. In the box, each food is rated by how well it satisfied subjects' hunger, compared to white bread.

BOX 6.2 THE SATIETY INDICES OF SOME COMMON FOODS COMPARED TO WHITE BREAD

Less satisfying	(-%)	More satisfying	(+%)
Ice-cream	4	Muesli	0
Crisps	9	Hot chips (French fries)	16
Yoghurt	12	Cornflakes	18
Peanuts	16	Jellybeans	18
Mars chocolate bar	30	Bananas	18
Doughnuts	32	White pasta	19
Cake	35	Cookie	20
Croissant	53	Crackers	27
		Brown rice	32
		Lentils	33
		White rice	38
		Cheese	46
		Eggs	50
		Grain bread	54
		Popcorn	54
		Wholemeal bread	57
		Grapes	62
		Baked beans	68
		Beef	76
		Brown pasta	88
		Apples	97
		Oranges	102
		Porridge/oatmeal	109
		Fish	125
		Potatoes	223

Source: Holt et al. (1995).

The rate of change of circulating insulin affects both cognitive efficacy and mood state: intakes of high glycaemic index foods release a lot of glucose and, thus, insulin in a short time (Benton 2002; Benton and Nabb 2003). Often this is associated with higher levels of cognitive processing and feelings of elation. The consumption of low glycaemic index foods like whole grains, fruit and most vegetables tends to release lesser amounts of glucose but for longer times (Benton and Parker 1998). Meals (e.g. breakfast) composed of such foods are more likely to be associated with happier mood states for several hours after the meal.

Relevance to nutrition promotion

The composition of foods and meals, and their frequency and timing, may be important influences on cognition, mood state and body fat storage. In the past, nutrition promotion has mainly been associated with the prevention of non-communicable diseases. To a large extent, it still is. However, the evidence from food psychology suggests that nutrition promotion could affect people's cognitive performance and mood state. These are important possibilities, since these psychological domains are important to everyone. In particular, depression is a highly prevalent condition which is a major disease in its own right and an independent risk factor for heart disease.

While glycaemic load is a fashionable and important concept, other food constituents such as n-3 fatty acids and B vitamins are also likely to play a role in the moderation of mood state and the prevention of depression (e.g. Hibbeln and Salem 1995; Bryan 2004). Clearly the scope of nutrition promotion may be more extensive than previously thought. It includes the effects of nutrients on present psychological states as well as long-term, abstract disease-prevention outcomes.

External environments—influences from outside the individual

Many of the environmental influences on human food consumption can be divided into social influences, and physical and 'geographic' influences (Furst et al. 1996), such as:

- family and social influences;
- cultural and societal influences and lifestyles;
- usage or purchasing situations;
- trends in the food system;
- organisational influences; and
- physical—climatic and geographical—influences.

Family and social influences are described below, followed by physical and organis-ational influences, and finally mixed models such as precede–proceed and the Ottawa Charter.

Family and small-group functioning—the home

Humans are fundamentally social beings. They live in small groups, whether at home in their families, at work, with friends, in leisure and sports pursuits, or even in internet chat rooms. Even hermits have to be social, if only to know when to avoid other people!

The domestic household (the 'home') is one of the key environments in which people spend a lot of time in the preparation and consumption of food, and in talking and thinking about it. Despite the dominance of food behaviours in our waking lives, we know very little about how people interact with food in their homes. Much of our knowledge is based on anecdotes, hearsay or occasional surveys. We know, for example, that in some families, children as young as eighteen months of age exert great power over the family's food choices (Worsley and Crawford 2005), demanding their own particular diets. In other families everyone eats the evening meal at the table with the TV turned off (Campbell et al. 2002), and in about a quarter of families there is an argument during the evening meal at least three nights a week (alas, we don't know what they argue about!). Clearly there is a lot of inter-action around food in the home. It is the setting where children learn most of their food habits and where adults maintain theirs. The home is the centre of food action, and nutrition promoters need to know much more about what goes on there and find ways to communicate with people there.

There are two sets of food-related domestic phenomena which are relevant to nutrition promotion. The first concerns the structure and economic status of current households (such as weekly food expenditures). The second concerns current socialisation practices (the family is traditionally regarded as the setting in which children learn the ways of their culture—socialisation).

Structure of households

In many countries, there has been a revolution in the structure of families and house-holds during the past 30 years. There are many definitions of families, from the traditional 'extended' and 'nuclear' families (e.g. male and female parents and children) through to single-person headed household (e.g. a mother and her child). High divorce and remarriage rates have seen the emergence of 'blended' families, composed of a woman and her biological children and a man and his children. The Australian Institute of Family Studies has summarised the types of families within which most people live during the early twenty-first century (see Figure 6.3). These new household structures are more complex than they appear, especially when people's work lives are also taken into account. For example, the distribution of child care in couples with children varies greatly from 'traditional families' in which the woman does most of this work to newer arrangements in which the man does most of the child care or in which child care is shared (Laws, personal communication, 2006).

FIGURE 6.3 TYPES OF FAMILIES IN CONTEMPORARY SOCIETY

Couples with dependent children	38.6%
Couples with non-dependent children	8.4%
Couples only 35.7%	
One parent families with dependent children	10.7%
One parent families with non-dependent children	4.7%
Other families	1.8%
Note: single people are not included here	

Source: ABS (2002).

These different types of household can have quite different incomes, financial expenditures and outside work commitments, which will affect the ways they deal with food. In some households, food expenditure takes up only minor proportions of the household income, while in others it takes up a substantial proportion. In the United States, food expenditure accounts for only 8 per cent of the average household budget (Jeffery, personal communication, 2006).

In turn, these factors are likely to relate to people's food security, their tastes in food (e.g. luxury foods or basic foods) and the time they are able to commit to activities like shopping and cooking. For example, Cole-Hamilton and Lang (1986), working in London, showed that financially deprived people have to spend more time searching for cheaper but nutritious food than better-off people. Cohabiting and non-cohabiting people tend to have quite different food consumption patterns (Krammer et al. 1999; Worsley 1988). Similarly, mothers in paid employment outside the home tend to eat different foods from those who are not in paid employment (Worsley 1991).

Family income and education may also have an impact on food consumption. For example, university-educated men and women appear to consume a wider variety of food regularly than less-educated people; they also appear less likely to be regular consumers of several types of meat products (Worsley et al. 2004). Similarly, people from socioeconomically disadvantaged backgrounds in Brisbane in Australia are less likely to purchase foods that are high in fibre and low in fat, salt and sugar (Turrell et al. 2002). People employed in blue-collar occupations (those living in low-income households) purchase fewer types of fruit and vegetables, and less regularly, than more educated, better-off people (Turrell et al. 2002).

Relevance to nutrition promotion

Today, household (family) types are more varied than in previous periods of history. Nutrition promotion can only work within the constraints of the environments in which people live. The structures of households affect the time and finance available for food activities. Therefore, nutrition promoters need to be able to characterise people's access to these resources so that they can offer programs which are feasible for households to use. If such analysis is not done, then there is a strong possibility that nutrition promoters will speak only to the converted, excluding those people who live in households which do not have the resources to adopt food and nutrition innovations.

Socialisation practices

Families often behave according to 'rules' or 'set ways of doing things', which parents and children create or adopt over time. Many of these family practices (eating styles such as not answering the telephone during a meal, or eating breakfast alone) form part of the socialisation of children. Educators have looked closely at parenting practices in socialisation, though there has been relatively little formal examination of the ways in which socialisation practices are related to food.

Birch (1999; Birch et al. 1995) is one of very few investigators to do so. She suggests that among Americans parenting styles vary according to whether they are *authoritarian* (the child has to follow inflexible rules), *democratic* (the child is allowed some freedom to choose their foods), *authoritative* (the parent guides the child's eating in a firm but flexible manner) or *laissez-faire* (little guidance or support is exerted over the child's food choices). There is mounting evidence to suggest that authoritative styles help children to develop healthy food choices.

Birch has found that in families in which there is close monitoring of eating behaviours, female children are more likely to develop eating disorders. In Australia, Campbell (see Campbell and Crawford 2001) has found similar styles. Russell's recent work with Australian toddlers and their parents suggests that children's food preferences are associated with *neophobia* (fussiness or rejection of new foods), the parent's views of the child's personality (e.g. whether the child is outgoing or anxious) and only to a small extent by the feeding strategies employed by parents. It is clear, however, that encouragement, repeated exposure to new foods, modelling of eating of healthy foods and positive reinforcement are used by many parents to develop healthy eating habits in their children (Russell 2007; Russell and Worsley 2007a, 2007b).

Relevance to nutrition promotion

Socialisation styles are likely to be of particular relevance to children's nutrition promotion. Sensitive issues are involved here. For example, should nutrition promoters work within the styles adopted by the family or should they attempt to alter styles which are counter to healthy eating? Usually, they follow the family's styles on ethical grounds (respecting the family's right to do what they wish in their own lives) and on grounds of practicality (it is very difficult to change whole sets of social practices), but there may be occasions when parents actually want to adopt new parenting styles. For example, a mother who uses a permissive parenting style with her infant and who finds the child frequently refuses food may wish to adopt a more authoritative style. Usually, however, the promoter has to work within the family's social practices.

Some parenting styles may be so opposed to healthy eating (for example, where the main caregiver feels they have no right to insist on which foods an eighteen-month-old eats) that it may be more cost effective to work with families whose styles are conducive to the adoption of healthy eating habits. However, 'giving up on difficult problems' only allows eating problems to develop further. More creative strategies may be required which allow some form of continuing communication with those families.

The broader food environment

Story and French (2004) have drawn attention to the broader environment within which consumers purchase foods in the United States. They note that during the past 30 years purchases of high-energy foods and beverages such as pizza, confectionery and soft drinks have all increased markedly because of more intensive marketing. In addition, the prevalence of eating out or eating foods prepared outside the home has also increased, as have portion sizes (e.g. of popcorn, cola drinks), which have effectively become cheaper since the 1970s. These findings have been echoed in other countries (e.g. Dalmeny et al. 2003; Hastings et al. 2003). Other societal trends affecting the demand for, and the supply of, food have been outlined in Chapter 2.

The key 'environmental factors' that influence food purchasing include:

- *price*—when prices of low-fat foods were lowered in vending machines, their sales increased;
- *proximity*—people buy foods and drinks that are near them when they want them (e.g. soft drink vending machines are conveniently placed in schools and sports venues);
- *location*—in rural areas, foods are often more expensive and of poorer quality than in cities, as are foods in poorer parts of cities and on the edges of cities. For example, poorer people without cars cannot shop at 'hypermarkets' where foods are cheaper or they have to hire expensive taxis or use infrequent public transport to get to them. The likely influence of geographic systems and town planning on human food selection is receiving increasing attention from researchers. Reidpath et al.'s (2002) study of the relationship between obesity prevalence and the density of fast food outlets is a good example (Chapter 3).

Organisational change theory

Outside the family, many people spend a lot of time in organisations such as their workplace, where foods and beverages may be served. We will look at some examples of organisational food provision in Chapter 9. However, as we saw in Chapter 2, many organisations in society produce, distribute or regulate foods. They can play constructive roles, neutral roles or downright anti-nutritional roles. So it is important to look at the ways in which organisational change can be fostered. We need to distinguish between the formal and informal characteristics of organisations.

Formal properties

Organisations can have hierarchical, flat or mixed management structures. Typically, those with flat levels of management—for example, where the managing director oversees departmental managers who manage the staff—tend to be more democratic in that they allow workers to have some input into decision-making. More hierarchical organisations (such as universities) tend to have several layers of management between the directors of the institution and those who perform its core functions; they tend to be top-down, authoritative, 'corporate' bodies. The main purpose of commercial organisations is to make a profit and to maximise profits to owners or shareholders. In trying to maximise profits, they have to pursue narrow corporate aims such as selling as much fast food as possible, irrespective of any negative consequences for their customers, so long as they are not legally responsible for such damages. Public organisations such as government departments have more disparate aims, not all of which may be clear to staff or customers. They usually include some formal commitment to 'service to all'—which often means they are vulnerable to the influence of lobby groups.

Informal characteristics

In reality, many companies and organisations are run by small groups (cliques). They operate according to the values of the organisation—values like 'working and playing hard', 'being tough', being 'open' (or closed) to new ideas, or being 'flexible and progressive'. If change is to occur within an organisation, those with power and influence have to become involved in the change process. Usually, change is successful because it is driven by a small number of influential champions. People are influential either because of their power (they can fire employees) or because of their social stature or prestige.

Relevance to nutrition promotion

Any proposed changes such as the establishment of a healthy food canteen, the creation of a corporate food and nutrition policy or the production of lower salt/fat products have to be in line with the ability of a company to continue to make its usual profits (or greater profits). It is imperative that proposed changes are taken up and advocated by powerful or influential company managers. For example, one UK supermarket chain introduced nutrition promotion programs simply because the managing director developed heart disease. Such 'converts', because of their powerful positions, were then able to persuade or direct other managers and junior staff that the nutrition promotion programs would probably enhance their profits.

It is important that the proposed change should be consistent with the organisation's values. If the organisation has a tough 'macho' culture, the nutrition promotion program will be more acceptable if it is couched in similar terms (e.g. with the potential to improve physique or sexual performance). The key strategy is to identify and persuade key decision-makers with simple proposals for change. Usually a lot of persistence is required over long time periods.

A problem in Anglo-American societies lies in the short-term orientation of government and commercial business. Governments are mainly interested in being re-elected within a few years, and companies have to make their usual profit within each financial year or else their shareholders will complain bitterly and may dismiss the managing director. This climate does not foster long-term innovations. In other countries, such as those in continental Europe and East Asia, longer time frames for profit-making tend to foster business innovation (Hofstede 2005).

Diffusion theories describe the factors which influence the ways innovations like washing machines, televisions, oral contraceptives, DVDs, clothing fashions and nutrition ideas and practices are taken up by the population. Evert Rogers (1995) popularised the concept in his book *The diffusion of innovations*. Not surprisingly, marketers and social scientists have studied diffusion intensively. Some innovations take a long time to move through society—for example, the washing machine took well over a century to be commonplace in Western countries. Others, such as low-fat products, took less than 20 years and compact disks took even less time to penetrate to most groups within society. Key concepts in diffusion theory are out-lined in Box 6.3.

BOX 6.3 KEY DIFFUSION THEORY CONCEPTS

- **Innovators:** venturesome, cosmopolite, risk-taking, information seeking, with a higher financial status.
- **Early adopters:** greatest degree of opinion leadership, respected by other members of social group. Strategies with a motivational emphasis may be most effective at getting them involved in the diffusion process.
- **Early majority:** deliberate, adopt new ideas just before the average member of a system.
- **Late majority:** sceptical, adopt new ideas just after the average member of a system. The pressure of peers is necessary to motivate adoption. Intervention strategies that help them to overcome barriers are needed to get them to take up the innovation.
- **Laggards:** traditional, last in a social system to adopt an innovation, pay little attention to the opinions of others.

Note: Innovations that are perceived by individuals as having greater relative advantage, compatibility, trialability and observability, and less complexity, will be adopted more rapidly than other innovations (Rogers 1995).

Stages of adoption

- **Awareness:** extent to which a target population is conscious of an innovation.
- **Interest:** personal intrigue on the innovation.
- **Trial:** experimenting with the innovation.
- **Decision:** adopter decides to continue, quit or recreate the innovation.
- **Adoption:** continuation or integration of the innovation into lifestyle.

Source: Calvo and Rahrig (1997).

These theories distinguish people according to their position on the 'diffusion curve'. So we can talk in terms of early adopters of a particular innovation, late adopters and laggards. It is believed that early adopters are always on the lookout for new things and new ideas ('neophiliacs'—lovers of new things), and that they influence late adopters and eventually the 'laggards'.

Some of the barriers to the adoption of innovations include price (of the product) or costs such as the loss of time in using the new product, negative comments from friends (e.g. when people become vegetarian), or physical risks (e.g. of the side-effects of oral contraceptives). Rogers (1995) and others have shown that, in general, people from high social status strata of society tend to be early adopters (they have greater spending power and are in a better position to bear any costs), and they also tend to be involved in social networks in which they have greater exposure to innovations.

Relevance to nutrition promotion

Diffusion theories are useful for nutrition promotion because they draw attention to the social nature of humans and to the reality that they are influenced by particular types of communication sources and groups of people. Many nutrition promoters have been entirely concerned with the accuracy of the nutrition messages that they want to communicate and have rarely considered the ways in which communication networks encompass the membership of populations. Particular sections of the population will take more notice of some sources of information than others. For example, teenagers are much more likely to take notice of a young TV star than a 'crusty'. If nutrition promoters are able to plot the flow of information in the community in which they are working, identifying 'opinion leaders' and 'followers', then they are likely to be successful in communicating their messages (Chapter 13).

Social marketing

Social marketing (Andreassen 1995; Miaback et al. 2002) uses the same methods as commercial marketing, but it aims to benefit the population or community rather than the marketer. It has been defined as 'the systematic application of marketing along with other concepts and techniques to achieve specific behavioural goals for a social good' (Wikipedia).

Marketing methods are complex but effective (as the high sales of fast foods demonstrate). They are concerned with placing the right product before a selected population. This involves determining the wants and needs of the target population, developing and selecting products that meet these needs and wants, communicating the benefits of the products, identifying the best points of access to that population (distribution channels), appropriately pricing the product, and determining the product's placement in relation to competing products. A useful example is the Food Cent scheme, which does not promote nutrition but instead promotes the dietary pyramid as a way of saving money—something that is highly valued by its target audience (Chapter 8).

Donovan et al. (1999) summarise the application of marketing segmentation techniques to public health under the mnemonic TARPARE. Any population can be divided into various groups or segments according to various criteria such as socio-demographics (e.g. age and sex) or psychographic criteria (such as attitudes to food):

TARPARE assesses previously identified segments on the following criteria:
T: The Total number of persons in the segment;
AR: The proportion of At Risk persons in the segment;
P: The Persuasibility of the target audience;
A: The Accessibility of the target audience;
R: Resources required to meet the needs of the target audience; and
E: Equity, social justice considerations.

Donovan et al. (1999) note that the assessment can be applied qualitatively or scores can be assigned to each segment. Targeting of high scoring may lead to greater program success than targeting of lower scoring segments.

Relevance to nutrition promotion

Social marketing approaches are valuable techniques for nutrition promotion. They help to identify groups of people in the population who are interested in particular nutrition topics ('market segmentation'), and they enable us to establish the effectiveness of particular approaches. Marketing techniques can be used for health promotion as well as for their more common commercial branding applications.

The precede–proceed model

This popular health promotion model (Green and Kreuter 1991; see Figure 6.4) provides a useful checklist of issues that any health promotion program needs to consider. It combines aspects of individualistic theories with broader community-based approaches.

FIGURE 6.4 THE PRECEDE–PROCEED MODEL FOR HEALTH PROMOTION PLANNING AND EVALUATION

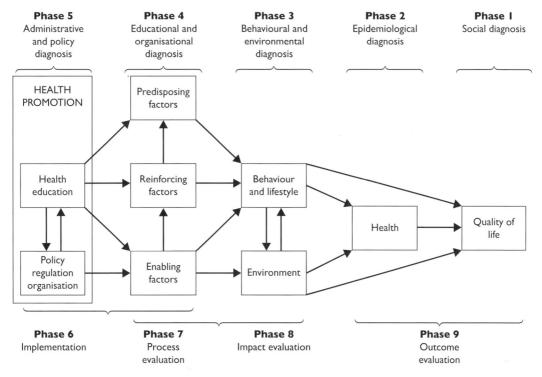

PRECEDE is composed of phases 1, 2, 3, 4, 5

| **Phase 5** Administrative and policy diagnosis | **Phase 4** Educational and organisational diagnosis | **Phase 3** Behavioural and environmental diagnosis | **Phase 2** Epidemiological diagnosis | **Phase 1** Social diagnosis |

Phase 6 Implementation   **Phase 7** Process evaluation   **Phase 8** Impact evaluation   **Phase 9** Outcome evaluation

PROCEED is composed of phases 6, 7, 8, 9

Source: Pomeroy (2007).

The model provides a useful framework which includes three sets of influences on people's behaviours: predisposing factors, enabling factors and reinforcing factors. *Predisposing factors* include attitudes, beliefs and values (guiding principles in people's lives, such as egalitarianism, achievement motivation, tradition and security). *Enabling factors* include skills (e.g. the person's cooking abilities), availability (e.g. the availability of fruit in the school canteen), accessibility (e.g. the position and pricing of fruit in the school canteen) and referrals (i.e. whether the person can be referred to sources of help to maintain or change their behaviours). *Reinforcing factors* include support from family, peers, teachers, employers and health providers, among other sources.

Health program planning occurs in a series of stages, or 'diagnoses', as outlined in Figure 6.4 (epidemiological diagnosis, behavioural diagnosis, etc.). From the point of view of nutrition promotion, an additional level should be included—*the food and nutrition diagnosis*—the state of the population's food consumption and nutrition status.

Reconciling individual, population and environmental approaches: The society–behaviour–biology model

The foci of the various theories reviewed so far differ substantially, from very broad influences to highly specific influences on individuals' behaviours. We have seen that these theories can be very useful for specific nutrition promotion challenges. Apart from the precede–proceed model, however, there has been no over-arching model which links the various environmental, behavioural and biological processes together within a population health context. Recently, Glass and McAtee (2006) have proposed the society–behaviour–biology model to reconcile the individual processes with broader environmental and biological influences. They use the analogy of a river to explain these relationships (see Figure 6.5). The temporal aspects of population health are represented on the horizontal axis, and the organisational levels are on the vertical axis. The various types of social and environmental influences ('uphill' influences) are above the water and the biological influences are 'underwater'. The various opportunities and constraints on human actions and behaviour are specified as various sets of population health 'risk regulators', such as material conditions (e.g. food insecurity), discriminatory practices, neighbourhood and community resources and influences, behavioural norms, conditions of work and laws and policies. All these factors in various settings and environments can influence the health and health behaviours of the population. As such, this model brings together much of the recent literature about the various environmental, social and biological influences on health (and food) behaviours.

The main contribution of this model is that it identifies new areas for research and application—for example, how do dietary norms actually affect the population's food consumption behaviours? Like several of its predecessors, the model provides a diagnostic tool for nutrition promoters in particular communities and settings. For example, in a kindergarten how can food policies be implanted to influence dietary norms and behaviours, given the kindergarten's material conditions and type of neighbourhood?

FIGURE 6.5 THE SOCIO–BEHAVIOURAL–BIOLOGICAL MODEL

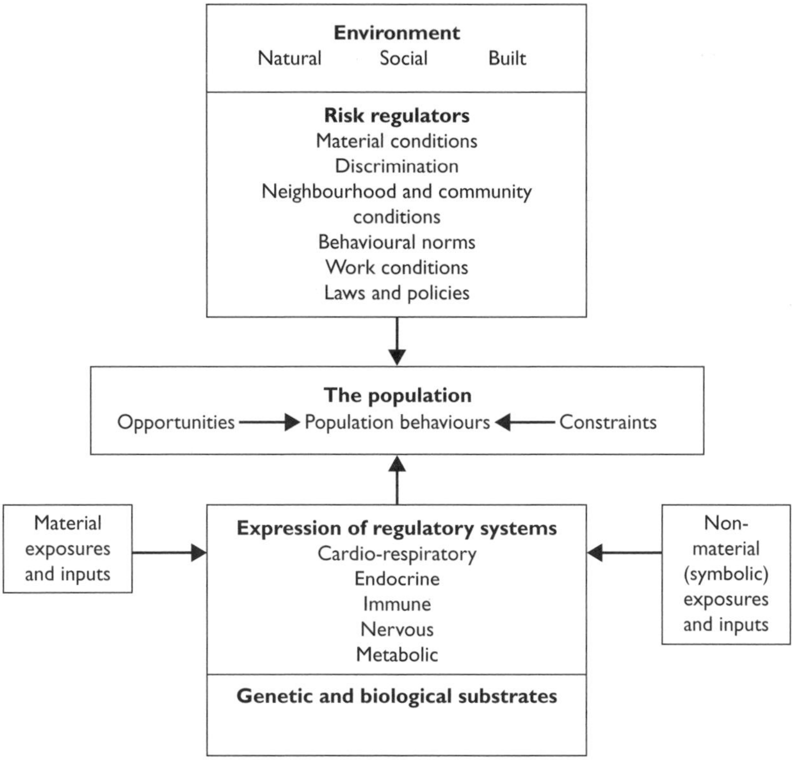

Environment

Natural   Social   Built

Risk regulators
Material conditions
Discrimination
Neighbourhood and community
conditions
Behavioural norms
Work conditions
Laws and policies

The population

Opportunities ——▶ Population behaviours ◀—— Constraints

Material
exposures
and inputs

Expression of regulatory systems
Cardio-respiratory
Endocrine
Immune
Nervous
Metabolic

Non-
material
(symbolic)
exposures
and inputs

Genetic and biological substrates

Source: Glass and McAtee (2006).

Conclusions

There are many models and theories which have been developed to explain health and food behaviours. In addition to the common individualist models which have been used to explain health behaviours, such as the transtheoretical model and the theory of planned behaviour, a number of environmental and systems models have been used to explain the broader influences on health behaviours. Further integration of environmental and individual approaches is likely in future, as well as a greater emphasis on *population* health and food behaviours.

Discussion questions

6.1 Discuss the advantages and disadvantages of expectancy value models of behaviour for nutrition promotion.

6.2 Compare and contrast the usefulness of the transtheoretical model and precede–proceed for nutrition promotion.

6.3 'Much nutrition promotion is flawed because it assumes conscious control over non-conscious processes.' Discuss the arguments for and against this proposition.

6.4 Why are sensory and satiety processes important for nutrition promotion?

6.5 Compare and contrast the roles of individual and environmental influences in nutrition promotion. Give examples to illustrate your points.

6.6 'Nutrition education is largely a waste of time.' Discuss.

6.7 Describe the main elements of: (a) the precede–proceed model; and (b) social marketing. Give food examples.

7
Change methods:
DESIGNS AND EVALUATION

Introduction

Nutrition promotion faces several major methodological issues. They include defining the types of behavioural change, the paradigms which underlie change activities, the criteria for successful change, the nature of the contrasting methodologies, particularly the intervention project, action research and systemic change, and, finally, issues of planning and evaluation.

The key questions to keep in mind in nutrition promotion are:

- Who or what do we want to influence?
- What do they want to do?
- What do we want them to do?

The scope for change

The answer to the first question can be complex. The question is about the *scope* of nutrition promotion. Nutrition promotion can occur at all levels of society in a wide range of settings. It can be directed towards:

- individuals—for example, in clinical settings between health professionals and patients or between fitness instructors and their clients;
- groups of people—for example, within families at home who undertake joint education programs such as in diabetic education, or between parents and children;
- people in workplaces;

- children at pre-school centres and at school or university;
- shoppers at supermarkets, or customers at restaurants, takeaway and fast food outlets;
- people in local communities, regions or countries;
- mass media consumers;
- managers of food manufacturing and retail companies;
- health and education professionals; and
- people in many other settings.

The scope or range of nutrition promotion is constrained by our views of society. We can think of society in many ways, in terms of individuals, or groupings and institutions such as companies, schools, hospitals and workplaces, or as categories of people like women, older people, people from non-English speaking backgrounds (now termed CALD—culturally and linguistically diverse backgrounds) and so on.

If we see society as only being made up of individuals and their immediate families, then we will try to reach people in settings where individuals abound, like doctors' surgeries or children in schools, where they are (hopefully) learning knowledge and skills which will prepare them for their future lives. However, if we see society as being made up of institutions, social networks, companies and other coalitions, as well as individuals, then there is far greater scope for nutrition promotion because we can communicate with those social groupings (and in particular their leaders) as well as individuals. This societal view will also affect the goals of any nutrition communications. If we focus on individuals as the 'prime movers', then we will probably aim to communicate knowledge and skills which will help them select healthier foods. However, if we adopt the view favoured in public health that settings, environments and *risk regulators* influence populations' behaviours and health status, we will put more effort into altering the settings and other factors which influence the population's food activities.

So the levels of organisation of society (individuals, social groups, categories) and the settings in which people are found can profoundly influence the ways in which nutrition promoters think about their goals and activities. We will go further into these distinctions later in this chapter, and in the settings chapters. Perhaps one initial point to make is that all of these social phenomena have a temporal aspect to them—that is, they change over time. Further, some institutions like health services and school systems can purposely be changed over time. For example, managers of school systems can make plans to increase the amount of healthy foods their children eat. Systematic change within settings is an important but frequently neglected opportunity for nutrition promotion.

Program aims must be consistent with settings and levels of organisation

The levels of social organisation and the settings in which food activities occur place practical limitations on nutrition promotion aims and the methods which can be employed to achieve them. However, a wide range of nutrition promotion and communication activities can be conducted across several levels of society using diverse methodologies:

- *The level of the individual* — Promotion is oriented to the actions an individual can take and to the settings in which individuals operate. So behaviours such as food shopping, dining out, cooking and eating can be influenced. This means that theoretical models which deal with individual behaviours are useful, such as the food-related lifestyle model (Grunert et al. 1997), choice and decision-making models, social cognitive models, learning theories and so on. Typical promotion methods include educational methods like didactic learning, discovery and group learning (especially for people in institutional settings like children at school), mass media messages aimed at individuals' food decision-making, point-of-sale programs in retail settings and more. In addition to bringing about changes in individual food consumption, nutrition promoters typically try to bring about increases in food and nutrition knowledge, changes in expectations about serving sizes and frequency of intakes of various food groups, and changes in beliefs and attitudes towards the consumption of food.
- *The level of the group or organisation* — Usually organisations and social institutions influence individuals' behaviours, but they may do so without the individuals being aware of their influence. For example, schools and hospitals can change the range of foods they provide, perhaps choosing not to supply high-energy snacks, favouring fruits and less processed foods and beverages. Some kindergartens do not allow parents to bring any pre-packaged foods, encouraging them to bring fresh, less processed food instead. Many childcare organisations and schools make their own food policies (sets of rules) to guide their clients' food behaviours. Supermarket and restaurant chains may make their own policies to assist customers to make healthy food choices — for example, Quinn's Supermarkets in Ireland (now SuperQuinn) was the first chain to stop the sale of confectionery products in checkout aisles so parents would not be harassed by their children's attempts to purchase them. The main methods used by these organisations are management rulings

communicated to staff, clients and customers by word of mouth and written messages.

- *The level of the locality or region* — Local governments may control food safety and environmental health and health services in their localities. They may decide, as Maribyrnong Council in Melbourne did, to devise their own local food policy which encourages food security for its poorest residents and the provision of bottled water and fresh fruit in its schools. Local council by-laws and regulations are used to enforce these policies (www.maribyrnong.vic. gov.au/Files/Food_Security_Policy.pdf).
- *The level of the nation and international community* — At this level, there are nationwide and international systems which affect governments, companies and individuals. Chief among them are the mass media (and the internet), which broadcast food and health-related messages in advertisements, on radio and TV programs and in print articles. Food advertising is often directed to children but can be regulated, if not banned, by national governments or limited through industry codes of practice.

Governments can also have a profound influence on the availability of food through their regulatory powers. If they wish, they can prevent or limit the import, manufacture and sale of foods on health grounds (e.g. because products have too much saturated fat or salt in them) or they can discourage consumption of some foods (like high-energy snacks and beverages) and encourage the consumption of other foods (like fresh fruit and vegetables) through taxation penalties and incentives. In Anglo-American countries such as the United Kingdom, Australia, New Zealand, Canada and the United States, such powers are rarely used mainly because of the dominance of neo-liberalism in these countries. However, other countries are more interventionist; Greece and Sweden, for example, limit the exposure of children to food advertising. (Neo-liberalism is a belief system which proposes that government should interfere as little as possible in the lives of individuals, preferring to let 'market forces' provide services and goods. It contrasts sharply with conservative or social democratic philosophies, which propose stronger roles for governments in the life of society.)

Governments also have the ability to communicate about nutrition, principally through regulations concerning the information which may be put on to food product labels (e.g. ingredients lists and nutrition information panels). Food Standards Australia and New Zealand, the European Community and British Food Standards Agencies and the Food and Drug Administration (USA) have these types of powers. Codex Alimentarius, an international body which helps regulate the international food trade, has provided the generic format for the food information on food product labels. Despite these opportunities, public education campaigns about the ways food labels can be used to promote healthier food choices have rarely been conducted by governments (the implementation of the US *Nutrition Labelling and Education Act* being a notable exception). However, from time to time non-government organisations and supermarket chains have used the food label to promote healthier food choices. For example, the National Heart Foundations in Australia and New Zealand have devised 'Pick the Tick' logos to guide consumers to products which contain reduced amounts of saturated fats and salt. Related schemes run by non-government organisations will be described in more detail in Chapters 12 and 13.

At the international level, notable examples of nutrition promotion include the World Health Organization's fruit and nutrition promotion programs (www.who.int/dietphysicalactivity/fruit/en/print.html), including the International Fruit and Vegetables Alliance (IFAVA) (http://ifava.org/), the World Bank (and UNESCO, WHO and FAO), the Health Promoting School scheme into the FRESH program (Focusing Resources on Effective School Health, www.fresh schools.org) and related programs (discussed in Chapters 8, 9 and 10). IFAVA's mission is to encourage and foster efforts to increase the consumption of fruit and vegetables globally for better health by supporting national initiatives, promoting efficiencies, facilitating collaboration on shared aims and providing global leader-ship — all of which is based on sound science.

Finding out what people need and want, and setting goals for nutrition pro-motion programs, requires thorough needs assessment methods, clear program planning and justification of program aims. Usually program designers are assisted by advisory groups made up of key stakeholders who meet regularly to assist in the design and management of the program. The promoters have to be able to justify their aims through a series of diagnoses (see Green and Kreuter 1991; see also precede–proceed model, Chapter 6), which they may present to the stakeholders. The views and needs of members of the population can be assessed through a formal needs assessment process (see below). Differences between the promoters' priorities and those of the population need to be resolved well before any implementation is undertaken. In community-based health promotion, the community's needs and views take precedence over those of the promoters — indeed, the main aim of this form of promotion is to respond to the community's priorities.

A typical dilemma faced by health promoters is illustrated in the example of an obesity prevention practitioner who is funded to begin the prevention of obesity in a locality but who soon finds that community representatives want to take action about other problems such as food insecurity issues or the presence of 'pesticide residues' in locally produced food. Further, they are outraged at the thought that many members of the community are overweight. What does the obesity prevention practitioner do? Well, there are several options: try to persuade the stakeholders of the importance of obesity prevention; find another more congenial group of advisers; or work with the stakeholders on *their* problems in exchange for help with the obesity issue. The last scenario would be sure to raise eyebrows in the funding agency.

Paradigms

Our views of society (and nutrition promotion) are examples of *world-views* or *paradigms*. This term was introduced by Thomas Kuhn in the 1960s in his attempts to understand the nature of science. He defines a paradigm in terms of the basic axioms or assumptions which scientists (and in effect all humans) use to build up their scientific explanations (Kuhn 1962). Euclid, for example, assumed that there were points, lines and angles—he didn't prove that they existed, but he showed that they were useful concepts for understanding geometric phenomena. From these basic axioms, he was able to build up a system of geometry which is still used today for many practical applications such as geometric positioning systems and building design.

The intervention project

There are at least three distinct 'world-views' which are encountered in health and nutrition promotion. The first is often described as a 'top-down' approach which suggests that nutrition promotion is something expert nutrition promoters do to people or institutions (see Chapter 8). The promoters are the people with access to expert (nutrition) knowledge, so they should know what are the healthiest things to do in a given situation, while the targets of the health promotion are essentially ignorant of, and unskilled in, nutrition issues. This is the 'expert' view of nutrition promotion. This world-view has led to the implementation of many interventions during the past century, some of which have been successful in changing people's behaviours in healthy ways. It is usually focused on the individual, usually on women and children; it is highly nutrio-centric rather than food oriented (i.e. targeting changes in *nutrient status* such as saturated fat reduction or increases in vitamin or mineral intakes). To a degree, it is a sensible approach since nutrition promoters usually have advanced knowledge of the latest findings in nutrition so one might expect them to lead change in the community. However, in other respects it is a very narrow view of nutrition promotion which tends to ignore people's needs, desires, habits and the influence of their social settings. The eradication of iodine deficiency (see Chapter 10) is a successful example of this approach.

The notion of the intervention project evolves from this paradigm. The main strength of this paradigm is its ability to examine how various factors influence nutrition promotion outcomes. For example, intervention projects can be designed to examine the effects of group learning among schoolchildren (compared with individual didactic learning) on nutrition-related outcomes. Or they might compare the relative effects of the school canteen versus individual knowledge acquisition on children's fruit consumption. Or they may wish to compare the responses of boys and girls to a standard healthy eating program in order to decide on its suitability for general release through a school system.

To make these comparisons, interventionists use experimental designs, often in the form of randomised control trials, so they can measure the effects of their interventions. Often exact controls are not feasible, so quasi-experimental designs may be used instead. For example, a supermarket chain may plan to launch a fruit and vegetables campaign so investigators may use another supermarket chain as a comparison group in an attempt to estimate the effectiveness of the first chain's nutrition promotion program. Both experimental and quasi-experimental designs have been well considered during the past century and several texts describe them (e.g. Shaddish et al. 2002). Their basic point is to allow the estimation of the effects of particular aspects of interventions.

The reasons for doing so vary. A good reason would be to establish the effectiveness of a demonstration project in order to raise awareness among others (e.g. other teachers, communities, companies) about the feasibility of making particular nutrition changes, or perhaps to show that a particular approach is both more effective in the long term and less expensive than another approach (e.g. that education of the whole family is more effective than education of an individual suffering from diabetes). This type of well-considered work can help develop better theoretical frameworks for nutrition promotion within particular settings (e.g. French's work on the use of vending machines in schools: see Chapter 8). However, judgment of the utility of any intervention also depends on a number of other factors such as sustainability, unintended negative consequences and generalisability.

The experimental (or 'counterfactual') approach in public health has come under intense scrutiny and debate in recent years. Many see it as inappropriate (Glass and McAtee 2006). For example, deliberate experimental changes in the siting of fast food outlets are likely to be impractical so their effects cannot be directly tested, yet there is compelling observational evidence that fast food outlets are part of a health-hostile food environment.

Action research

In several ways, action research is the antithesis of the intervention project. It places great emphasis on a 'bottom-up' approach and on widespread consultation with 'stake-holders'—those groups and individuals who have some form of involvement or 'vested interest' in the promotion program (see Box 7.1). Indeed, participation by stake-holders in the design, implementation and evaluation of the promotion program is usually actively encouraged. Most strikingly, action research tends to eschew experi-mental designs and hypothesis testing in favour of a strong goal orientation combined with the use of any combination of processes which will enable the attainment (or modification) of the program goals. In this sense, action research sacrifices scientific explanation for gains in flexibility, relevance and outcomes. It is the approach favoured by economic and community developers, and by many health promoters. The approach can be fairly narrow and focused on clearly defined, specific goals or it can be more diffuse, concentrating more on the development of harmonious commu-nity relationships and using a series of goals to do so. The Health and Social Development Councils which ran in South Australia during the 1980s were good examples of this latter approach. Various community-initiated goal-driven projects, such as the control of local air pollution, were used to disseminate the advocacy skills base among under-privileged people (Worsley and Shannon 1991). Several definitions have been coined for action research, as shown in Box 7.1.

BOX 7.1 DEFINITIONS AND STAGES IN ACTION RESEARCH

Action research is a three-step spiral process of: (1) planning which involves reconnaissance; (2) taking actions; and (3) fact-finding about the results of the action. (Lewin 1947a and b)

Action research is the process by which practitioners attempt to study their problems scientifically in order to guide, correct and evaluate their decisions and actions. (Corey 1953)

Kemmis (1988) suggests that action research should have four cyclical phases:

1. **Planning**—or definition of the problem and organisation of research practices.
2. **Acting**—or implementation.
3. **Observing**—or action and collection of data. Observation is followed by subsequent reflection and action.
4. **Reflecting**—and developing revised action derived from what has been learned.

Action research can be used in practically any setting and for any purpose. It is commonly used by health promoters and reflects the primary values of public health—for example, the high valuation of equity and democracy being expressed in terms such as 'participation', 'empowerment' and 'consultation'. Nutrition promotion differs from health promotion in that, for much of its long history, action research has not been as dominant, intervention projects being more prominent.

Total quality management—systemic change in organisations

This paradigm can be seen as a derivative of action research. It is mainly concerned with the ways in which the functioning of organisations like businesses and government departments can be made to change over time. Typically managers of businesses wish to improve their profitability or to maintain or increase their market share. To do so, they often resort to making changes in the running of the company—for example, increasing their expenditure on advertising, merging marketing and technical divisions into innovation centres, putting more sales staff 'on the road' and much more. Astute managers record the consequences of these decisions so they are able to discern successful decisions from unsuccessful decisions. The successful decisions become 'institutionalised' or incorporated into the normal running of the company. That is, the system changes in the long term.

Some of the settings in which food and nutrition are relevant are systems rather like these businesses—for example, hospitals, schools and supermarkets and the mass media. To date, however, there have been relatively few nutrition-related systems which have attempted to permanently include nutrition communication and promotion into their day-to-day activities. One could imagine a school system attempting to make its food purchases reflect the dietary guidelines, regularly monitoring the extent to which it did so and developing ways to ensure that the gap between the guidelines' ideal and actual practice was narrowed. We will examine some examples of systemic change programs in several settings along with opportunities for future systemic change in the settings chapters, especially Chapters 11 (health systems) and 12 (food industry systems). A useful description of the application of Total Quality Management approaches to health systems is given by Ader et al. (2001).

Changing systems', organisations' and professions' operations

Glasgow et al. (1999) proposed the RE-AIM framework for health promotion program planning. It has several methodological dimensions including:

- *Reach*—the proportion of the at-risk population that are involved in the program. Many health-promotion programs target those at high or moderate risk but Rose (1985) suggests that even those at low risk should be targeted.
- *Efficacy*—the degree to which the program brings about positive (or negative) effects. Unwanted negative 'side effects' such as increase in social stigmatisation are quite common and can undermine programs. Promoters have to be certain that the harm caused by a program is 'acceptable' and does not outweigh the benefits of the program.
- *Adoption*—the proportion of individuals and settings (like worksites) that take up a program. Identifying and overcoming barriers to adoption are major health-promotion priorities.
- *Implementation*—the extent to which the intended objectives and goals of a program are delivered.
- *Effectiveness*—is about the to degree to which intended effects are brought about in 'real life' settings by (non researcher) employees, such as teachers, managers, nurses. (Effectiveness can be defined as the product of the degree of Implementation multiplied by the degree of Efficacy.)
- *Maintenance*—the duration that the changes brought about by the program last. As a program becomes a routine part of the organisation's practices it is more likely to be maintained.
- *The Public health impact*—of a program depends on how well the above themes are put into practice. Some of them are discussed below as criteria for change.

Criteria for change

There are several criteria which can be used to judge the adequacy of any nutrition promotion program. The various paradigms differ in the ways they meet these criteria. The criteria are based upon general health promotion principles and on the values espoused by public health.

Effectiveness (and efficacy)

These criteria are about the extent to which the program meets its aims and goals. There are broad aims of nutrition promotion which include the promotion of healthy eating and high nutrition status among as many people as possible within a population group. However, there are also narrow, program-specific goals such as 'increasing the number of pieces of fruit consumed to three pieces per day' or 'to raise haemoglobin levels to greater than 12 mg/dl'. These appear straightforward but, as we saw in Chapter 1, the definitions of 'healthy eating' and 'adequate' or 'high' nutrition status can be problematic. When examining reports of health promotion programs, we need to question the choice of project aims and goals carefully. For example, why was fruit intake (or saturated fat intake) chosen as a goal? Were there other aspects of diet which might have changed as a result of the pursuit of this aim—for example, in their quest to eat three pieces of fruit per day, did participants alter another part of their usual diet such as their vegetable intake and if so was this nutritionally desirable or not? Was the chosen goal feasible in the lifestyle context in which the participants lived? For example, if fruit was not available at reasonable prices at certain times of the year (as in much of northern Europe), was this a feasible or 'sensible' aim?

Specificity of program goals

Goals should be as specific and as concrete as possible. They should be expressed in terms of the behaviours the participants should perform in order to achieve the general aims involved. So if the aim is to increase the consumption of fruit among a population, a specific goal might be the consumption of a piece of fruit at 11.00 a.m. every day (as happens in classrooms in Footscray Primary School, described in Chapter 8).

The assessment of effectiveness or efficacy is often more difficult than it may at first seem. Beware of average or mean changes. While fruit intake may have risen by a mean of one piece a day (from, say, a mean of 0.5 pieces per day), a considerable proportion of participants may not have increased at all or even reduced their intake. Arithmetic means hide a lot! The strength of the effect of a program is usually calculated by the effect size, which can be expressed as either the difference between the mean effects observed in two groups (e.g. treatment and control groups) or as a proportion of the standard deviation associated with the program. Effect sizes are useful because they allow comparisons to be made between different programs or studies; meta-analyses of the literature are based on effect sizes (see Rosenthal et al. 2000). Many different indices have been used to calculate effect size.

Validity of results

Observed changes may or may not be due to the program—that is, the *validity* of the program may be unclear. How do we know that the apparent effects of a nutrition promotion program are in fact due to the program and not to some artefact? There are several types of factors which threaten the validity of conclusions from (nutrition) promotion activities (Shadish et al. 2002; see also Box 7.2). For example, the whole community may change its food intake at the same time as a program is run, perhaps because of seasonal conditions (e.g. more fruit being available in spring and summer), or there may be rising consciousness of the importance of fruit. This is where experimental and quasi-experimental designs (which often incorporate comparison groups) are useful, though they are often expensive.

BOX 7.2 THREATS TO THE VALIDITY OF RESEARCH DESIGNS

Internal validity refers to the extent to which it is possible to make an inference that a relationship is causal (the experimental manipulation resulted in the observed differences). According to Cook and Campbell (1979), there are thirteen threats to internal validity:

1. *history*—events take place between the pre-test and the post-test that are not the treatment of research interest;
2. *selection*—difference between kinds of people in one experimental group as opposed to another;
3. *maturation*—observed effect is due to respondent growing older and wiser between the pre-test and the post-test, when this maturation is not of research interest;
4. *testing*—familiarity with a test where items and error responses can be remembered at a later testing;
5. *mortality*—different kinds of people drop out and the experimental group are composed of different kinds of persons at the post-test;
6. *instrumentation*—when the effect might be a change in the measuring instrument between pre-test and post-test and not to the treatment's differential impact at each time interval;
7. *statistical regression*—movement of extreme scores toward the mean in pre-test/post-test designs where the treatment may have not been the cause;
8. *interactions with selection*—selection–history, selection–maturation, selection–instrumentation;

Box 7.2 continued

9. *ambiguity about the direction of causal inference*—not sure whether A caused B or B caused A or if A and B interacted in a non-causal way;

10. *diffusion or imitation of treatments*—the control group gains access to the treatment;

11. *compensatory equalisation of treatments*—may insist that the control group receive the same treatment;

12. *compensatory rivalry of respondents receiving less desirable treatments*—attempt to reduce or reverse the expected treatment effect;

13. *resentful demoralisation of respondents receiving less desirable treatments*—effects may be due to reactions rather than the treatment.

External validity refers to the ability to generalise to particular target populations, settings and times, and generalising across particular target populations, settings and times. There are three threats to external validity:

1. interaction of selection and treatment—those who volunteer and decline;

2. interaction of setting and treatment—bias in the settings or the organisations which participate;

3. interaction of history and treatment—circumstances under which the study is conducted.

Source: Shaddish, Cook and Campbell (2002).

There are two important types of validity. *Internal validity* refers to the correspondence between a measure and what it was intended to measure (e.g. the degree to which self-reported body weight actually corresponds to true body weight). *External validity* refers to the generalisability of a study's findings. For example, an experiment may show that a parent-led approach in one school may boost children's fruit and vegetable consumption. The external validity would be the extent to which these findings would also be found if the approach were used in other schools.

Feasibility

Program goals may be achieved, but that may take a long time and a lot of money; other equally valuable goals might have been achieved within less time for a smaller cost. In our experience, we have found that many people find it much more difficult to increase their vegetable intakes than their fruit intakes (Nowson et al. 2004).

Cost benefits (cost effectiveness)

A key aspect of feasibility is cost effectiveness. A program may bring about some desired benefits but at what costs? The costs of a program can be assessed in several ways:

- the financial expense outlaid on the program, including the estimated costs of volunteer labour;
- the amount of time taken to achieve the goal—could other goals have been achieved more quickly?
- opportunity costs—while a particular program was being pursued could other programs have been undertaken which would have been more effective?
- negative consequences and the precautionary principle (the 'do no harm' principle of public health). Often projects and programs have unanticipated negative consequences. For example, in their attempts to reduce saturated fat intakes, participants may reduce their consumption of dairy and meat products with associated reductions in iron, calcium and vitamin intakes. Another possibility is the risk that obesity prevention programs may encourage the development or worsening of eating disorders.

These unintended consequences may be judged to be so small that they do not matter or they may be regarded as serious negative problems. A big problem in program evaluation is that such negative side-effects may not be anticipated because the program designers have not systematically considered all the likely consequences of the program. As a result, negative effects are often unobserved and unreported.

Unwanted effects may occur outside the food and nutrition domain—for example, participants may gain the false idea that so long as they eat healthily, other aspects of their lifestyle such as physical activity and substance abuse do not matter.

Clearly estimation of the costs of a program can be complex. Health economists have developed several ways of calculating cost benefits and related indices (see Box 7.3).

Equity is a major value in public health. This means that the needs of all participants, especially those who have poor access to society's resources, should be met. This raises two issues. First, the program planners have to make sure that the program meets the food and nutrition needs of the participants; usually this is done through a needs assessment process. It is quite easy to design a program with only the designers' needs in mind. Several of the major American heart disease prevention programs of the 1970s and 1980s suffered from the overwhelming desire of the

BOX 7.3 INDICES OF THE COSTS AND BENEFITS OF PROGRAMS

- **Cost-benefit analysis (CBA)** is a type of analysis that measures costs and benefits in monetary units and computes a net monetary gain/loss in terms of a cost-benefit ratio. This analysis may be helpful and necessary in setting priorities when choices must be made in the face of limited resources. This analysis is used in determining the degree of access to, or benefits of, the health care to be provided.
- **Cost-effectiveness analysis (CEA)** is a type of analysis that compares interventions or programs having a common health outcome (e.g. reduction of blood pressure; life-years saved) in a situation where, for a given level of resources, the decision-maker wishes to maximise the health benefits conferred to the population of concern.
- **Cost-utility analysis (CUA)** is a type of analysis that measures benefits in utility-weighted life-years (QALYs) and that computes a cost per utility-measure ratio for comparison between programs.

Source: The World Bank, http://web.worldbank.org.

investigators to reduce the population's risk of heart disease. Unfortunately, the communities involved did not prioritise heart disease as highly. This is a real dilemma for 'experts'. On the basis of their expertise, they may believe that the population has a need to do something (e.g. reduce its intake of saturated fats) but the members of the population may not appreciate the importance of this expert-defined problem. Instead, the community may see a need for other types of initiatives, such as programs to reduce body weight or school breakfast programs for impoverished children.

While the first example might appeal to cardiologists, the second probably will not. Should the experts engage in awareness-raising—for example, using the mass media to show the evil effects of high saturated fat intakes—or is another approach indicated? In community health promotion, the emphasis is on consultation with all the stakeholders so as to arrive at some consensus about the (nutrition) needs of the population, or at least to arrive at some workable compromise. The danger with well-funded external experts engaging in awareness-raising campaigns is that the population may simply reject their ideas, or they may see the campaign as outside interference and a waste of public money. On the other hand, if the expert-defined problem is a really serious one which actually does affect many in the population, then awareness-raising is likely to be successful (see Thompson et al. 2003 for further discussion). Campaigns about smoking cessation, car seatbelt wearing, safe sex practices and sunscreen use probably have been successful in part because they tapped concerns held by many in the population.

A second equity issue is related to the reach of the program into the population. There is a great danger, especially with nutrition promotion programs, that they mainly reach people who already have sound nutrition knowledge and habits while people in need of help are not reached by the program. That is, they may 'preach to the converted'. People from lower socioeconomic status backgrounds, those with poorer financial resources, those from less educated backgrounds or non-English speaking people may be missed by programs, or they may find them less useful than others. This may be partly due to the lesser availability of most societal resources for poorer people, and partly because their cultures may use different communication channels—for example, only viewing TV programs which are in their first language (e.g. Arabic or Spanish in the United States) or spending more time in settings like churches which may be ignored by program implementers. It is important to ensure that the program is adequately promoted through use of the communication channels preferred by the population. A good example is the American Black Churches program, which communicates in several languages via distinct channels (mass media and church groups) with African Americans and Hispanic Americans (McClelland et al. 2001; for a comprehensive review of healthy eating programs among people of colour, see Yancey et al. 2004). Another is the Verb advertising campaign, which produces separate advertisements for Native American youth, black Americans and Caucasian Americans (Huhman et al. 2005).

Lack of source credibility may also reduce 'reach'. Many people's lifestyles have little in common with those led by the program planners. That is, they may belong to cultures and subcultures which find the 'middle-class' appearance of the pro-moters irrelevant or offensive. For example, teenagers may view messages from adult health professionals as less important or credible than the same messages communicated by their peers (Story et al. 2002). These difficulties often require additional special sub-programs which attempt to communicate with disadvantaged groups in the population.

Recently Tugwell's group in Canada have proposed the 'equity effectiveness loop' which proposes a five part action cycle to ensure that programs reach all parts of the population. They suggest that program efficacy is modified by at least four modifiers: access, diagnostic accuracy, provider compliance (in health service applications) and consumer adherence (Tugwell et al. 2006). Mark Boyd (2008) provides a clearer picture of equity which is useful when trying to understand the ways in which pro-grams can be effective. He proposes the 'equity triangle', which describes three principal aspects of equity (Figure 7.1). It acts as a prompt to consider how to improve the targeting of projects, programs and services to reduce inequalities.

FIGURE 7.1 THE EQUITY TRIANGLE

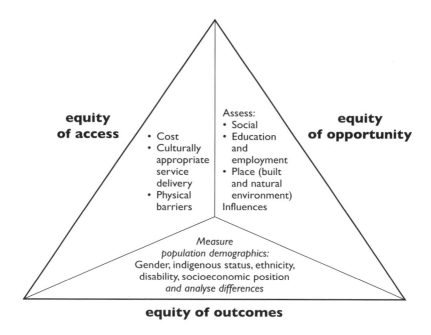

Source: Boyd (2008); VicHealth (2008).

Boyd notes:

The aim of the triangle is to introduce an equity dimension to program planning and delivery without this becoming a cumbersome process.

As a minimum in the short-term, services and programs should address equity of access and equity of outcomes, and create the partnerships (such as with local government or state advocacy organisations) to assist with addressing equity of opportunities. (Boyd 2008)

Sustainability is a key issue in nutrition promotion. Basically it is about how long a program, or a healthy food consumption style, can be maintained over time. It has several important aspects.

Economic sustainability

Often funding or labour is provided to establish a program but the real test is whether these resources can be maintained for the duration of a program or on a permanent basis. Externally funded intervention projects are notorious for short-term funding. For example, a six month-long intervention may be funded from an external research agency to provide nutrition promotion services for a primary school. The funding is usually minimal but does provide the required nutrition promotion expertise. Usually the host institution makes considerable changes to the ways it operates—for example, by providing teachers who take on extra work, canteen staff who attend classes in nutrition, clerical help in accessing child and parental permission and much more. The project usually continues until its designated end date, when the funding ceases. The researchers write their reports, and the host institution is left with unmet raised expectations and with the problem of getting back to normal operations. Ideally, success in short-term projects should be used to advocate for further, prolonged funding. If such advocacy is not envisaged from the start, the logistic costs imposed on the host institution may be too great to justify the immediate health gains of the program participants and the external nutrition promoters' scientific publications and reports. The ethics of such unsustainable projects are questionable.

Sustainability of internal communications—administrative sustainability

Suppose a nutrition program brings about benefits desired by all, and suppose the host institution adopts it on an ongoing, funded basis. Sustainability may still be an issue if adequate management systems are not in place. Typically, the first wave of workers in the host institution are well trained in the procedures designed by the nutrition promoters. However, over time trained staff leave and are replaced by new untrained staff. For example, hospitals which have a 'baby-friendly pro-breast feeding policy' depend on nursing and other staff trained in the principles of breast-feeding promotion. If they leave, the replacement staff are unlikely to have the same knowledge and skills, so they may not consider breastfeeding to be important or they may not know how to promote it to new mothers. The original initiative will soon dissipate unless the hospital management has a clear policy which allocates resources to staff orientation and training. Such a policy will be greatly enhanced if managers are provided with information about the efficacy of the program (e.g. the percentage of mothers leaving the hospital who have initiated breastfeeding.). That is, feedback about the working of the system is essential for its continuance. Some of the factors which may affect the sustainability of health promotion programs are outlined in Box 7.4.

BOX 7.4 WHAT MAKES A PROGRAM SUSTAINABLE?

O'Loughlin et al. (1998) investigated 189 local heart health promotion programs across Canada in an attempt to discover the likely factors associated with their long-term sustainability. They found that key factors appeared to be:

- **No paid staff**—Interventions like walking groups, grocery store tours and health fairs that were run by volunteers only and did not employ staff tended to be more permanent. (This may not apply in other countries and contexts.)
- **Interventions that were modified during their implementation tended to last longer**—The authors believe that 'modifiability' of a program is a key factor (e.g. interventions might be subdivided into their component parts or have alternative content forms). Obviously, any changes need to be monitored to make sure the program still serves its original general aim.
- **The nature of the intervention–provider fit is important**—The program should fit well with the host organisation's mission and practices. Other research suggests that the fit between the organisation and the intervention ('institutionalisation'), in terms of their objectives and values, is important.
- **Long-term programs usually have a 'champion' who works to keep it alive**—The champion's efforts may be easier if other key individuals like managers and community leaders are also involved in the intervention.

Source: O'Loughlin et al. (1998).

Environmental sustainability

Sound nutrition promotion requires that everyone is fed in a healthy manner throughout their lifetimes and throughout their children's and grandchildren's lifetimes (Gussow and Contento 1984). It is not sufficient only to provide a population with health-sustaining food for a limited period of time. Many people today, especially young people, are vitally interested in the long-term sustainability of food supplies. Therefore, nutrition promoters have a responsibility to ensure that the habits they promote are consistent with long-term environmental sustainability. The use of polluting technologies or those which use unnecessarily large amounts of fossil food energy in the production and distribution of food, or the use of excessive amounts of packaging, are issues that nutrition promoters have to consider.

In a sense, environmental sustainability overlaps several other criteria. For example, it can be regarded as a form of equity through time and place. Nutrition promoters have to ensure that local practices help or do not harm other people around the world, and that current practices do not restrict the availability of food for future generations. This is not a major feature of nutrition promotion at present but it is an essential goal. The reduction of fossil fuel use is an increasing concern of both the community and industry. The Carbon Disclosure Project, for example, links future investment decisions to large companies' plans to reduce fossil fuel use (www.cdproject.net).

Cultural appropriateness, fairness and feasibility— go with the flow!

Whatever is promoted should be feasible for participants to adopt without much cost to themselves. Bad habits are not broken, they are replaced by healthier habits. Promoters have to examine carefully the contexts in which people live (i.e. their lifestyles). So, if adolescents view eating fruit as undesirable, it would be unfeasible and unfair to teenagers to expect them to consume fruit in front of their friends— perhaps it would be more realistic to get them to eat it at home or to consume it in the form of a dessert or a fruit drink. The proposed changes have to be consistent with that lifestyle—that is, they have to be culturally appropriate. Another example would be the promotion of foods to raise the iron status of 'vegetarian' teenagers. The nutrition promoter could point out that red meat is the richest source of iron but if the participants are against eating meat because it conflicts with their personal values, then it would be unfair and unfeasible to insist on this strategy. Instead, it might be wiser to help them become healthy vegetarians through the promotion of vegetarian principles and cuisines. As an anonymous marketer once noted: 'Public opinion is like a lava flow, only silly people would think they could alter its course but smart people can anticipate where the flow is moving to and place their products there.'

Generalisability

One of the maxims in health promotion is 'give it away'. Health promoters should always be willing to give their findings and techniques to others so that the population's health is maintained and improved. Therefore, this is an important criterion, particularly for intervention projects. A major potential strength of intervention projects is that they can serve as demonstrations which can be adopted by other groups. Therefore, the methods used and the results gained should be generalisable.

This usually means that communication materials, methods and evaluations are straightforward to use. The generalisability of action research may be much lower than that of intervention projects because continuous attempts are made to match the particular needs posed by local circumstances, but even here elements of the program may be generalisable. Features of good programs are likely to be copied by others.

Feedback and communication

The continuing flow of communication inside a program, and between it and the community, is essential for its effectiveness and for its sustainability. If the program implementers do not have regular information about the progress of the program towards its goals, then they are unable to take further action to support the attainment of those goals. Similarly, if there is no information flowing to and from key stakeholders, the program is likely to lose support from them. Reports of the program's success or otherwise require wide dissemination so that others can learn from its activities. Too often, project reports have been made to a few funding bureaucrats and to no one else. All participants and stakeholders should be informed, at reasonable intervals, of the state of play within the pro-motion program: many of them have contributed their time (and perhaps money) so there is an ethical responsibility for program managers to ensure they are adequately informed.

Advocacy is a special form of communication. This can have many end-points, but the basic aim is to bring about the release of resources to promote the health (or food-related nutrition status) of a particular population group. This is related to the common injunction that the job of public health workers is to create community outrage where it is justified by empirical evidence. The sorts of tech-niques involved are letter writing campaigns, seminars, media communications, lobbying of politicians and government officials, and networking among pro-fessional organisations which are organised to achieve the one aim. Recent examples include advocacy for fortification of foods with folic acid, reductions of salt content in processed foods and advocacy for funding for children's obesity prevention. Data collected from demonstration interventions and other research can provide persuasive evidence for lobby groups. Not all advocacy efforts may be in the public interest, and many under-privileged groups such as the severely disabled do not have effective advocates.

The conduct of nutrition promotion

Many health promotion textbooks describe the ways in which programs are designed and implemented (e.g. Wass 1994). The main stages in program development include the following.

Identifying and contacting stakeholders

The main players (stakeholders) who influence the population group's food consumption should be identified and contacted. Depending on population group, these stakeholders may be children, parents, teachers, health professionals, manufacturers and retail managers, canteen staff and local food vendors.

Raising support from the organisation and community, and raising awareness

The nutrition promoter may perceive a problem with the eating habits of the population group but community members may not. It is usually necessary to communicate with the stakeholders about the importance of healthy eating and nutrition, and to negotiate with them on the need for specific health promotion actions. It is important to explore and understand the stakeholders' views. For example, while everyone might agree that the school or company canteen sells a poor range of foods, there may be little agreement on whether anything can be done about it ('we will lose money') or what steps might be taken to improve the situation.

The promoter will usually elicit support from a group of influential stakeholders who are concerned about the population group's food and who want to play an active role in improving the situation. This process will take many meetings and much discussion and negotiation until everyone agrees on a set of practical aims. Many people do not realise the important role of eating habits in population health and nutrition, so the nutrition promoter is often in a good position to raise their awareness of the issues and to suggest ways in which to translate their concern into concrete actions.

Setting up advisory groups—communicating with representatives of the population group

It is essential to communicate with representatives of the population group who inform the nutrition promotion managers about their concerns and about the outcomes of their attempts to change their food behaviours. Such communication allows the program to explain its aims and distribute other information. Strategies to communicate with the population group may include the creation of local advisory groups, the use of school newsletters and local newspapers, and the recruitment of local pharmacies, general practitioners and retail food outlets to keep the population informed.

Many indigenous communities have well established ways of enlisting this type of support, often in the form of long, democratic meetings (e.g. meetings on the Maori *marae*).

Formative evaluation

Formative evaluation is about scoping ideas for the goals and implementation of the project—it is essential.

Setting goals and making plans—towards a food policy

One component of good communication with stakeholders is establishment of a clear set of behavioural goals and plans to achieve those goals. The nutrition promoter should specify exactly what the members of the target group (e.g. children and their families) are being asked to do—for example, 'eat one extra piece of fruit every day'. Goals may also be set for people in different settings within a community—for example, the school canteen may have its own goals (perhaps to increase sales of salad rolls by 10 per cent in the next three months). Goals should specify what the desired behaviour change is, who should do it and in which context (at home, at school, in the supermarket and so on). Examples of the activities in particular settings are given in Table 7.1. They are essential for the conduct of the project and for its evaluation. The planning of goals is also essential, because it will clearly state the sequence of activities and the timelines for goal achievement.

TABLE 7.1 SUMMARY OF NUTRITION PROMOTION METHODS USED IN COMMON SETTINGS		
Method	**Aspects**	**Settings—examples of activities**
Learning approaches	Food exposures—tasting and making food Modelling Reward Goal-setting and feedback	**Home** **Education/child care** **Health services** **Retail** **Mass media—modelling** **Workplace** **Leisure sector**
Education and communication	Goals Feedback	**Home**—goal-setting, feedback, nutrition communication, needs responsive

Table 7.1 continued

	Sound food and health information schema Facts Discovery and group learning	**Education**—development of schema, needs responsive **Health services**—brief messages, needs responsive **Workplace, leisure sector**—brief messages **Retail**—Point of sale: brief messages, needs responsive **Mass media**—detailed and short messages **Local government**—information services **Primary and secondary producers**—promotion of healthy foods
Social support	Empathy Material support Information support	**Home** **Education** **Workplace and leisure** **Health services**—group support **Local government**—provision of facilities for social support
Environmental change	Change in food availability Change in price Differential promotion	**Home**—change in buying and storage patterns **Education, health services, workplace and leisure**—change in food service **Retail**—labelling of foods, differential promotion at point of sale, in-store nutrition promotion **Mass media**—pro- or anti-healthy food advertising **Local government**—support for farmers' markets, healthy food networks **State/national government, agriculture**—promotion of sustainable food production
Food policies	Informal and formal rules and regulations which reduce consumption of unhealthy foods and promote healthy foods and associated information and facilities	**Home**—family rules re eating, cooking, shopping **Education, health, workplace leisure**—policies prohibiting/allowing consumption of foods and beverages **Mass media**—regulation of food advertising **Local government**—regulation of sales of foods and environmental health **State/national government, agriculture**—encouragement of production, distribution and consumption of healthy foods, discouragement of unhealthy foods

Identifying barriers and opportunities (including costs)

The precede–proceed and social–behavioural–biological models are useful at this stage of planning. When the goals are being planned, members of the planning group are likely to identify opportunities for behaviour change, as well as barriers to change. The planning group should mainly comprise local stakeholders because they will be in a good position to identify these opportunities and barriers. The group may wish to conduct a preliminary survey to identify these factors more precisely (formative evaluation). Typical opportunities are access to local supplies of fruit and vegetables, and the availability of willing volunteers. Barriers may be related to ignorance of the issues, lack of family support and poor business skills in the running of school food services. The planning group has to suggest ways of overcoming the barriers and using the opportunities. It will also begin to consider how the project is to be evaluated (see below).

Establishing strategies and activities

The planning group will soon recognise ways of meeting the project goals. The Ottawa Charter and the precede–proceed and social–behavioural–biological models in particular may be useful in identifying strategies and associated activities. One of the first decisions will be about the scope of the project: is it going to be fairly narrow, dealing with the children and their families in a single context (such as the classroom or a clinic), or is it going to be a broad, community-wide project? If the former, then individual focused models (such as the theory of planned behaviour or social cognitive theory) might be useful; if the latter, then broader models are required. The strategies adopted by individual oriented approaches might include self-monitoring of eating behaviours or the use of social (group) reinforcement via experiential learning (e.g. in food tastings). Strategies used in broader approaches might focus more on the food environment, such as fruit and vegetable promotions in supermarkets or reform of the school canteen. A combination of individual and environmental change approaches is valuable in appropriate contexts.

Conducting an evaluation (measurement tools)

Evaluation of project activities is invaluable. It is a key project goal, and planning for evaluation must be undertaken from the inception of the project. Evaluation need not be expensive or unduly onerous, but it should be appropriate to the extent and intensity of the activities undertaken. Evaluation provides evidence of efficacy and can be used to defend the program if it is questioned. Many authoritative members of the community are sceptical of healthy eating programs, so evidence of efficacy is an important way of bringing them on side.

Process evaluation is essential. It involves keeping a record of events during the project's implementation (from the first meeting until the end). This account helps identify the barriers to implementation and the actions that were taken to overcome or avoid those barriers. It also acts as a 'group memory': people often forget what they have decided, so a record helps keep everyone task focused. Above all, the process evaluation allows transparency so outsiders can be shown what the project has considered and done. It is especially useful for recording events and processes which are not easily measured. So, if it happens, write it down!

Outcome evaluation is often the only form of evaluation considered. It is often perceived to be difficult to do because it usually involves some form of quantitative measurement, but it can be quite qualitative (e.g. interviews with key stakeholders before and after the intervention to gauge how they have changed). More usually, measurements are made of participants (and often of a comparative group of similar non-participants) before and after various stages of the intervention. Such measurements are undertaken to assess the size and type of the changes that occur. The types of variable that can be assessed depend on the goals of the project. They might be people's body mass, the range of foods that people eat each day, parents' involvement in cooking with their children, the sales of different types of foods in canteens, or even the number of times that a general practitioner provides healthy eating advice to parents. Every variable, and even every participant, does not have to be measured; rather, the aim of evaluation is to check whether the intervention approaches have the desired effects and, if not, to identify ways in which to improve the program.

Reporting of findings is an important part of any program. Program leaders have an ethical responsibility to report back to participants and stakeholders (because it is their project, to which they have given time and other resources). There is also a 'political' imperative to report back: the reporting of success or failure will keep people involved and keep healthy eating on the local agenda.

Reports do not have to be long, but they should cover the issues that the project was designed to promote or remedy. They should be written in the plain language(s) of the community. They can be in a range of forms (e.g. written reports on paper or on websites, CDs, posters, audio reports or videos), depending on the target audience's preferred communication channels. Ideally, reports about all projects should be stored in a publicly accessible location so everyone may benefit from the information. One of the main findings of the Review of Children's Healthy Eating Interventions was that many healthy eating promotions are not reported, so no one can learn from them.

Institutionalising the program and training new staff

A one-off intervention is not going to help the target group for long. Unfortunately, many interventions have been temporary. Under the injunctions of the Ottawa Charter, health and education services should intervene continually to promote healthy eating to children through changing circumstances. Such promotion should be a part of the job. Once an intervention has been shown to be effective, therefore, we need to find ways of making it (or features of it) permanent. This effort involves changing the work duties of health and education staff so they will continue aspects of the intervention.

Putting it together: Program logic

Several helpful instruments which are freely available have been developed to aid in project planning. These deal with the flow of proposed actions which are carried out to attain project (or program) goals and objectives. Figure 7.2 shows one of these instruments. A recent development is intervention mapping, which is a planning and evaluation protocol which includes needs assessment, intervention design and evaluation (Kok et al. 2004). Major resources for program planning and evaluation are available on the internet. They include:

> The Univerity of Wisconsin website:
> http://www.uwex.edu/ces/pdande/evaluation/evallogicmodel.html

> The US Centers for Disease Control:
> http://www.cdc.gov/eval/resources.htm#logic%20model
> and DoView software:
> http://www.cdc.gov/eval/resources.htm#logic%20model

Methods used in nutrition promotion

Many methods are used in nutrition promotion and food communication to influence people to change their food habits. They fall into several groups which have more or less applicability according to the settings in which people are found. Table 7.1 summarises them and their associated settings. The methods often overlap with each other. For example, methods based on learning theories which encourage modelling and reinforcement actually require supportive environments in which

FIGURE 7.2 PROGRAM PLANNING AND EVALUATION

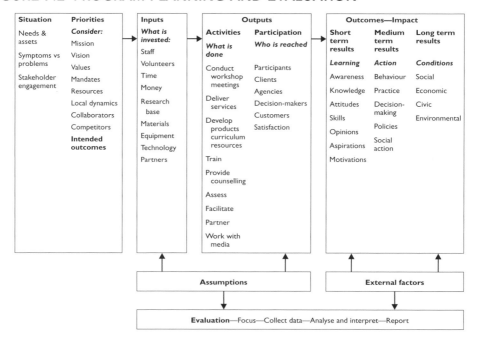

Source: http://www.uwex.edu/ces/pdande/evaluation/evallogicmodel.html.

someone such as a parent models and rewards the desired food behaviours. Most of the methods have been designed to change individual's food behaviours but they can be used selectively to change settings. For example, Hazard Analysis Critical Control Points (HACCP) methods with appropriate performance incentives can be used to modify the behaviours of manufacturing and retail managers to make healthy foods more available for their customers.

Approaches based on learning theories

Typically, these are used with children in pre-schools and schools, though aspects of them can be used in many other settings, such as modelling of behaviours in the mass media. They include ensuring exposure to healthy foods through tasting, food preparation and cooking sessions, modelling of behaviours such as choosing fruit over confectionery and non-food rewards for doing so, usually in the form of social reinforcers such as smiles or praise, 'gold stars' or similar symbols of social approval. Goal-setting methods (cybernetic models such as self-monitoring—see Chapter 3) can be used to extend these learning methods into long-term food behaviour change plans for individuals and organisations. The Food Dudes program in the United Kingdom developed video instruction to show parents how to shape their young children's eating utilising learning principles (Lowe et al. 2001).

Educational and communication methods

These methods incorporate the above learning principles but tend to do so in the context of formal education and communication settings. Educationalists have developed teaching and learning strategies which have been employed by nutrition promoters, such as didactic instruction, discovery or problem-based learning, and group learning. These methods are well understood and have been used successfully to impart nutrition knowledge and to bring about attitudinal and behaviour change (Johnson and Johnson 1985a). Piagetian and other schema theories are used to enable learners to assimilate knowledge frameworks which are appropriate to their cognitive development. These enable learners to interpret nutrition 'facts' during their lives.

These forms of education have been adapted to mass communication. In particular, the notions of the framing of information and the use of persuasive messages employing fear arousal have been studied extensively. They will be discussed in Chapter 13.

Social support

This is now recognised as essential for most people to make any behavioural changes. Humans are social beings, and for most of us the attention and responses of other people (especially family and friends or 'significant others') are crucial in our daily lives. Social support provides us with emotional support and empathy as well as problem-solving information and access to material resources. For example, if a new food is unpleasant a parent or friend can provide support to 'try again' and they can explain why the behaviour is so valuable (it will make you healthier), they can show the learner the best ways to eat the new food (e.g. cut it up), and in the case of parents they actually supply the food—the material resource. Adequate social support plays a crucial role in the working of the immune system and the prevention of illness (Kawachi and Berkman 2000, 2003; Cohen and Syme 1985).

Most settings can provide social support. One of the aims of nutrition promotion is to recruit such support. For example, Burke and colleagues (2003) showed that nutrition and physical activity promotion among married and de facto couples was more effective than among singles.

The mass media cannot provide social support directly, but they often use the concept of the *reference group* to show learners how people they admire might react to their changing (or maintaining) their behaviours. Various cola advertisers are expert at this, frequently showing their products in the presence of groups of happy friends.

Environmental change

Environments have long been recognised as major influences over health behaviours; they are directly alluded to in both the precede–proceed (Green and Kreuter 1991) and social cognitive theories (Bandura 1986). However, there has been a renewal of interest recently, especially in social epidemiology (Berkman and Kawachi 2002). This includes detailed analyses of food availability, pricing effects and differential promotion in various settings. The social–behavioural–biological model put forward by Glass and McAtee (2006) represents an advance in the conceptualisation of environments with its specification of *risk regulators* which are likely to provide new foci for health promoters. Thompson et al. (2003) and Dooris (2005) have emphasised the interactions which can occur between environments at different levels of organisation, and thus the need for sophisticated analytic methods such as nested designs.

Food policies

Institutional food policies are a useful way of assimilating the lessons learned from interventions. This assimilation has to be done in a deliberate manner. Too often, the effects of interventions are assumed to continue long after they have ceased. Given the competing influences on people's eating behaviours, this is a naïve assumption. For this reason, all food-related settings such as hospitals, schools and pre-schools, retail outlets and local government should have active staff-training programs that explain to staff the importance of following the institution's food policies in specific ways.

The importance of food policies

Food policies are part of the institutionalisation of change. They are sets of rules and social norms about the supply and consumption of food in nations, organisations and settings. They may be broad reaching (as in international food policies) or very local (as in the local kindergarten). Examples are given in the settings chapters. They are best implemented by key stakeholders after thorough discussions between them. They convey key advantages, including the prevention of conflict between stakeholders (e.g. school food policies often prevent conflict between groups of parents) and the induction and training of new staff and other stakeholders ('our policy is to . . .'), and they enable the provision of healthy food to be a 'normal' part of the life of the setting (e.g. in schools parents do not have to argue for a supply of healthy foods).

Conclusions

As our thinking shifts from a predominant focus on individuals to populations, and from change within individuals towards change in environments, settings and risk moderators, so do the methods used. Methodology is becoming far more sophisticated as health promoters realise that health and health behaviours are influenced by many factors which interact with individuals' actions and environments over long time spans.

Recently, the distinction between top-down and bottom-up approaches has become blurred. There is now more emphasis on formative and process evaluation and the use of a wider range of theories to guide interventions. There is a strong focus on how best to deliver sufficient 'dosages' of interventions. This requires more detailed consideration of environments, settings and risk regulators, as well as a clearer focus on the health behaviours of *populations* (as distinct from *individuals'* behaviours). Consequently, the newer methodologies employ multiple methods with multiple components at multiple societal levels (as outlined in the socio–behavioural–biological model). There is now more optimism in health promotion, with the dawning realisation that small changes in habitual population health behaviours can have major, long-term cumulative effects, as Rose (1985) indicated long ago.

Discussion questions

7.1 Compare and contrast the advantages and disadvantages of action research approaches with those of the experimental project tradition (as they relate to nutrition promotion).

7.2 Consider methods to reduce the prevalence of heart disease in the general adult population. What are the likely costs and benefits of possible preventative approaches?

7.3 Describe the methods appropriate to particular settings within three levels of society.

7.4 Discuss the various threats to validity which face nutrition promotion. Give a nutrition promotion example of each threat.

7.5 What do you consider to be the main factors which promote the sustainability of nutrition promotion programs?

7.6 Select an example of a real or imaginary nutrition promotion project. Describe the key steps in the conduct of the project.

7.7 Briefly outline the main steps in the evaluation of a primary children's nutrition promotion program.

8

Community nutrition promotion programs

Introduction

The idea of working in the community is a potent dream for many, especially for some health professionals locked away in clinical specialities and biochemistry laboratories. Community nutrition is an elusive term which conveys many differing connotations. Broadly, it emphasises that nutrition promotion needs to get out from the confines of institutions to work with members of the community on their own terms. The term 'community nutrition' represents the integration of nutrition promotion initiatives in all the settings in which people live. This chapter is the first of several dealing with nutrition promotion in the settings in which people spend their lives, including health care settings, schools, worksites, retail food outlets and the mass media.

What is 'community'?

This is a highly charged philosophical and political question. To neo-liberals like former UK prime minister Margaret Thatcher, there simply is no community—there are only individuals and their families. Others would say that 'community' is a figment of the urban imagination, something that we may believe existed in the 'good old days' when life was supposedly simpler and people were more caring about each other. Still others believe that there is 'the' community—an abstraction for which we pay taxes to maintain basic services like water, sewerage, transport, electricity, education and health systems. In many countries, this is synonymous with the 'nation'. And then there are post-modernists and post-structuralists who note that there are many sorts of community, such as communities of ideas and communities of discourse, as well as communities based on place and relationships between people (Cooke 1990).

As usual, public health finds itself somewhere 'in the uncomfortable middle'—and clearly not in fashion in many English-speaking countries. It proposes that the notion of community is important and real, so it is 'communitarian' rather than 'individualist'. It sees community at various levels of abstraction, from the local street or neighbourhood community, through communities derived from membership of institutions like small business, hospitals and local regions, through to the 'general community'. This last term can mean 'the population' of unrelated individuals who share some things in common—like poverty—or even the 'global community', which emphasises the challenges faced by all humans such as threats to ecological sustainability. The term 'population' is not synonymous with 'the community' because the latter conveys some sense of shared resources, relationships and identity.

Thompson et al. (2003) have summarised the various definitions of community proposed by health promoters as follows:

- a group of people sharing locality, interdependence, values and institutions (Warren 1958);
- a social system (Thompson and Kinne 1999);
- a place and a 'sense' (as in a 'sense of belonging') (Colombo et al. 1991).

Communities may be small (e.g. religious organisations, workplaces, schools) or large (cities, counties, states), where people 'gather together on a regular basis'. Internet communities (e.g. regular participants in chat rooms) and mass media audiences also fit these definitions in that they are socially mediated spaces where people regularly gather and share a sense of interdependence and belonging.

Supply and demand side

Activities within communities can be influenced by social forces which can act in two opposing directions: first, 'upstream' or 'supply-side' influences such as government and big business initiatives like national food policies, taxation, rules and regulations (see Chapter 7); and second, 'downstream' or 'demand-side' influences—the demands made by individuals about what they want in their lives, like the demand for sweet beverages, healthy foods or convenience foods. Nutrition promoters can work with either set of influences; however, until the past decade most of their attention has been on demand-side influences. Nutrition promotion programs have focused on downstream influences, often resorting to education and cognitive models in a bid to get individuals to change some aspects of their lifestyles voluntarily. It is worth noting that, despite our own personal, values-driven preferences for either of these approaches, there is no compelling evidence that one approach is better than the other. Rather, some combination of both is usually more effective than either approach alone.

Top down and bottom up

A closely related set of terms, possibly derived more from community development theorists, is 'top down' and 'bottom up'. These relate to the ways health promoters work. 'Top down' places the emphasis on 'expert knowledge' guiding the actions of mere mortals in the community. Many intervention programs are 'top down' in that a predetermined set of actions and goals is imposed upon 'people in need of enlightenment'—for example, the goal of reducing cholesterol levels in the community. Not surprisingly, such attempts are often rejected and viewed as paternalistic or condescending. However, there is little doubt that well-informed top-down interventions can be crucial—for example, seat belt legislation is a 'top-down' approach that in many countries is viewed by auto drivers as 'reasonable'.

'Top-down' and 'bottom-up' approaches can often be combined to produce more effective programs than either alone. For example, attempts to promote fruit and vegetable consumption at worksites require the education of the workforce (to value and eat fruit and vegetables), but they can be strengthened by governments which provide tax incentives for employers to provide these foods at low prices.

Characteristics of community programs

Community programs have several characteristics which are increasingly seen by health promoters as definitive. They centre on the prefix 'multi':

- *Multi-level*—They usually operate on several levels—for example, at the level of the individual, small groups (like families), local institutions (like worksites) and at the levels of regional, state, national and international policies. This is very much a feature of the social ecological and social–behavioural–biological model (see Chapter 6).
- *Multi-component or multi-facet*—They often include several components—for example, a communication program (to inform individuals), structural change such as the provision of lower-cost fresh foods at worksites, and policy changes such as regulations governing the sales of fast foods in school canteens.

- *Multi-method* — To achieve the outcomes associated with the various levels and components, different methods may be employed. For example, archival methods may be used to assess the changes over time in the content of newspaper and magazine articles about food and health to gauge the long-term impact of a program. Recent community intervention programs have emphasised formative evaluation and process evaluation. The Working Well Trial (Biener et al. 1999; Patterson et al. 1998) is a good example. It used intensive process evaluation to assess the extent to which program initiatives had actually been delivered (and thus the 'dose' of the intervention).

Dooris (2005) has provided a useful illustration of the multi-faceted nature of whole-of-community programs as applied to a university health promotion program. This example draws attention to the ecological nature of current community health promotion, the emphasis being on the interdependence between the various components of programs (see Figure 8.1).

FIGURE 8.1 SETTINGS AS SYSTEMS: THE EXAMPLE OF A UNIVERSITY

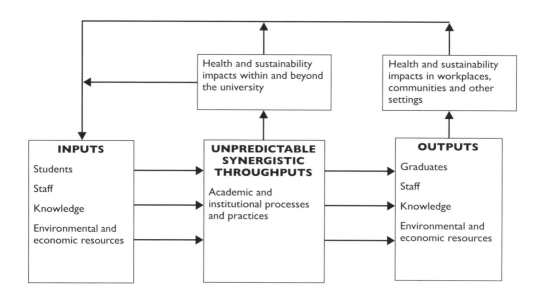

Source: Dooris (2005).

The inclusion of environmental and ecological perspectives: The importance of food policies

Until very recently most community nutrition programs were based in the downstream, aiming to bring about changes in individuals' behaviours through persuasion and education. However, there is an increasing realisation (especially in relation to obesity prevention) that 'upstream' or environmental factors have important effects on behaviours and that public health policies are required to modify these influences. Nutrition promoters like Sorensen et al. (2004), Seymour et al. (2004) and others have drawn attention to the importance of food policies. These integrate top-down and bottom-up approaches because they can be designed and implemented at local levels (e.g. in a school or a local government area), at the regional or state level and at the national and international levels. National food policies, for example, can provide over-arching sets of rules, regulations and codes of practice which allow full participation in the adoption of policy themes at the local level. Fruit and vegetable promotion might be encouraged at the national and state level through upstream measures like tax subsidies, whereas at the local level education and individual behaviour-change strategies can be used to promote the same themes according to local needs and circumstances.

Many health and nutrition promoters see 'community nutrition' in very simple and practical terms. They view members of the population as living in a series of different settings, so if they can make each setting support 'healthy choices' and integrate these settings towards some common goal, then the community's health (or nutrition) will be promoted. For example, fruit consumption can be promoted in pre-schools and schools, at worksites, in hospitals, at sports events, at transport interchanges, in the mass media and in many other settings. A community-wide nutrition strategy might evaluate the feasibility of promoting fruit and vegetables at each of these settings and put resources into those that are most likely to influence the population's consumption of these foods. The highly successful Danish 6 a Day program (www.6aday.com/abstract6adayDenmark2004.doc), California's 5 a Day program (Foerster et al. 1995) and its successor, the American 5 a Day program (Beresford et al. 2001), integrate 'top-down' and 'bottom-up' approaches.

The definitions of 'community' and 'community interventions' are important for nutrition promoters because they help to direct their activities towards certain problems and away from others. So if they view the 'community' as being the same as the 'population', they may be likely to focus on general interventions for the 'person in the street', ignoring the particular problems faced by certain groups such as those living with disabilities. In turn, this can lead nutrition promoters to use methods which are suited to individuals only and cause them to fail to use more social or environmental approaches. For example, Timperio et al. (2005) have shown that there are fewer food stores which sell fresh foods in the poorer parts of Melbourne than in the richer parts of Melbourne. Similar observations have been made in the United States and the United Kingdom, though McIntyre points out that this is not a universal phenomenon (Sorensen et al. 1998; McIntyre 2005). We could try to educate people living in poorer areas to value fresh foods more, but to buy them they would need to travel to the richer areas, which is difficult since many of them have no car and no access to safe, convenient public transport. So, at the very least, a combination of individual and structural approaches is usually required to bring about changes in health behaviours and status.

Social hierarchy and physical environments

Social and physical environments have important roles in creating unhappiness and happiness, ill-health and health. Kawachi and Berkman (2000), Marmot and Wilkinson (1999) and other social epidemiologists have drawn attention to the 'social hierarchy' (the relative power of people to influence outcomes in their lives, and thus their health) and to physical conditions like the layout of streets and towns, the positioning of shops and small businesses, and the ways in which they can influence many aspects of the human condition such as physical activity, food consumption and drug abuse. It is difficult, and perhaps unwise, to try to separate the influence of the purely physical environment from that of the social environment. In a public lecture in 2005, McIntyre noted that 'people make places and places make people'. This is in accordance with Giddens' (1993) structuration theory, which suggests that structures (in time and space) are intimately related to people's activities and behaviours—you can't have one without the other. McIntyre notes how, over several hundred years, the city of Glasgow has developed a geographic distribution of social disadvantage and ill-health— 'systematic disadvantage'. Recently the World Health Organization set up a commission on the Social Determinants of Health which has amassed evidence about the importance of factors which influence social inequities in health and has made proposals to change them (www.who.int/social_determinants/en; see also Baum 2007).

Social capital

The concept of 'social capital' refers to the fact that only some of the resources in a local community are material structures, like housing, roads, hospitals and businesses. In addition, people form helping relationships and coalitions for various causes (e.g. babysitting circles, domestic unpaid work in the household, car pools, community groups, clubs and societies), all of which contribute to the social resources of the community (Baum et al. 2000). The value of such 'social capital' in financial terms is immense. Marilyn Waring (1998), the New Zealand social scientist, first drew attention to the financial value of 'women's domestic work' in the 1980s—if it had to be paid for, it would represent a major part of the Gross National Product.

Relevant examples of community intervention programs

Most current community health intervention programs are multi-faceted, and nutritional impacts are usually only one of several desired outcomes. However, in addition to setting specific programs, there are at least three sets of 'whole-of-community' interventions which have had strong nutritional or healthy eating themes. They include heart health promotion and other disease- and nutrient-centred programs, fruit and vegetable promotion, and community participatory health promotion programs.

Heart health intervention programs

For much of the past half-century, Western nutrition has been dominated by cardiovascular disease, particularly by the saturated fat–serum cholesterol–atherosclerosis pathway. This concern led university-based physicians and their colleagues to develop community intervention programs. These attempted to persuade people to change their lifestyles in terms of stopping smoking, reductions in fat and salt consumption and increases in physical activity. The first programs started in the early 1970s and their successors continue today. It has been estimated that about 25 per cent of the reduction in ischaemic heart disease mortality rates in North America between 1980 and 1990 was due to primary prevention efforts along the lines developed by these programs (Hunink et al. 1997). A useful summary of community-based nutrition and lifestyle programs is provided by Verheijden and Kok (2005).

There have been three generations of these heart health interventions (Huot et al. 2004), as follows:

First-generation programs

These were largely based on individual behaviour models such as social learning theory and the theory of planned behaviour (or its earlier version, the theory of reasoned action). The prime example was the Stanford Three City Program led by Jack Farquhar and Nathan Maccoby (Maccoby and Wood 1990). It used a wealth of methods (newsletters, TV advertising, restaurant menu labelling changes, counselling sessions) in two towns in California to get members of these communities to change their lifestyles compared with a third 'comparison' town where little promotion activity was undertaken. The overall effects were small. There were reductions in saturated dietary fat intakes and serum cholesterol levels which were sustained for at least one year after the conclusion of the study. Despite its apparently limited impact, the project served as a model which was emulated throughout North America and elsewhere.

Members of the Stanford team were invited by cardiovascular physicians to North Karelia in Finland. These physicians were alarmed at the high rates of cardiovascular mortality in that county and wanted to adopt some of the methods used in the Stanford Three City Program. So the North Karelia Program was set up using similar social learning principles and using Vassa county as a 'comparison'. An interesting feature of the North Karelia study is that it was largely funded from local government funds, since in Finland hospitals and other health services are administered by local government. The North Karelia administration decided to divert funding from its health services into 'prevention' activities. This was an important difference from the North American university research-funded interventions since local people were responsible for the prevention program, which was therefore closer to the perceived needs of local people. Perhaps as a result, the North Karelia Program was more 'flowing' than its American predecessor and it appears to have been more successful in bringing about changes in behaviours, and in serum cholesterol levels and associated death rates. It was later generalised to a Finland-wide prevention initiative.

The North Karelia study lasted for a long time. After ten years, there were significant reductions in smoking, and in serum cholesterol concentrations and systolic and diastolic blood pressure among middle-aged men and women (Puska et al. 1983). In the 20-year period between 1972 and 1992, there was a 55 per cent reduction in cardiovascular disease risk among men (68 per cent among women), which has been attributed to dietary changes similar to those promoted by the North Karelia Program (Puska et al. 1983; Puska et al. 1985). It did demonstrate quite impressive dietary changes, such as reductions in the use of high-fat products.

Heartbeat Wales was a first-generation, community-based demonstration intervention conducted between 1985 and 1990. The intervention area was Wales and the comparison community was Northeast England. Public education campaigns, policy and infrastructure changes were conducted throughout Wales with the aim of altering the population's heart disease risk behaviours. Fifteen self-report instruments were used to assess behavioural indicators relating to smoking, exercise, dietary habits and body weight. Compared with the comparison area, there were reductions in the prevalence of smoking and a number of dietary improvements, though there was no overall intervention effect between Wales and the reference area. In part, this was due to an unexpectedly high uptake of heart health promotion activities in the reference area (Tudor Smith et al. 1998).

Second-generation programs

These were more focused on community processes such as diffusion processes and community organisation, but they retained a strong emphasis on the saturated fat–serum cholesterol pathway, which led to a narrow focus on sources of fats in general—especially severe restrictions on the use of animal products like red meat and whole milk. Other aspects of the diet were largely ignored. This was very much a prohibition ('diet police') approach to nutrition promotion. Key studies included the Stanford Five City Project, the Minnesota Heart Health Program and the Pawtucket Heart Health Program. The Stanford Five City Project again demonstrated similar changes in cardiovascular risk factors in the intervention towns to the Stanford Three City Study, but in addition showed that some of the changes could be sustained (Winkleby et al. 1996). The Minnesota Heart Health Project and Pawtucket Heart Health Project showed few changes, but increases in body mass indices were smaller in the Pawtucket intervention communities than in the control community (Carleton et al. 1995).

The results of these studies varied, but in the main they had minimal overall effects on their outcome (impact) variables. This was partly because secular (country-wide) trends were also changing. For example, in Canada between 1970 and 1998, adults reduced their fat intake by 40 per cent of total calories. Similar changes also occurred in the United States (Sorensen et al. 1998). The anti-fat message was strong in North America and certainly not confined to 'intervention' communities. Another problem was the uneven delivery of the interventions—the 'dosage' of the interventions was poorly controlled. In some communities, few people showed up to the intervention meetings or did not attend to TV advertisements (Huot et al. 2004).

Generally, these programs found it difficult to reach all the population in their communities. Instead, the results were 'patchy' though promising. For example, in the Pawtucket and Stanford Five City Programs, members of the intervention communities appear to have either not gained weight (Pawtucket) or gained less weight (Stanford) than members of the comparison communities. In the Minnesota Heart Health Program, however, people who had high serum cholesterol levels seem not to have gained as much weight as others. While point-of-sale programs *overall* in the Stanford, Pawtucket and Minnesota Programs did not affect dietary variables, the Pawtucket Program did (Hunt et al. 1990). Between 1984 and 1988, the percentage of shoppers who could correctly identify shelf labels indicating low-fat, low-energy and low-salt contents increased from 11 per cent to 24 per cent. Similarly, the percentage of shoppers who felt encouraged to buy these types of products rose from 36 to 54 per cent.

Third-generation programs

Designers of the third generation programs have tried to learn from the experiences of the first two generations. They target hard-to-reach sub-populations (e.g. blue-collar workers and members of disadvantaged ethnic communities) by adapting to local circumstances and constraints and using empowerment approaches. Examples include the Quebec Heart Health Program (Huot et al. 2004), which operates in three urban, suburban and rural communities, the Washington Heights Heart Health Program (Shea et al. 1990) and the Kahnawake Schools Diabetes Prevention Project (Mohawk Nation; Paradis et al. 2005; Levesque et al. 2005); and the Finnmark Heart Health Program (Norway: Strand and Tverdal 2004).

The Canadian Heart Foundation Program is an interesting derivative of heart health programs. It utilises the autonomous nature of the ten Canadian provinces in the Canadian Federation. Each provincial Heart Foundation has set up its own heart health initiatives within a fifteen-year 'federal' program which aims to build capacity in public health and to implement it (Canadian Heart Health Surveys Research Group 1992; Conference of Principal Investigators of Heart Health 2002). In the first five-year phase of the Quebec Heart Health Program, regional public health departments conducted nutrition promotion activities in urban suburban and rural areas, each of which included experimental and control groups. Examples of the activities included hypertension and hypercholesterolemia screening sessions for hypertension, supermarket tours, distribution of healthy recipe booklets, healthy food tastings, cooking classes and walking clubs. A Global Dietary Index (GDI) was used to assess dietary change. The GDI improved in the urban and suburban sites in both the exposed and the non-exposed groups, but deteriorated in the rural area. The investigators concluded that the efficacy of this type of program could be improved through greater attention being given to public policy and to the influence of social and physical environments (Huot et al. 2004).

Examples of nutrient-specific programs

Several programs have been conducted in the past two decades which certainly reach into a large number of communities and utilise community support, but which are 'top down' and systems oriented rather than purely community-based interventions. They are also clearly nutrient specific. Three prominent examples are the International Council for the Control of Iodine Deficiency Diseases' worldwide program, the American Opportunities for Micronutrient Interventions, and the recent World Action on Salt and Health. All three aim at structural upstream change.

Eradication of iodine deficiency

Over one billion people are at risk of iodine deficiency diseases, which present primarily as goitre, cretinism and intellectual disabilities. The International Council for the Control of Iodine Deficiency Disease (ICCIDD) network reaches into over 100 countries. Its aim is to provide adequate levels of iodine for populations — especially for pregnant women, since iodine deficiency causes long-term neural damage during the first trimester of pregnancy. The primary mode of intervention is the iodisation of salt, though injections of iodised oil are sometimes given to women of reproductive age. Much of the ICCIDD's activity is devoted to lobbying for salt iodisation policies in individual countries and to monitoring the effectiveness and durability of the salt iodisation process under real-life conditions and the effectiveness of the distribution of iodised salt (Hetzel and Pandav 1997).

The ICCIDD program has prevented iodine deficiency among hundreds of millions of people. However, in Western populations iodine concentrations continue to fall with unfortunate consequences. Iodisation of salt may be required in affluent Western countries though their high salt consumption is a major cause of hypertension, stroke and heart disease. However, it appears that even halving the salt content of processed foods in a country like the United Kingdom would still deliver sufficient iodine if all the salt was iodised.

Opportunities for Micronutrient Interventions (OMNI)

This program was a form of American aid to the Third World during the 1990s. It recognised that, although the Green Revolution of the 1960s and 1970s provided many people around the world with sufficient daily energy, it did nothing to relive micronutrient deficiencies—particularly deficiencies of iron, vitamin A and iodine. OMNI provided funding and policy initiatives for a variety of projects around the world, usually in conjunction with national and international health agencies. For example, vitamin A and iron were added to soy sauces used in many Southeast Asian countries or delivered in serving-size portions for children. Some of these initiatives were more enduring than others, but they illustrate the value of upstream approaches which do not require major educational efforts. However, their fragility and narrow nutrient orientation suggest that more food-based approaches are likely to be more feasible in the long term. OMNI's work is being continued by the Micronutrient Initiative Canada (www.micronutrient.org).

World Action on Salt and Health

This is a recent development which has emerged out of the British government's initiative to encourage the food industry to halve the amount of salt in processed foods (UK Food Standards Agency, www.salt.gov.uk/index.html). Directives to reduce population salt intake (to around 5 or 6 grams per person per day) have been included in national dietary guidelines for almost 30 years (Sharp 2004; NHF 2006). Most attempts to implement the guideline have focused on individuals' use of table and cooking salt. This is difficult to do (Beard 1997) and this route accounts for only 15 per cent of the salt in the population's diet. In countries like the United Kingdom, United States and Australia, salt in bread accounts for around one-quarter of the population's salt intake. There is now strong evidence that the salt content of white bread can be reduced by up to 25 per cent before most people notice any change, and that it is also technically feasible (NHF 2006). However, with the notable exception of the United Kingdom, little has been done to reduce the salt content of manufactured foods, so advocacy, lobbying and government incentives are required. WASH is the lobby group which aims to bring about these changes worldwide (www.worldactiononsalt.com); the Australian branch of this organisation (AWASH) can be found at www.thegeorgeinstitute.org/media-and-publications. A core recommendation of WASH is that foods containing more than 120 mg salt/100 g should be avoided where possible (Beard et al. 1997).

Fruit and vegetable promotion in the community

Most of the community interventions during the decades from the 1960s to 1980s, particularly in the United States, were oriented towards reductions in the consumption of saturated fats. Since that time, some reconsideration has occurred concerning the wisdom of this narrow approach. There are several reasons for the subsequent change of emphasis. First, the emphasis on fat reduction was not accompanied by reductions in body weight. Worse still, the incidences of obesity and type 2 diabetes have increased alarmingly. Second, studies have shown that heart disease aetiology is related to a number of other dietary factors; the homocysteine hypothesis, for example, proposes that atherosclerosis is an inflammatory process which is partly related to deficits in folate and thus intakes of fruits and vegetables. Studies like the Lyons Heart Study (de Lorgeril et al. 1999) have shown that it is not only excessive intakes of saturated fats but also deficits in the consumption of polyunsaturated and short-chain fatty acids (plant and marine fats) that are crucial in cardiac pathology. Third, large bodies of evidence concerning the aetiology and prevention of other diseases such as colon cancer (and other cancers) have shown that consumption of whole groups of foods is important. For example, there have been several major reports about the health benefits of higher than conventional American intakes of fruits and vegetables (e.g. WHO 2002; World Cancer Research Fund 1997).

In a systematic review of interventions designed to influence adult fruit and vegetable consumption, Pomerleau et al. (2005) have shown that approaches like face-to-face counselling, telephone contacts, computer-tailored information and community-based multi-component interventions can be effective ways to change healthy individuals' dietary behaviours. Similarly, Sahay et al. (2006) showed that ten of fifteen controlled nutrition interventions they reviewed showed significant intervention effects. They found that key elements of successful interventions included a theoretical basis, family involvement, participatory planning and implementation models, clear messages, and adequate training and ongoing support for intervention staff.

The World Health Organization website has a detailed description of its fruit and vegetable promotion initiative (www.who.int/entity/dietphysicalactivity/media/en/gs_fv_report.pdf), and the Report of the Fourth International 5 a Day Symposium held in Christchurch in 2004 also provides practical updates on fruit and vegetable promotion (www.5aday.co.nz/symposium).

All of these developments have changed the emphasis of community nutrition promotion to a more positive approach: people are now encouraged to consume more of certain foods like fruits, vegetables, cereal grains and fish while trying to reduce their energy and saturated fat intakes. The World Health Organization Global Strategy on diet and physical activity emphasises the generation of stronger evidence for policy concerning the links between diet, physical activity and chronic disease, advocacy for policy change through informing decision-makers and interventions, greater stakeholder involvement in the promotion of the strategy, and the development of a strategic framework for action within and between countries (www.who.int/dietphysicalactivity/goals/en).

The 5+ a Day Programs

The US National Cancer Institute set up a series of nine promotion programs in the mid-1990s which aimed to increase Americans' consumption of fruit and vegetables (to five servings a day). As we have seen, these were in various settings such as health services, supermarkets, restaurants, worksites, churches and schools. A lot of experience has been gained as a result. Most of these programs were successful in that they increased the consumption of fruit and vegetables, usually by small but aetiologically significant amounts. The early American models have now been formalised and extended into a 5 a Day Movement which is supported by the WHO and various health foundations around the world (e.g. Anti-cancer and Heart Foundations). Useful American examples are California's 5 a Day Program (Foerster et al. 1995) and the Seattle 5 a Day Worksite Program (2001).

Australian examples include the annual Fruit and Vegetable Campaign conducted by the Western Australian Department of Health (see Box 8.1) and the Coles Supermarkets-Dietitians Association of Australia 'Two Fruit and Five Veg' point-of-sale program, among others. Links to the World Health Organization's Global Strategy are given in Box 8.1.

The *Danish 6 a Day Program* commenced in 1998 as a multi-sectorial fruit and vegetable promotion program. It was preceded by a systematic review of the literature and is accompanied by a regular dietary survey of children and adults. To date, the children's target has been achieved and the adult target is on track to be achieved by 2010. The worksite and mass media components of the program have been particularly successful. Further details are available from www.6aday.com/abstract6adayDenmark2004.doc. The program is part of the European platform for action on diet, physical activity and health (www.eurocoop.coop/events/en/seminarobesity/default.asp).

The *British School Fruit and Vegetable Scheme* is part of the UK 5 a Day Program. Under the scheme, all four- to six-year-old children in local education authority infant, primary and special schools are given a free piece of fruit or vegetable each school day. Early evaluations of the program suggested that it may have increased consumption of fruit and an improvement in healthy eating practices in children from both affluent and deprived areas (www.dh.gov.uk/PolicyAndGuidance/HealthAndSocialCareTopics/FiveADay; www.Scotland.gov.uk/Publications/2005/12/1395341/53422).

BOX 8.1 WHO GLOBAL STRATEGY ON DIET, PHYSICAL ACTIVITY
AND HEALTH

In May 2004, the 57th World Health Assembly (WHA) endorsed the WHO Global Strategy on Diet, Physical Activity and Health. This strategy provides a comprehensive set of policy options for member states and other stakeholders, which if fully implemented will lead to a significant reduction in chronic diseases and their common risk factors, including low consumption of fruits and vegetables.

With this mandate from the 57th WHA, and the known benefits of fruit and vegetable consumption, WHO aims to actively promote an increase in fruit and vegetable intake worldwide, especially in developing countries. Incorporation of fruit and vegetable consumption as part of national chronic disease prevention and school health programs is a central aim.

Fruit, vegetables and NCD disease-prevention

- Up to 2.7 million lives could be saved annually with sufficient fruit and vegetable consumption.
- Low fruit and vegetable intake is among the top ten selected risk factors for global mortality.
- Worldwide, low intake of fruits and vegetables is estimated to cause about 19 per cent of gastrointestinal cancer, about 31 per cent of ischaemic heart disease and 11 per cent of strokes.

Source: www.who.int/dietphysicalactivity/media/en/gsfs_fv.pdf; www.who.int/dietphysicalactivity/fruit/en/index2.html.

Community-relevant dietary change studies

Several studies have shown that increases in nutrient-rich, moderate or low-energy food, especially when combined with moderate levels of daily physical activity, can have major effects on disease risk:

- The Dietary Approaches to Stop Hypertension Studies (DASH) (Appel et al. 1997) and the PREMIER Study (Elmer et al. 2006; Appel et al. 2003) represent a landmark in nutrition promotion trials. These studies were not 'community' programs; instead, they were tightly controlled experiments.

However, they attempted to show the benefits of 'whole-of-diet' interventions among participants from impoverished American minorities, including Hispanics and Afro-Americans. They showed that increased intakes of fruits, vegetables and low-fat dairy products markedly reduced the participants' systolic and diastolic blood pressure, whether or not it was in the hypertensive range. Reduction of salt intake increased the beneficial effects.

In our own work on the extension of the DASH approach among free-living community volunteers in Melbourne (the OZDASH and OZWELL studies—Nowson et al. 2004; Nowson et al. 2005), we confirmed the DASH findings but found that the volunteers initially had difficulties consuming more servings of vegetables and low-fat dairy products. However, in the OZWELL study, participants were fed back information about their food consumption and were able to achieve the project goals relating to vegetables, low-fat products and other foods. When combined with a mild exercise program, substantial improvements' in participants health status were observed (e.g. in terms of triglycerides, blood pressure, body weight and serum calcium concentrations—see Figure 8.2), with persons taking anti-hypertensive medication such as ACE inhibitors improving about twice as much as other participants. These studies show that if community volunteers make small changes to their dietary and physical activity patterns, major improvements in health status are likely to follow. The key next step is to find ways to diffuse these change techniques throughout the community.

FIGURE 8.2 SYSTOLIC AND DIASTOLIC PRESSURE OVER TWELVE WEEKS OF INTERVENTION IN THE OZWELL STUDY

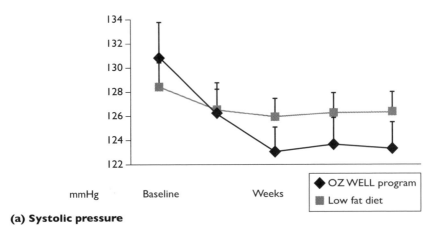

(a) Systolic pressure

Figure 8.2 continued

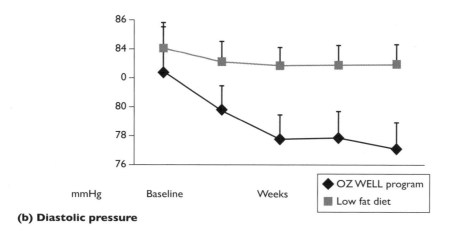

mmHg   Baseline   Weeks

◆ OZ WELL program
■ Low fat diet

(b) Diastolic pressure

- The Western Australian Heart Disease Prevention studies (e.g. Bao et al. 1998) showed that reductions in alochol consumption, moderate exercise and fish consumption reduced hypertensive volunteers' blood pressure.
- The Trial of Nonpharmacologic Interventions in the Elderly (TONE) Study (Whelton et al. 1998) showed that improvements in dietary quality and moderate exercise reduced elderly patients' blood pressure and improved their health status.
- The Coronary Artery Risk Development in Young Adults (CARDIA) Study is a longitudinal study of aetiology of cardiovascular disease in 5115 black and white American men and women who were aged 18–30 years at their initial examination in 1985–86. Five further examinations were held in 1987–88, 1990–91, 1992–93 and 1995–96. A wide variety of dietary, biomedical, behavioural and psychological variables were measured and many findings established. In particular, the positive association between fast food consumption and weight gain and insulin resistance has been established (Pereira et al. 2005). A description of the study and the results from the first examination are provided in Cutter et al. 1991.

- The European Prospective Investigation into Cancer and Nutrition (EPIC) (Bingham et al. 2003; Bingham 2006) is a major population study of the antecedents of bowel cancer which has yielded many findings, including an inverse relationship between dietary fibre intake and bowel cancer.
- The Survey in Europe on Nutrition and the Elderly (SENECA), a Concerted Action (Buijsse et al. 2005), is a prospective population study which has shown several relationships between health indices and food and nutrient consumption, supporting the view that a varied diet is conducive to good health in old age.
- Similarly, the Food Habits in Later Life study (Wahlqvist 1995; Wahlqvist et al. 2005) showed that dietary variety and regular moderate physical activity were positively associated with duration of survival among people over 70 years of age in several cultures.
- Comprehensive lifestyle modification programs in inpatient settings in Sweden (Kaati et al. 2005) have shown that small changes in dietary habits (such as the inclusion of more fruit and vegetables), along with increased moderate activity, resulted in reductions in blood pressure and other chronic disease indices in hypertensive and type 2 diabetic patients for up to five years after an initial one-month lifestyle change program.
- In a systematic review, He et al. (2006) have shown that high fruit and veg-etable consumption is posively associated with potassium intakes and reduced risk of hypertension.
- The Multiple Risk Factor Intervention Trial (MRFIT) (Gorder et al. 1986) was one of the early intervention trials which showed that increases in dietary quality were associated with reductions in chronic disease indices.
- The Diabetes Prevention Program (Diabetes Prevention Program Research Group 2002) showed that dietary and lifestyle change was a feasible and more effective alternative to type 2 diabetes prevention than metformin treatment.

Action research and community participatory health promotion

Action research in the community ('community development', 'community partici-patory health promotion') starts from different premises to those of intervention research. Its focus is more on social and physical environmental barriers to health. As a result, in English-speaking countries (and elsewhere) this form of activity is seen as inherently more 'political' than interventions since it implicitly or explicitly challenges the socio-political 'status quo' of these highly individualist societies. Community participatory health promotion (CPHP) tends to operate among disadvantaged groups in the community and is certainly a 'bottom-up' approach starting out from the daily lives of men and women who experience disadvantages such as poor access to health services and to healthy foods, as well as other systemic problems associated with unemployment, violence and racism (Thomas et al. 2003). However, this bottom-up approach does not prevent collaboration between government, universities and the community, as will be seen in some of the examples below. Indeed, blurring of the distinction between bottom up and top down is a recent trend, as can be seen in the Kahnawake Schools Diabetes Prevention Project (Mohawk Nation—Paradis et al. 2005; Levesque et al. 2005) and the Washington Heights Heart Health Program (Shea et al. 1990).

The complex issues which people have to deal with in daily life were well illustrated by Griffiths (2002). She showed that nutrition and healthy eating are just a few of the 'issues' experienced by Melbourne parents. For example, they may suffer from inadequate resources such as finance, housing and food; family members may have behavioural problems associated with mental and physical health and drug abuse; the parents may have poor parenting skills and knowledge; emotional and material support structures may be inadequate; and there may be many external pressures associated with long working hours, poor pay, inadequate transport and poor access to facilities.

This illustrates a key difference between nutrition intervention projects and CPHP. While top-down project designers may know the importance of nutrition for population health (and be suitably obsessed about it), for many members of the community, nutrition and healthy eating may not be important issues—or even issues at all. True enough, they may suffer from various forms of malnutrition, but they may not recognise food and eating as being important contributors to their ill-health. Their problem may be more to do with getting enough food to eat than with the health attributes of the food. This raises the need to negotiate between experts' knowledge and needs (e.g. university-based nutrition promoters and their funding agencies) and the felt needs of members of the community. It is quite possible that the nutrition promoter who raises funds for a community-based obesity prevention program may find their proposals 'watered down' or opposed by the very communities they intended to help. Instead, they may want to spend the 'obesity prevention' funds on the provision of foods for hungry people, or on non-nutritional goals such as the prevention of domestic violence or the treatment of mental illness. A great deal of trust and long-term commitment by all parties is required to work out such difficult issues.

Recent examples of community nutrition programs include the following:

- The Black Churches Projects (Resnicow et al. 2000) and the REACH Coalition of the African American Building a Legacy of Health Project's. Los Angeles Community Participatory Health Promotion Project (Sloane et al. 2003) utilised participatory approaches within existing networks in the black American community and achieved major population-wide increases in consumption of fruit and vegetables.
- The Barwon Sentinel Site Program (Bell et al. 2004) is one of the first obesity-prevention programs to be conducted in Australia. It uses a variety of community strategies in the rural Victorian town of Colac, including school nutrition education, changes to retail food outlets, and various community involvement strategies. Early evaluation suggests that family attitudes to obesogenic foods have become more negative and children's age-adjusted body weight has reduced.
- The Australian National Child Nutrition Program was conducted in numerous settings, mainly in pre-schools and primary schools, following the introduction of the Goods and Service Tax (GST) in the late 1990s and early 2000s. Unfortunately, little evaluation was designed into the program. A post hoc attempt to evaluate some of the regional interventions in rural Victoria suggested that there may have been some local impacts (www.health.vic.gov.au/archive/archive2005/nutrition/forums/downloads/pe_ncnp_prelimeval.ppt).
- The German Children's Obesity Prevention Program (Contest) is one of 24 programs funded by the German Ministry of Health which are aiming to prevent or reduce overweight and obesity among children up to ten years of age during a three-year period from 2006. Each program is using its own approach, though a uniform evaluation protocol is being used to monitor key dietary, behavioural and biomedical variables (www.besseressenmehrbe wegen.de/index.php?id=425).

Disadvantaged groups, communities and populations

Unfortunately, in every society there are groups of people who are disadvantaged in various ways. There are various forms of disadvantage, though poverty or lack of money and/or material resources is a major component which can influence food consumption and nutrition status. Several groups of people can be identified in many countries, including the following:

- *People with insufficient material resources* include homeless people, people on pensions that are insufficient for their needs (e.g. single mothers with children) and working people whose incomes are insufficient for their needs. Mental illness and infectious illness are highly prevalent among the homeless, and low incomes and limited education characterise many who have insufficient incomes. Newly arrived refugees and members of indigenous minorities are also over-represented in this group.
- *Older people in the community and in care* frequently have poor nutrition status, especially widowed women living on their own. Meals on Wheels and other services have been devised to support them, but many remain malnourished—even in residential care homes.
- *People living with some form of disability* make up about one-fifth of the population. Those with severe neurological, physical or intellectual disabilities represent the 'hidden minority' of most societies. They frequently have very poor health and nutrition, with minimal access to fresh foods and health services. Little is known about their living conditions, and they present a major, urgent challenge for health promotion—one which remains largely unanswered.
- *People living in total institutions* such as prisons and the military represent another large group living in restricted settings whose nutrition status is likely to be problematic.

Examples of programs that focus on disadvantaged people

Often the role of the nutrition promoter is to help the people affected to gain access to food in order to maintain or improve their nutrition and health status. Ideally, the ultimate aim is to prevent the occurrence of such problems or at least to find a permanent solution to the problem. There are many examples of nutrition promotion among these special groups, particularly in developing countries. Several prominent examples which may be adapted for use in other settings include the following.

The Women, Infants and Children Program
This American federal program ameliorates the effects of poverty on the food supplies of women, infants and children. It has been operational for several decades, offering food stamps for the purchase of foods as well as numerous community

extension activities (see Box 8.2). There is some coordination with centralised surveillance such as the Pediatric Nutrition Surveillance System (see Box 8.3) and the National Health and Nutrition Examination Surveys (NHANES) (www.cdc.gov/nchs/nhanes.htm).

BOX 8.2 THE WOMEN, INFANTS AND CHILDREN PROGRAM (USA)

The WIC target population is low-income, nutritionally at-risk:

- pregnant women (through pregnancy and up to six weeks after birth or after the pregnancy ends);
- breastfeeding women (up to the infant's first birthday);
- non-breastfeeding post-partum women (up to six months after the birth of an infant or after the pregnancy ends);
- infants (up to their first birthday); WIC serves 45 per cent of all infants born in the United States;
- children up to their fifth birthday.

WIC participants receive supplemental nutritious foods, nutrition education and counselling at WIC clinics, as well as screening and referrals to other health, welfare and social services.

WIC is not an entitlement program, as Congress does not set aside funds to allow every eligible individual to participate in the program. WIC is a federal grant program for which Congress authorises a specific amount of funds each year. WIC is administered at the federal level by Food and Nutrition Service and by 90 WIC state agencies, through approximately 46 000 authorised retailers, and it operates through 2000 local agencies at 10 000 clinic sites, in 50 state health departments, 34 Indian Tribal Organisations, the District of Columbia and five territories (Northern Mariana, American Samoa, Guam, Puerto Rico and the Virgin Islands).

Examples of where WIC services are provided include county health departments, hospitals, mobile clinics (vans), community centres, schools, public housing sites, migrant health centres and camps, and Indian Health Service facilities.

Source: www.fns.usda.gov/wic/aboutwic/wicataglance.htm.

BOX 8.3 THE US PEDIATRIC NUTRITION SURVEILLANCE SYSTEM (PEDNSS)

PedNSS is a child-based public health surveillance system that describes the nutritional status of low-income US children who attend federally funded maternal and child health and nutrition programs. PedNSS provides data on the prevalence and trends of nutrition-related indicators.

PedNSS uses existing data from the following public health programs for nutrition surveillance:

- the Special Supplemental Nutrition Program for Women, Infants and Children (WIC);
- the Early and Periodic Screening, Diagnosis and Treatment (EPSDT) Program; and
- the Title V Maternal and Child Health Program (MCH).

A majority of the data are from the WIC program that serves children up to age five. Data on birthweight, short stature, underweight, overweight, anaemia and breastfeeding are collected for infants, children and adolescents from birth to 20 years of age who go to public health clinics for routine care, nutrition education and supplemental foods.

The PedNSS provides nutrition surveillance reports for the nation defined as 'all participating contributors' as well as for each contributor. A contributor may be a state, US territory or tribal government. Each contributor can receive more specific reports by clinic, county, local agency, region or metropolitan area.

The goal of PedNSS is to collect, analyse and disseminate data to guide public health policy and action. PedNSS information is used for priority-setting and the planning, implementing, monitoring and evaluation of specific public health programs.

The Pregnancy Nutrition Surveillance System (PNSS) is a similar program-based public health surveillance system that monitors risk factors associated with infant mortality and poor birth outcomes among low-income pregnant women who participate in federally funded public health programs.

Source: www.cdc.gov.

The *FoodCents program* for unemployed, less-educated women was described in Chapter 2. It has been successful in teaching basic food skills which these women continued to used for up to four years after the brief intervention (Foley 1998; Foley and Pollard 1998).

Indigenous groups' programs

Many indigenous groups are impoverished minorities in many countries, so specific nutrition promotion programs have been developed for them. Examples include the Mohawk nation in Canada, Maori in New Zealand and the Aboriginal Nutrition Network in Australia. (A comprehensive description of Australian indigenous nutrition promotion programs is given in *FoodChain*, September 2007: www.nphp.gov.au/workprog/signal/foodchain.htm.)

Foodbank programs are common in affluent societies. They are often organised as cooperatives formed by consumers and charity agencies. They frequently take healthy food products (e.g. non-standard size fresh eggs) and redistribute them to impoverished groups of people. Prominent examples include the North Carolina Foodbank (www.foodbankcenc.org) and other similar programs in many countries. A typical foodbank is described in Chapter 11. They are often large undertakings. For example, in Britain in 2005, some 2000 tonnes of food were redistributed through the community food network of 300 organisations to 12 000 disadvantaged people each day in 34 cities and towns (http://fareshare.org.uk/about/index.html).

Breakfast programs for children are also commonplace. They try to ensure that schoolchildren have a substantial breakfast. A good example of a participatory form of a breakfast program is the Red Cross breakfast program run in Sydney (Miller and Lennie 2005). Williams et al. (2003) have criticised such programs on the grounds that they can lead to further stigmatisation of poor children and allow government authorities to avoid their responsibilities to provide adequate care for all children, though of course governments may contribute to such programs (Miller and Lennie 2005). Nevertheless, the demand for these programs is very high. However, the Primary School Free Breakfast Initiative in Wales has demonstrated several benefits for children (G.F. Moore et al. 2007; L. Moore et al. 2007; Tapper et al. 2007).

The new consciousness in community health promotion

Community heath promotion is a rapidly developing area. Several themes have emerged in the past decade, including the following:

- *Increased methodological sophistication* (discussed in Chapter 7).
- *Recognition of the validity of alternative approaches to RCT designs* — the coming together of methods and approaches. Increasingly, 'intervention project'

workers are incorporating methods from the social sciences, especially those relating to formative and process evaluation such as the use of qualitative, archival and observational methods as well as the use of time series, quasi-experimental and other designs. Community (and Employee) Advisory Boards are now regarded as mandatory, though opinion about their roles varies (Thompson et al. 2003).

- *Recognition of the importance of societal norms and the cumulative impact of health promotion on the general population and secular trends.* There has been a realisation (Thompson et al. 2003; Sorensen et al. 1998) that the small effects brought about during the 1980s and 1990s have had widespread cumulative influence in society. They have brought about long-lasting changes in normative attitudes towards the importance of preventative health behaviours. One of the main purposes of community health promotion is to change societal normative structures (Thompson et al. 2003).

- *Recognition that public health significance is quite different to clinical significance and that small individual changes can have major impact at the population level*, as suggested by Rose (1985) a quarter of a century ago. While fractional increases in daily servings of fruit and vegetables may not sound much to an individual, changes of this magnitude across the population can have major effects on disease prevalence. This realisation has been accompanied by increased optimism about the utility of dietary change—for example, reductions in the salt content of foods will have profound effects on the prevalence of several diseases without the public having to exert much personal effort. Accompanying this trend is a renewed focus on the importance of changes in dietary and other health practices for their effects on disease and health prevalence.

- *The importance of theory and formative and process evaluation* for the delivery of effective health promotion (theory-based evaluation, such as Auspos and Kubisch 2004). Health and nutrition promoters are now using a much wider range of theories and theory-building approaches such as ecological theories (Dooris 2005). Appropriate theories are very useful guides to project implementation. They suggest the key factors which must be modified (or accounted for) in order to bring about project goals. There is a reciprocal relationship between process evaluation and theory construction—each process contributes to the other.

- *Ways to reconcile community needs with those of the research community.* The COMMIT study (Thompson et al. 1991) assessed the power balance between the program researchers and community representatives. Apart from some equality at the beginning of the program, the researchers exerted the most influence over the conduct of the program through its various phases, most probably because they were the main conduit for external funding.

Intervention programs are alive and well, but have changed since their inception during the 1970s to promote a variety of nutritional messages among a broader range of settings and community groups. Most have been concerned with the provision of information and individual skills within micro-settings such as supermarkets, schools and worksites. Issues relating to program focus and effect measurement (especially of dietary change) remain serious problems, as does their failure to come to terms with the implicit food and economic policies which influence the population's dietary habits. Such 'upstream' issues (such as the location and pricing of fresh food) have become quite obvious in recent action research and community participatory health promotion. Further examples will be given in the following chapters.

Conclusions

Nutrition promotion in the community has evolved over the past 30 years from an individualist 'top-down' orientation to a more sophisticated, broader, multi-method approach which incorporates social, environmental and individual influences on population health. The advantages and power of the combination of nutritional and other aspects of health are becoming more apparent. The inclusion of nutritional themes in community health programs tends to prevent several diseases simultaneously—unlike pharmaceutical treatments. The evaluation of community health programs is also becoming much more sophisticated in terms of formative and process evaluation, Novel designs attempt to account for the complexities of societal forms and also assess the *public health* significance of programs. The net effect of this evolution has been greater realisation of the importance of the institutionalisation of programs, advocacy and policy implementation within local, national and international contexts.

Discussion questions

8.1 Define 'community' and think of three examples.

8.2 Compare and contrast 'top-down' and 'bottom-up' approaches to nutrition promotion, providing examples.

8.3 How have heart disease prevention programs evolved in the past 40 years? What are their strengths and disadvantages?

8.4 How has the focus of nutrition promotion changed in the past 40 years?

8.5 What are the advantages and disadvantages of nutrient-specific programs?

8.6 What are the relative strengths and weaknesses of community participatory health promotion programs compared with community intervention approaches?

9

Nutrition promotion for children and young adults

Introduction

Children vary in many ways—for example, some are shy while others are outgoing; some live in traditional families with a mother and a father while others live in blended families or with one parent. They do, however, share three characteristics. First, they are dependent on others for their care. As they age, their dependency usually decreases. Most often, but not exclusively, this care is provided by their parents or other 'caregivers'. Second, they grow physically, mentally and socially. They require nutritious food and stimulating, safe and highly social environments. Third, human offspring mature much more slowly than the offspring of most species. Partly as a consequence of this, children are excellent learners. They actively explore their environments and learn rapidly. Most societies provide special opportunities for them to learn from previous generations about the world, though they usually perceive the world quite differently from their parents' generation. Various forms of food and nutrition education are prescribed for children in most societies because these are generally regarded as important life skills.[3]

Children's nutrition promotion issues span several broad age groups or life stages: pregnancy and the peri-natal period; toddlers and pre-schoolers; primary children; and adolescents and young adults. This messy classification recognises that the issues occurring within the various life stages are dissimilar—for example, sixteen-year-olds face very different challenges from month-old babies and their families.

---

3 Much of this chapter is based on Worsley and Crawford (2005b).

Within each life stage, children live in one or more settings or environments. These settings frequently overlap. Even unborn foetuses in their third trimester in the well-defined setting of their mothers' bodies are exposed to a wider environment in which they hear the sounds made by other family members or taste the foods consumed by their mothers. As children become less dependent on their immediate families, they spend more time in settings outside the home.

The main settings in which children live may influence their food consumption. They include:

- the child's home—this is usually a long-term setting, yet it is one of the least studied;
- antenatal clinics;
- maternity hospitals and birthing centres;
- maternal and child health centres;
- long day-care centres, preschools and kindergartens;
- primary and secondary schools;
- community settings such as scouting and guides clubs, or sports clubs;
- the workplace;
- retail outlets such as supermarkets and fruit and vegetable markets;
- nutrition advisory systems run by Departments of Health, professional associations (e.g. dietitians' associations) and food companies' information 'hotlines' and information programs;
- the mass media—children (and parents) encounter nutritionally relevant communications in the mass media via program content and in food advertising.

Children present opportunities

Many nutrition promoters and food marketers see children as a great opportunity. Marketers increasingly see them as having considerable financial power (Box 9.1). This includes their own purchasing power as well as their power to 'pester' their parents to buy advertised food products and toys. There is a continuing debate over the benefits, disadvantages and ethics of such commercial exploitation (see Chapter 13). On the other hand, many nutrition promoters—especially nutrition educators—see the opportunity to instil knowledge, skills and habits which will steer children on to the 'good nutrition road' for life. There is a common assumption that things learned early in life will last throughout the lifespan. Although there is some evidence for this view, there is contrary evidence which suggests that personal knowledge and skills can be overwhelmed by changing circumstances and by the influence of powerful settings. For example, a teenager may strongly believe that milk drinks are nutritious but may be reluctant to drink them in front of disapproving friends. Nutrition promotion is required at all life stages and in all relevant settings because knowledge acquired early in life may not transfer to different settings and circumstances in later life.

BOX 9.1 ADVERTISING TO CHILDREN

The marketing of food to children is a massive enterprise. Since the 1960s, when children were singled out as a lucrative market, corporations have sought to reach children early to create lifelong consumers. A more immediate objective is to capture the economic power wielded by children that includes their influence over parental purchases ($200–$500 billion) and their own subsequent independent spending ($20 billion).

Food and beverage industries spend more on food marketing each year. Food marketing expenditures in the United States were estimated at $361 billion in 1991 and $538 billion in 2000, not including imports and certain food categories. Marketing accounts for 81 per cent of food costs, up from 78 per cent a decade ago. With few exceptions, industry is free to market calorie-dense, nutrient-poor foods to children of any age, in any demographic group, in any amount and in any way. And there is little to compete with food industry messages. In the year that funding peaked at $3 million for the main government nutrition education program (5-a-Day), McDonald's spent $500 million dollars on a single advertising campaign ('We Love to See You Smile').

Businesses spent $15 billion marketing products to children in 2004 and were rewarded with $200 billion in sales; one-third to one-half of those advertising dollars are spent on food advertising, the overwhelming majority of which is for sugared breakfast cereals, fast food, soft drinks, snacks and candy and gum. Advertising's overall message to children is clear—eat a lot of food, snack between meals, lobby parents to buy certain products (what the industry calls 'pester power'), forget the distinction between treats and meals (e.g. Cookie Crisp and Reese's Puffs breakfast cereals), and aspire to the attributes portrayed in the marketing (fame, fun, friends) by eating the advertised foods. The effect that this advertising has on childhood obesity is robust and clear.

Source: http://yaleruddcenter.org/what/advertising.

Casey and Rozin (1989) suggest that children's food choices may play three important roles for the developing child:

- The child's health and growth are partly determined by the amount and quality of nutrients consumed.
- Early food choices may establish habitual patterns of relating to foods and might influence food choice in later childhood or adulthood.
- Food choice is one of the critical arenas for conflict between parents and children—for example, refusal to eat foods prepared by parents.

What is children's healthy eating?

'Healthy eating'[4] is not easy to define, though healthy eating guides and dietary guidelines (NHMRC 1998, 2003b) have provided de facto definitions. Typically, healthy eating is regarded as the consumption of a wide variety of fresh fruit, vegetables, legumes and wholegrain cereal foods, as well as dairy and animal foods (or other protein-rich foods). However, this is a fairly limited definition because 'healthy eating' can also refer to:

- eating only to satisfy appetite or hunger (so-called 'intuitive eating');
- enjoying a variety of different foods and flavours (e.g. see Ehrlich and Murkies 2001);
- uncoerced eating—that is, not being forced to eat particular foods; and
- having regular meals and snacks, not skipping breakfast and so on.

Nevertheless, various countries' dietary guidelines do reflect nutritionists' views of healthy eating and they are useful guides for the promotion of healthy eating among children. They form the basis of most public health nutrition approaches because they summarise the basic principles of human nutrition as they relate to the population's health. For this reason, they are essential reading for all practitioners who are trying to promote healthy eating among children because they define 'healthy eating'. Examples are *The Australian dietary guidelines for children and adolescents* (NHMRC 2003b; see Box 9.2) and *The Australian guide to healthy eating* (NHMRC 1998; see Chapter 4).

---

4 Described in more detail in Worsley and Crawford (2005b).

BOX 9.2 EXTRACT FROM *THE AUSTRALIAN DIETARY GUIDELINES FOR CHILDREN AND ADOLESCENTS*

Encourage and support breastfeeding

Children and adolescents need sufficient nutritious food to grow and develop normally

- Growth should be checked regularly for young children.
- Physical activity is important for all children and adolescents.

Enjoy a wide variety of nutritious foods

Children and adolescents should be encouraged to:

- eat plenty of vegetables, legumes and fruit;
- eat plenty of cereals (including breads, rice, pasta and noodles), preferably wholegrain;
- include lean meat, fish, poultry and/or alternatives;
- include milks, yoghurts, cheese and/or alternatives;
- avoid reduced fat milks, as these are not suitable for young children under two years because of their high energy needs, but reduced fat varieties should be encouraged for older children and adolescents;
- choose water as a drink;
- avoid alcohol, as it is not recommended for children.

Care should also be taken to:

- limit saturated fat and moderate total fat intake;
- avoid low-fat diets, which are not suitable for infants;
- choose foods low in salt;
- consume only moderate amounts of sugars and foods containing added sugars.

Care for your child's food: Prepare and store it safely

These guidelines are not in order of importance.

Each one deals with an issue that is key for optimal health. Two relate to the quantity and quality of the food we eat—getting the right types of food in the right amounts to meet the body's nutrient needs and reduce the risk of chronic disease. Given the epidemic of obesity we are currently experiencing in Australia, one of these guidelines specifically relates to the need to be active and to avoid over-eating. Another guideline stresses the need to be vigilant about food safety and, in view of the increasing awareness of the importance of early nutrition, there is a further guideline that encourages everyone to support and promote breastfeeding.

Source: NHMRC (2003b).

Problems with children's eating

These vary according to the age and social situations of children and adolescents. Several general concerns have been expressed, as follows.

Narrow preferences for a small number of foods

This occurs among children of all ages (Magarey et al. 2001). Findings from the 1995 National Nutrition Survey (ABS 1999; see Box 9.3) indicate that many children consume foods containing large amounts of energy in the form of fats and sugars, along with salt, and less than 'optimal' (as recommended by children's dietary guidelines) amounts of nutrient-dense foods such as fruits, vegetables and wholegrain foods.

BOX 9.3 A SNAPSHOT OF CHILDREN'S EATING HABITS

On the day of the 1995 National Nutrition Survey, the following were the findings among 5–8-year-olds:

- About 40 per cent of children ate no fruit.
- Almost 30 per cent of children ate no vegetables.
- Potato comprised half of the vegetables eaten.
- Seventy-five per cent of potatoes were fried or mashed with added fat.
- Other than potato, children consumed only 1.5 types of other vegetables.
- As many children ate confectionery (60 per cent), while 80 per cent of children ate foods such as cakes, biscuits and pastries.
- One-third ate 'snack' foods such as crisps, Twisties and Cheezels.
- Only 25 per cent of children reported drinking water.
- Seventy per cent reported drinking fruit or vegetable juice.
- Thirty-eight per cent reported drinking soft drinks.
- Twenty-five per cent used beverage flavours (e.g. cordial).

Source: Based on the analysis of CURF data from the 1995 National Nutrition Survey (ABS 1999).

Difficulties with meals

Many infants and young children (less than four years of age) tend to prefer familiar foods (Birch 1999). Community dietitians have reported that some children as young as two years of age present problems with food refusal and fussy eating that appear to be related to narrow preferences for high-salt, high-fat and high-sugar foods (Worsley and Crawford 2005a). Some nutritionists have also expressed concern about the timing and content of meals among primary and secondary students. Skipping breakfast or lunch, for example, has been linked with cognitive and mood deficits (Pollitt and Mathews 1998).

Impact of working life

About 35 per cent of Melbourne parents with children of pre-school or primary school age have work or leisure schedules that make it difficult for them to eat the evening meal with their children (Campbell et al. 2002). Thirty-five per cent of mothers find it difficult to find time to prepare the evening meal; 15 per cent of them do not consider the evening meal to be a pleasant family time; and 30 per cent of families have the television on during the evening meal on most nights (Campbell et al. 2002).

One possible effect of the increased duration and intensity of parental working lives (Pusey 2003) has been the rapid increase in the provision of food for children outside traditional school hours. The number of out-of-school care programs in Melbourne, for example, trebled in the first five years of this century (NOSHA website, www.nosha.org.au), and over 50 per cent of high schools now provide breakfast programs (Maddock et al. 2005).

Excessive energy consumption

Between 1985 and 1995, there was relatively little change in the *amount* of foods consumed by children (Magarey, personal communication 2003; Magarey 2000; see Figure 9.1), but there was a major increase in the *energy* that children consumed. This was associated with large increases in children's consumption of the following types of food:

- a 40–56 per cent increase in confectionery intake (lollies and chocolates);
- a 29–48 per cent increase in beverage intake (includes soft drinks);
- a 46 per cent increase in the intake of cereal-based products and dishes (e.g. cakes and sweet biscuits);
- a 59–136 per cent increase in the intake of sugar products and dishes.

FIGURE 9.1 CHANGE IN THE AMOUNT AND ENERGY DENSITY OF FOOD CONSUMED, 1985–95

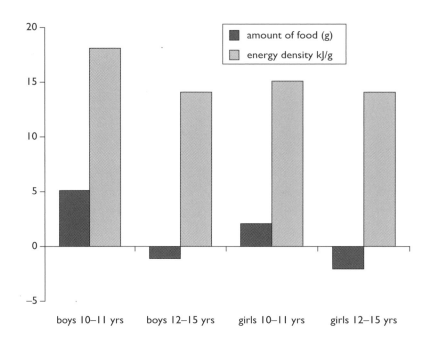

Source: Magarey, personal communication (2003).

Overweight and obesity

Table 9.1 shows the results of three surveys conducted during the past fifteen years. About 20 per cent of children are now overweight and about 5 per cent are obese (Booth et al. 2001). Obesity is a serious chronic health condition (WHO 2000), and obesity in childhood is associated with increased risk factors for heart disease and other conditions in later life (Baur 2001; Must et al. 1992). Disturbingly, type 2 diabetes (or adult-onset diabetes) has begun to appear among adolescents (AIHW 2003; Callahan and Mansfield 2000). In addition to its effects on physical health, obesity in children is also associated with reduced psycho-social health (WHO 2000); obese children often experience social isolation and poor self-esteem.

TABLE 9.1 OBESITY IN AUSTRALIAN CHILDREN (FROM SURVEYS CONDUCTED 1995–97)

	New South Wales Schools Fitness and Physical Activity Survey		National Nutrition Survey		Health of Young Victorians Survey	
	Boys (%)	Girls (%)	Boys (%)	Girls (%)	Boys (%)	Girls (%)
Acceptable	79.9	78.7	80.7	77.7	78.9	76.4
Overweight	14.9	16.3	14.4	16.9	15.8	17.8
Obese	5.2	4.9	4.9	5.4	5.3	5.7
Overweight/ Obese	20.1	21.3	19.3	22.3	21.1	23.5

Source: Booth et al. (2001).

The effectiveness of children's nutrition promotion

Nutrition promotion among children is one of the most common activities occurring in education and health systems around the world. Antenatal clinics, birthing centres, community health clinics, long day-care centres, pre-schools and kindergartens and primary and secondary schools routinely communicate about nutrition and healthy eating. Unfortunately, it is one of the least reported and evaluated activities. This may be why nutrition promotion is often thought to be either non-existent or ineffective in influencing children's food consumption.

Fortunately, several reviews of the area (e.g. Glanz et al. 2005; Huon et al. 1996; Setter et al. 2000; Thomas et al. 2003; Worsley and Crawford 2005a) have shown that nutrition promotion activities can be highly effective in changing children's attitudes, beliefs and eating behaviours. Thomas et al.'s (2003) findings are particularly instructive. They show that interventions are commonly associated with small increases in fruit and vegetable consumption, though it is easier to increase fruit consumption than vegetable consumption. They found six contextual issues which can inhibit or facilitate behavioural change:

1. Children do not see it as their role to be interested in health.
2. Children do not see messages about future health as personally relevant or credible.
3. Fruit, vegetables and confectionery have very different meanings for children.
4. Children actively seek ways to exercise their own choices with regard to food.
5. Children value eating as a social occasion.
6. Children see the contradiction between what is promoted in theory and what adults provide in practice (Thomas et al. 2003).

The authors suggest a number of strategies based on their findings, including 'branding fruit and vegetables as tasty rather than healthy', 'do not promote fruit and vegetables in the same way', 'make health messages relevant and credible to children' and 'create situations for children to have ownership over their food choices'. In a more positive vein, O'Dea (2003) found that children and adolescents saw the major benefits of healthy eating as: improvements in cognitive and physical performance; greater fitness and endurance; feeling good physically; and being more energetic. Barriers included the greater convenience and preferred taste of unhealthy foods, as well as opposition from friends and parents. She suggests that healthy eating and physical activity programs require more involvement of parents and school staff.

The Review of Children's Healthy Eating Interventions (Worsley and Crawford 2005a; see Box 9.4) clearly showed that interventions can be effective in changing children's eating behaviours, though many studies have methodological limitations. However, most interventions have been restricted to maternity hospitals and school settings. There is little information about the effectiveness of programs in family and community settings.

The promotion of children's healthy eating and nutrition in different life stages and settings

There have been thousands of healthy eating and nutrition programs for children. In this chapter, it is only possible to give the barest glimpse of them and to provide a few indicative examples.

BOX 9.4 SUMMARY OF FINDINGS FROM *THE REVIEW OF CHILDREN'S HEALTHY EATING INTERVENTIONS*

Children's healthy eating interventions can be highly effective. The research literature suggests interventions can improve the quality of children's food intake, both in the short term and possibly for several years after the intervention.

The review found 115 publications that reported the results of evaluated interventions, with over one-third of those interventions found to be successful in meeting their stated aims of changing children's eating behaviours. However, the review identified several weaknesses in this area:

- Many interventions have not been evaluated, so the efficacy of particular approaches is impossible to judge.
- Most evaluations have been conducted immediately after (within three months of) the completion of the intervention.
- Most reported interventions have been conducted for only short durations (typically three months or less), and long-term interventions (over a year or more) have been rare (Resnicow et al. 1993).
- Many interventions have had nutritional end-points—for example, the investigators aimed to reduce serum cholesterol levels or saturated fat consumption. Fewer studies have focused on eating and food (and beverage) consumption, such as increases in the consumption of fruit and vegetables (for example, Tooty Fruity).
- Most interventions among school-aged children have been conducted by external researchers with limited funding from grant agencies. Few intervention programs have been conducted within health and education systems by professionals employed in those systems (e.g. teachers, community workers and pre-school educators), although there were some useful exceptions—for example, pre-school programs in Western Australia (Pollard et al. 2001). As a result, healthy eating programs have had limited lifespans and have rarely been institutionalised within health and education systems.
- Most interventions have been conducted in primary schools and maternity hospitals. These settings, along with pre-school day-care centres, offer major opportunities for practitioners to influence children's food and beverage consumption. Wider community settings and family settings have rarely been used to focus on the promotion of children's healthy eating. The impact of the North Karelia (Puska et al. 1985), Stanford Heart Health (Farquhar et al. 1990) and similar programs on children and their families is uncertain.
- Often, only the briefest details have been given about the mode of intervention, such as 'classroom education' or 'family counselling', or about the changes observed in food consumption. Greater emphasis has been given to changes in biomedical status (such as body weight or serum cholesterol concentrations).
- Interventions that use multi-method approaches (such as classroom instruction combined with improved food services and parent involvement) may be more successful than single-method approaches (Cliska et al. 2000), although the quality of input into each method is crucial.

Source: Worsley and Crawford (2005a).

The early years

The early period of life offers many opportunities for nutrition and health promoters and food communicators. It has several advantages. First, parents are really interested in the health and well-being of their children, and are generally receptive to health and nutrition communication and assistance. Second, with the entrenchment of women in paid work outside the home and the occurrence of double-income families, more children are being placed into forms of day care and early childhood education. This makes the feeding of children necessary in public settings in which 'quality' can be assessed and monitored. The professionalisation of child care means that carers have to meet community expectations about their performance, which usually requires training and continuing assessment. These trends will create stronger demand for nutrition promotion in these settings.

Antenatal clinics and care of pregnant women

The nutritional quality of the pregnant mother's diet is an important lifelong influence over the child's health (Barker 1994; Moore and Davies 2001). For example, the protein status (especially milk proteins) of the mother's diet appears to influence placental size and function and birth outcomes (Moore and Davies 2001). Adequate intake is important in the prevention of neural tube defects (NHMRC 2005). Current Australian guidelines recommend that all women of reproductive age have 400 micrograms of folate (or Dietary Folate Equivs) each day (NHMRC 2005). Foods such as citrus fruit and beans are important sources. In some countries, such as the United States and Australia, breads and other foods have been fortified with folate to prevent the adverse consequences of folate deficiency (for example, neural tube defects such as spina bifida) among newborns. It is especially important to encourage a varied diet during pregnancy (NHMRC 2003a).

Many women become more interested in nutrition during pregnancy and often commence the use of dietary supplements and 'healthy' diets then (Devince and Edstrom 2001). Their increased concern about nutrition may present opportunities to guide them (and their partners) towards healthy eating. For example, expectant mothers can be reminded of the value of fruits, vegetables, lean meats and dairy foods in providing these nutrients. Suggestions about convenient ways in which to prepare these foods are also likely to be well received. Pregnancy and the months after childbirth offer many opportunities (or 'teachable moments') for nutrition promoters and other health practitioners to advise women and their families about food and health issues. Unfortunately, access to healthy foods during this stage of life is a problem for some women.

Maternity hospitals and birthing centres

Counselling and education in hospital before and immediately after birth can help mothers establish breastfeeding (Worsley and Crawford 2005a). The Australian Department of Health and Aged Care has suggested ten steps to successful breast-feeding, based upon the international Baby Friendly Hospital initiative (Chapter 11).

Australian research suggests that support from fathers, other adult family members and the general community is required if mothers are to breastfeed for at least six months after birth (McIntyre et al. 1999). Many breastfeeding mothers find that the environment outside their homes is not conducive to breastfeeding. Many workplaces, for example, do not provide quiet, private rooms for breastfeeding, and some employers actively oppose it. Similarly, shops, restaurants, bus and rail depots and airports often do not provide adequate facilities. This lack of conducive environments often convinces mothers that it is more practical to cease breastfeeding relatively soon after birth.

The National Breastfeeding Strategy (see Box 9.5) includes several initiatives, such as the identification of 'baby-friendly' shops and restaurants through simple signs that inform mothers that they can breastfeed on the premises, and the distribution of information kits for mothers (McIntyre et al. 1999).

BOX 9.5 THE AUSTRALIAN BREASTFEEDING STRATEGY

The 'Goals and Targets for Australian Health in the Year 2000 and Beyond' include recommendations for breastfeeding. The recommended goal for breastfeeding initiation in Australia is 90 per cent of new mothers. The policy also aims to ensure that 80 per cent of mothers should be breastfeeding six months after birth. It has been estimated that 85 per cent of Australian babies are breastfed when they are discharged from hospital, falling to 60 per cent at six months. The Australian Breastfeeding Association's five-year plan to protect and promote the initiation and increased duration of breastfeeding aims to remove barriers to the initiation and extended duration of breastfeeding. Strategies to achieve these aims include:

- a national collection of breastfeeding statistics;
- the incorporation of breastfeeding into the relevant National Priority Areas;
- the development of appropriate public policies;
- the establishment of a panel and/or coordinator to promote and protect breastfeeding initiatives to encourage maternity hospitals to attain Baby-friendly Hospital Accreditation;
- the inclusion of breastmilk into the National Food Accounts—illustrating the economic value of breastfeeding.

Strategies to educate women about the benefits of breastfeeding include media campaigns, school programs, breastfeeding education classes and education in the workplace, as well as provision of telephone services, home visits and peer support programs.

The Australian Breastfeeding Association website is at www.breastfeeding.asu.au.

Source: Department of Human Services, Victoria, 2005.

The strategy takes a multi-faceted approach to encourage breastfeeding, incorporating family education, national accreditation standards for maternal and infant care services, data collection and antenatal educators. Special attention is also given to employer support (via the Workplace Support Project), health professional education (via relevant guideline documents) and Indigenous health.

An information kit for businesses about ways in which they can provide supportive environments for women to breastfeed their babies is available from the Australian Department of Health and Aged Care. The kit also includes signs that can identify premises as 'baby friendly'. Such environmental changes involve major adjustments to societal norms about the acceptability of breastfeeding in public life. Strong advocacy and awareness-raising programs are required, one example being the mass breastfeeding occasions that are staged from time to time.

Maternal and child health centres

Maternal and child health nurses provide community-based education, counselling and support for families, and health surveillance of infants and young children. For example, in Victoria parents may be seen by nurses at birth, then when babies are aged two weeks, four weeks, eight weeks, four months, eight months, twelve months, eighteen months, two years and three + years, with possible extra visits according to need and resources. Maternal and child health nurses play a key role in nutrition promotion because they often have first-hand experience of parents' problems in feeding their children.

New parents may be uncertain about how to feed and care for their babies, and many young parents live away from the support of their families. Strong encouragement and emotional support can help young parents feed and care for their infants, and overcome problems such as food refusal and fussy eating. Maternal and child health services provide advice and support to help parents overcome these sorts of problems.

Long day-care and family day-care centres

Many parents place their infants and pre-school children in the hands of carers who operate from their own homes (family day care) or from commercial or community premises. These carers feed the children during the day and they may influence the development of food and taste preferences, given the importance of early experiences (Temple et al. 2002).

Workers in child-care centres are often keen to teach the principles of healthy eating to children and their parents (Montague 2004). The Start Right, Eat Right program offers basic nutrition and healthy eating training for child-care workers to enable them to provide children with healthier food choices and to advise parents about healthy eating (Lewis and Pollard 2002; Pollard 2001; Pollard, Lewis, Barker et al. 2002; Pollard, Lewis and Miller 1999, 2001; see Box 9.6). The course includes training in nutrition and child feeding principles, and the serving of healthy menus that include whole foods and discourage highly processed foods. It appears to have developed the roles of carers as healthy eating advisers (Lewis, personal communication, 2002).

BOX 9.6 WHAT IS THE START RIGHT—EAT RIGHT AWARD SCHEME?

The Start Right—Eat Right award scheme encourages and recognises long day-care centres that provide nutritious and varied food for children. These centres serve food in ways that meet children's social, cultural and educational needs. To receive the award, centres must meet specific standards of excellence for safe, nutritious and varied food. The Start Right—Eat Right award scheme was developed and trialled as part of the Western Australian Health Promotion Foundation's (Healthway) Cent$ible Food Service Project (1996–99).

Implementation of the scheme was a collaboration between:

- the Department of Health's Nutrition and Physical Activity and Environmental Health Branches;
- Family and Children's Services Childcare Licensing Board;
- key childcare training organisations, including Gowrie Inc.;
- Family Day Care Schemes of WA;
- Australian Institute of Environmental Health (WA branch);
- Care West; and
- the Independent Childcare Association.

The scheme aims to motivate and encourage long day-care centres to provide:

- at least 50 per cent of the recommended dietary intake (RDI) of nutrients for children while in care;
- a FoodSafe environment;
- a supportive and enjoyable healthy eating environment.

To achieve the award, long day-care centres must have:

- a cook and coordinator who have completed recognised nutrition training such as the Food Service Planning for Childcare Centres or the Community Service Unit's Prepare Nutritionally Balanced Food in a Safe and Hygienic Manner qualifications;
- the centre menus assessed to show that they provide at least 50 per cent of the recommended daily intake for children in their care;
- recognition of current FoodSafe practices by the local council;
- a suitable nutrition policy to support these standards;
- the centre assessed by a trained assessor.

Between its inception in 1997 and 2004, over half of Western Australian long day-care centres had received the Start Right—Eat Right award and over 1200 individuals had attended nutrition training. Many of these people are still working in the industry.

Source: http://www.population.health.wa.gov.au/promotion/srer.cfm.

Pre-schools and kindergartens

The education provided for three- and four-year-olds in these centres includes socialisation and education about food and eating. The centres offer important opportunities to influence children's food learning and eating practices. *Food Facts for Preschoolers* is a useful education guide produced by Kindergarten Parents Victoria (www.kpu.org.au).

The child's home as a setting for nutrition promotion

Relatively little is known about the ways in which families buy, prepare and consume foods, despite them being the main source of children's food and the main influence over their food choices. Many families have 'rules' or ways of eating. For example, low-SES families in Melbourne tend to eat at the table (in contrast to higher SES families), though they tend to serve more soft drinks; some ban television watching or telephone calls during meals, and about one in four have arguments during the main evening meal at least three times a week (Campbell et al. 2002). Permissive and authoritarian parenting styles can have negative effects on the quality of foods consumed (Davison and Birch 2001).

Parents are difficult for nutrition promoters to reach. The Food Dudes studies in Britain (Tapper et al. 2003; see Box 9.7) show how the home can be used to promote healthy eating. Other ways to reach families include cooking and healthy eating demonstrations in supermarkets, as well as at child-care centres, primary and secondary schools and worksites, direct mailing (e.g. of a series of nutrition communication postcards) as well as radio and television programs and internet sites.

BOX 9.7 THE FOOD DUDES STUDIES

These studies into increasing children's fruit and vegetable consumption were first carried out in the home environment with a small group of five- to six-year-old children (identified by their parents as 'fussy eaters') who ate little fruit and vegetables (Dowey 1996; Horne et al. 1995; Lowe et al. 1998). The studies employed a multiple baseline research design (Kazdin 1982) in which, following baselines of varying duration, the start of the intervention was staggered over time across foods, being introduced first for fruit and then for vegetable

Box 9.7 continued

consumption. The studies evaluated the effects of four different procedures on children's consumption of a range of fruit and vegetables presented to them. The procedures were as follows: fruit and vegetable presentation only; rewarded taste exposure; peer modelling; and rewarded taste exposure combined with peer modelling.

The peer modelling element consisted of a video featuring the heroic 'Food Dudes', a group of four slightly older children who gain superpowers from eating fruit and vegetables. The Food Dudes do battle against evil 'Junk Punks' who threaten to take over the planet by destroying all the fruit and vegetables, thereby depriving humans of their 'Life Force' foods. Throughout the video, the Food Dudes eat and enjoy a variety of fruit and vegetables. The reward consisted of items such as Food Dude stickers, pens and erasers, awarded to the children for eating target amounts of fruit and vegetables.

The results showed that the combination of peer modelling and rewards was very effective at increasing children's consumption of both fruit and vegetables. Prior to the introduction of the intervention, the children were consuming an average of 4 per cent of the fruit presented to them at home by their parents, and just 1 per cent of the vegetables. However, upon the introduction by their parents of the video and rewards, fruit consumption increased to 100 per cent and vegetable consumption to 83 per cent.

Follow-up measures taken six months later showed that not only were the increases large, they were also maintained over time. The children were still eating 100 per cent of the fruit presented to them and 58 per cent of the vegetables, even though they were no longer receiving the rewards or watching the video. In addition, there was evidence to show that the effects were not simply restricted to the fruit and vegetables that the children had been rewarded for eating, but also occurred for other items children were able to name as fruit or vegetables.

By way of contrast, the results also showed that continued presentation of fruit and vegetables alone had no effect on children's consumption. Likewise, the effects of the peer-modelling video without the rewards were minimal. There were some effects when the rewards were used without the video (especially with fruit), but by far the greatest increases in consumption were achieved when the video and rewards were combined. We believe that because the rewards are labelled as 'Food Dude' items, they acquire considerable potency through their association with the characters on the video, and that for this reason the effects of the combined elements are greater.

Sources: Tapper et al. (2003); Food Dudes website, www.fooddudes.co.uk.

The primary years: Primary schools as settings for nutrition promotion

There are thousands of primary schools in most countries. The vast majority of 5–12-year-olds attend them (see Gibbons 2002). They are often the hub of the local community—that is, they 'reach' deeply into the local population affecting children, their families and their friends and relations. School-based nutrition promotion programs have the potential to influence the lifelong food consumption of children and their families.

Nutrition promotion activities that involve parents or that change the food school supply are probably more effective than those based solely on classroom activities (Worsley and Crawford 2005a). The use of group-based, experiential learning techniques within well-designed curricula can influence children's food attitudes and beliefs (Johnson and Johnson 1985a). There have been many nutrition education/promotion projects and programs in primary schools. In Australia, several educational initiatives are pertinent to primary (and secondary) schoolchildren's eating habits.

The model national nutrition education curriculum

The national curriculum recommends the content and pedagogy regarding food and nutrition (Reynolds 2000a, 2000b, 2000c). The educational activities carried out under the framework are likely to influence children's and families' eating behaviours. Unfortunately, these effects have not been evaluated, and the ways in which teachers develop and reinforce healthy eating habits are unclear. There is a major need to evaluate the effects of day-to-day classroom nutrition education. For example, to what extent (if at all) do school nutrition curricula affect the food choices of children and their families?

The health-promoting schools network

The 'health-promoting schools' movement has been developed in Australia over the past two decades. It emphasises the community- and health-building roles of schools (Nutbeam and St Ledger 1997) and promotes relationships between the curriculum, teaching and learning, the school organisation and the community. Three points illustrate the approach.

First, the quality of school food services should be consistent with what is taught in the curriculum. Children should not be taught about the nutritional benefits of fruit and vegetables only to find they are not on sale in the school canteen. Second, the school organisation should be inclusive. Teachers, administrators, parents and students should be involved in menu planning. The school food service caters to the needs of students from diverse sociocultural and socioeconomic backgrounds. Third, in developing links with the community, students should visit local shops and food markets and community organisations, and their schools should develop links with farming and environmental groups. A model charter for a health-promoting school is shown in Box 9.8.

BOX 9.8 A MODEL CHARTER FOR A HEALTH-PROMOTING SCHOOL

Our school aims, through all our activities and structures, to assist students, staff and other members of our school community to experience physical, mental and emotional wellbeing. We are committed to:

- ensuring our physical surroundings are safe, pleasant and stimulating;
- effectively teaching skills for health in the classroom;
- relating and communicating well with members of our school community;
- creating school policies and procedures that promote health;
- participating with staff, students and their families in planning and carrying out health-promoting initiatives;
- inviting local organisations to work with us to make our school community more healthy.

Our school will:

- provide personal development, health and physical education programs that are integrated with student welfare;
- provide at least three supervised sessions of vigorous physical activity per week for all students;
- involve the local community in the review, implementation or evaluation of at least one health-promoting program per year;
- provide the canteen with a policy of selling health-promoting foods;
- provide a fully equipped and well-maintained first aid area, staffed by a qualified person, and ensure careful attention to practices and medications;
- recycle paper, aluminium and glass, and use environmentally friendly products where possible;
- address safety in all school activities, including sport, playground, practical lessons and school traffic environments;
- provide an environment that minimises health risks, with particular regard to air and noise pollution;
- provide programs that address major public health issues such as road safety and drug education with community participation in planning and implementation;
- establish links with local health services on issues relating to the health of students and staff.

Source: WHO (1995), at www.deakin.edu.au/faculty/education/math_sci_enviro/hpsschools/model.htm.

The World Health Organization originally endorsed the 'health-promoting schools' concept in the 1990s. The concept has now evolved, internationally, into the FRESH framework (Focus on Resources for Education in School Health), which recommends the provision of portable water, sanitation, hygiene and food and nutrition services in all schools around the world (see Table 9.2). In the United Kingdom, it is now mandatory for all schools to be health-promoting schools and to be able to demonstrate their commitment to Office for Standards in Education (OFSTED) inspectors.

The FRESH framework for school health interventions includes health-related school policies, provision of water and sanitation, skills-based health education and health and nutrition services in schools.

TABLE 9.2 THE FRESH FRAMEWORK

Objectives	Core interventions	Beneficiaries Target groups	Indicators
Health-related school policies			
Increase the number of schools with adequate water and sanitation facilities	Clear policies to ensure the provision of water and sanitation in all schools	The school population	• Percentage of schools with safe water and adequate facilities
Increase access to sanitation facilities for teachers, boys and girls	Policies on basic hygiene education in curriculum that lead to increased demand and responsiveness from children, parents' groups and the community for well-maintained facilities	The school population The community Adolescent girls	• Percentage of schools with well-maintained sanitation facilities • Increased gross enrolment rates of girls, when adequate toilets are available in schools
Increase family life education and access to family planning services	Clear policy to include family life education and family planning in secondary school curriculum	Adolescents	• Percentage of schools with family life education and contraceptive/STD counselling

Table 9.2 continued

Objectives	Core interventions	Beneficiaries Target groups	Indicators
Reduce dropouts because of pregnancy	Clear policy that pregnant girls can stay in school and continue in school after delivery	Focus on women's rights to education Adolescent girls	• Increased gross enrolment rates of girls • Reduced dropout of adolescent girls
Reduce tobacco and substance use	Policies that prohibit smoking and substance use in schools	The school population	• Percentage of schools tobacco free
Reduce discrimination against people with HIV/AIDS and their families	Policies to avoid discrimination against people with HIV/AIDS and their families	The school population	• Reduce number of children with HIV/AIDS and HIV/AIDS orphans excluded from schools

Provision of safe water and adequate sanitation in schools

Increase the number of schools with adequate water and sanitation facilities, well maintained and with separate facilities for boys and girls	School construction norms that include adequate water and sanitation, with separate facilities for boys and girls	The school population, especially adolescent girls	• Percentage of schools with safe water and adequate well-maintained sanitation facilities • Increased gross enrolment rates of girls
Reduce incidence of diaorrhea and intestinal infections among schoolchildren	Safe water available in schools	The school population and the community	• Reduced absenteeism and repetition rates

Table 9.2 continued

Objectives	Core interventions	Beneficiaries Target groups	Indicators
Skills-based health education			
Reduce the number of unwanted pregnancies and dropouts from school	Skills-based health education, including family life education	Adolescents	• Increased gross enrolment rates of girls • Reduced number of girls who drop out due to pregnancy and discrimination
Reduce risk behaviours and address lack of knowledge on HIV/AIDS	Skills-based health education and HIV/AIDS/STD prevention included in school curriculum	All schoolchildren	• Percentage of schoolchildren with life skills to prevent HIV/AIDS/STD transmission
Reduce short-term hunger and improve nutrition	Skills-based nutrition education	All schoolchildren	• Percentage of children that have food before they go to school
Reduce tobacco and substance abuse	Skills-based health education	All schoolchildren	• Percentage of schoolchildren who have used any tobacco product in the previous 30 days

Source: http://siteresources.worldbank.org/INTPHAAG/Resources/AAGSchoolHealth-FRESH.pdf.

School fruit and vegetable programs

Fruit and vegetable awareness-raising programs encourage children to taste and prepare fruit and vegetables. The Western Australian School Fruit and Vegetable

Program, conducted during the past decade, has increased children's expectations about the number of servings of fruit and vegetables that should be consumed each day. An interesting extension of this approach is the British Free Fruit for Schoolchildren Program which appears to have increased fruit consumption (see Box 9.9). Interestingly, a large number of schools in the United Kingdom have been implementing healthy eating programs for the past decade. A very useful summary of practical 'how to do it' examples is provided by Wheelock (2006). These programs were further supported by the social marketing approach taken by Jamie Oliver in his television series on school food. This campaign, which was not based on massive amounts of scientific evidence, helped to bring about major changes in formal UK children's food policy.

BOX 9.9 BIG CHANGES IN UK SCHOOLS

Several new programs have been launched by the UK government to combat the rising prevalence of overweight and obesity, poor nutrition and physical inactivity among children.

- **The 5 a Day Program** aims to increase fruit and vegetable consumption by raising awareness of the health benefits and improving access to fruit and vegetables through targeted action.

  The 5 a Day program has five strands which are underpinned by an evaluation and monitoring program: the National School Fruit Scheme, local 5 a Day initiatives, national/local partners—government health consumer groups, a communications program including 5 a Day logo, and work with industry—producers, caterers and retailers.

- **The National School Fruit Scheme** is part of the 5 a Day program to increase fruit and vegetable consumption. Under the scheme, all four- to six-year-old children in state schools will be entitled to a free piece of fruit or vegetable each school day (currently an apple, banana, pear or satsuma). This will eventually entail distributing around 440 million pieces of fruit to over 2 million four- to seven-year-olds in some 18 000 schools across the United Kingdom each year.

- **The Food in Schools (FiS) Program** is a joint venture between the Department of Health (DH) and the Department for Education and Skills (DfES). The program is developing a whole range of nutrition-related activities and projects in schools to complement and add value to the wide range of other initiatives in schools. The DH-led strand of the FiS Program comprised eight pilot projects which followed the child through the school day. It complemented, but was outside of, the formal curriculum. The projects were distributed across UK regions. They included healthier breakfast clubs, tuckshops, vending machines, lunchboxes and cookery clubs, as well as water provision, growing clubs and the dining room environment.

Sources: www.wiredforhealth.gov.uk/cat.php?catid=888&docid=7555; www.intute.ac.uk/social sciences/cgi-bin/search.pl?term1=fruit&limit=0; www.schoolfoodtrust.org.uk/index.asp.

Fruit in New Zealand schools

Fruit in Schools aims to encourage healthy eating and lifestyle choices among children in socially disadvantaged areas. The program encourages participating schools to become health-promoting schools, supporting healthy eating, physical activity, a smoke free environment and sun protection. In high-needs schools, children will receive a free piece of fruit each day for up to three years (www.moh.govt.nz/fruit in schools).

The New South Wales Fresh Taste Program

This program commenced as a result of the New South Wales government's Children's Obesity Summit in 2004. The Minister of Education requires government school food services to use a 'traffic light' system based on the energy density of foods. Fruits and vegetables are usually 'green' foods which can be served without restriction, 'amber' foods are those which have higher energy densities, and red foods are those with energy densities of 1000 kj/100 g or more. These foods can only be served twice per term. The program appears to be working well. In part, this is because the senior levels of government discussed the issues involved thoroughly before embarking on it. For example, involvement of students in decision-making is a cardinal feature of the program. There was also a requirement to consult widely with key stakeholders such as the State Schools Parents' Association and the Catholic and Independent Schools systems. In addition, a lot of supporting groundwork was done—for example, short courses on the provision of healthy food for school canteen staff were provided by local technical colleges, and web resources, including meal planners and energy analyses of common foods, were provided plus a great deal of additional support. The result is that parents, teachers and children appear happy with the improved service, few children go out of the school to buy high-energy foods like fizzy drinks, pies and chips, and most industry suppliers appear to be satisfied because, without the opportunity to supply cheap fat- and sugar-rich foods, it is possible for caterers and suppliers to provide more novel healthy products (www.health.nsw.gov.au/obesity/canteens).

A detailed demonstration of the holistic approach to school nutrition promotion is provided by the Tooty Fruity Vegie Project, a two-year intervention to increase fruit and vegetable consumption among primary children in northern New South Wales (see Box 9.10). Several valuable lessons were learned from this program, which apply to most nutrition promotion projects—especially the complexity involved in school nutrition promotion and the high level of management skills required.

BOX 9.10 THE TOOTY FRUITY VEGIE PROJECT

The Tooty Fruity Vegie (TFV) Project was a two-year, multi-strategic health promotion program aimed at increasing fruit and vegetable consumption among primary schoolchildren in the Northern Rivers region of New South Wales. The project aimed to achieve this increase by improving:

- children's fruit and vegetable knowledge, attitudes, access and preparation skills;
- parents' fruit and vegetable knowledge and preparation skills, and their involvement in fruit and vegetable-promoting activities in the schools and elsewhere;
- teachers' attitudes towards teaching about fruits and vegetables in schools and their skills and confidence in relation to teaching about fruits and vegetables.

Intervention

In late 1998, ten volunteer primary schools (1174 students in total) were recruited as intervention schools, and another six local primary schools (992 students in total) were recruited to act as demographically and geographically matched controls. The project, which ran during the 1999 and 2000 school years, promoted a whole-of-school approach to implementing a range of classroom, canteen, family-oriented and community-based strategies promoting fruits and vegetables. The strategies were developed from the evidence available at the time and were designed to create a supportive environment by developing, and helping schools to implement, fruit and vegetable-promoting educational resources and activities for children, their parents, teachers, schools, school canteens and the broader community.

Schools were encouraged to form project management teams to oversee the project's implementation in their school. Membership varied between schools but could include teacher, principal, child, parent, canteen, community nutritionist and Aboriginal education assistant representatives. These teams, assisted by a TFV project officer, were responsible for choosing the TFV strategies to be implemented in their school, organising their implementation, monitoring the response to them and modifying them as necessary. They also often initiated new, innovative strategies they found or developed themselves. Small grants of A$270 to A$750 per year were made available to schools, based on need, to assist with implementing TFV strategies. In line with the aim of creating a self-sustaining program, all intervention schools were encouraged and helped to recruit and train volunteers (mainly parents) to help with implementing many TFV strategies. The TFV project officers ensured that information about successful strategies was communicated between intervention schools.

Box 9.10 continued

Evaluation

The TFV project had a comprehensive process, impact and outcome evaluation plan, of which only the first two are presented here. The latter involved prospective 24-hour food records at the beginning, middle and end of the project; these are currently being analysed and will be reported separately.

In order to evaluate the quality of the project's implementation and its success in relation to its broad range of impact indicators, we drafted, pilot tested, revised and administered surveys to all the children, parents, teachers, principals, volunteers and other health professionals involved in or exposed to the TFV project. In addition, a 'participation index' was completed by each intervention school's project management team to indicate the reach, frequency and quality of implementation for each key TFV strategy.

Results

The results showed that the TFV project was well implemented, reached the vast majority of all target groups and was overwhelmingly positively received by them. The project enhanced the quality, diversity and frequency of classroom fruit and vegetable-promoting activities, substantially increasing children's involvement in and enjoyment of such activities. It also increased the amount, range and use of fruit and vegetable-promoting materials distributed to parents, as well as increasing parental interest and involvement in, and enjoyment of, fruit and vegetable-promoting activities in schools and beyond. The fun, practical and hands-on nature of many of the TFV strategies, and the parental involvement, seem to have been key factors in the project's success.

The TFV project improved children's fruit and vegetable-related knowledge, attitudes and preparation skills, and their access to fruits and vegetables at home and in school settings; it also may have improved their fruit and vegetable-eating intentions and actions.

Source: www.nrahs.nsw.gov.au/population/promotion/tooty_fruity.

A major problem with many school programs is that they can be complex and difficult to sustain in the long term. For example, new staff may experience difficulties in knowing how to implement a program. Simple interventions are more likely to be sustainable in the long term. Two examples of this genre are scheduled fruit breaks and the use of bottled water at school (Muller 2003). At Footscray Primary School, the percentage of children eating fruit at school rose from 4 per cent to over 70 per cent when scheduled in-class fruit eating breaks were introduced (see Figure 9.2). Similar increases in the consumption of water were observed when children were allowed to drink in class from their own water bottles.

FIGURE 9.2 CHANGES IN STUDENTS' FRUIT INTAKES AT FOOTSCRAY PRIMARY SCHOOL

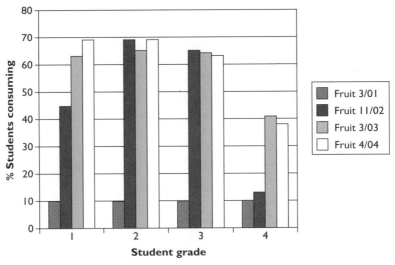

Source: Muller (2003).

Other examples of school fruit and vegetable promotion include:

- *visits to fresh fruit and vegetable markets* by pre-school and primary children to expose them to new taste and food experiences in vibrant settings;
- *community garden and school garden schemes*, which allow children to acquire production experience and skills and positive attitudes towards whole foods—for example, Collingwood College's vegetable garden, the Royal Children's Hospital (Melbourne) roof garden, local government city farms;
- *electronic resources*—The www.human-race.org.au system allows children to record their performance in physical activity tasks and some aspects of food preparation. These recordings can provide teachers with feedback. Such programs need to be further developed, perhaps to evaluate the effects of self-monitoring of food intake. A similar website that can be used to promote children's healthy eating is www.kidsfoodclub.org.

Other types of healthy eating programs in primary schools

Older intervention projects, especially those conducted in the United States, were less focused on the establishment of healthy eating habits than today's programs. They aimed to bring about changes in children's nutrition status, such as reductions in their consumption of saturated fats. Unintended adverse consequences are likely in such 'nutrio-centric' programs, such as reductions in children's micronutrient status (through avoidance of meat associated with reductions in saturated fat intakes). The CATCH study is a good example of this type of program (see Box 9.11).

BOX 9.11 THE CATCH (CHILD AND ADOLESCENT TRIAL FOR CARDIOVASCULAR HEALTH) STUDY

The study involved 5106 third to fifth graders in four American states. The intervention, conducted in 56 schools, included a combination of school food service modifications, enhanced physical education and classroom health curricula. Students in 40 schools acted as controls. A wide variety of health indices were measured before (1991), during and at the completion of the program (1994).

The percentage of energy intake from fat fell significantly more in the intervention school lunches (down from 38.7 per cent to 31.9 per cent) than in the control school lunches (down from 38.9 per cent to 36.2 per cent—$p < 0.001$). Self-reported daily energy intake from fat among students was significantly reduced in the intervention schools (down from 32.7 per cent to 30.3 per cent) compared with control schools (down from 32.6 per cent to 32.2 per cent—$p < 0.001$).

Other findings reported from this large-scale trial included: significantly greater response scores for dietary knowledge; dietary intentions and self-reported food choice changes for the intervention schools compared with control schools; significantly higher perceived social reinforcement for healthful eating patterns in the intervention groups; significantly reduced dietary cholesterol among children in the intervention groups (down from 223 milligrams to 206 milligrams) compared with controls (up from 218 milligrams to 225 milligrams); a significant increase in the intensity of physical activity in physical education classes in the intervention schools compared with the control schools; and significantly more self-reported daily vigorous activity in intervention students compared with controls. The study showed that combinations of intervention approaches can be effective in bringing about dietary and health behaviour changes in large school systems.

A three-year post-intervention follow-up included 73 per cent of the initial CATCH cohort (when students were in Grades 6–8). At Grade 8, self-reported daily energy intake from fat was significantly different for the intervention group compared with the control group (31.6 per cent versus 30.6 per cent—$p = 0.01$). There were also significant differences in dietary knowledge and dietary intentions, but not in social support or physical activity in the intervention students compared with controls at Grade 8.

Source: Luepker et al. (1996); Nader et al. (1999).

The Pro Children Program

Recently, this program has been set up in several European countries to assess children's fruit and vegetable consumption and to develop and test strategies for the promotion of fruit and vegetables among children and their parents. Detailed surveys of children's and parents' dietary habits, other behaviours and likely determinants have been conducted. These were based on intervention mapping (Kok et al. 2004; see Chapter 5). In the various countries, schools have randomly been allocated to an intervention arm or to a delayed intervention arm (Klepp et al. 2005). Preliminary findings suggest that key determinants of children's fruit and vegetable consumption include age, gender, socioeconomic status, food preferences and accessibility and availability of these foods at home (Rasmussen et al. 2006).

School food services: What should we do about them?

Children spend a long time at school, so they need to eat while they are there. In Australia and New Zealand, parents are generally assumed to be responsible for supplying food to their children at school, either by preparing the food children take to school or by giving them money to purchase food from canteens, tuckshops, corner dairies, fast food outlets and so on. Parents and adults are quite divided about this issue (see Table 9.3). Until recently, schools have been allowed to develop their own ways of regulating the consumption of foods on their premises. However, the childhood obesity crisis has forced some governments (e.g. in the United Kingdom, New South Wales and Victoria) to ban certain products from sale in schools, such as soft drinks and confectionery (see Box 9.12).

TABLE 9.3 VICTORIAN ADULTS' VIEWS OF SCHOOL FOOD POLICY OPTIONS n=420		
	% Total	
	(n=420)	
View	**Agree**	**Disagree**
Chocolate fundraisers should be banned	23	56
Sponsorship of school activities by companies like Cadbury's and McDonald's should be banned	29	50
Snack foods for children are okay in moderation	86	5
All soft drink vending machines should be banned from schools	71	18
Food companies should not be allowed to market high-energy and high-fat products at school	77	10
School canteens should not sell hot chips	45	34
Foods high in fat, salt and sugar should not be sold at school canteens (i.e. chocolate, lollies, potato chips)	64	22
Governments should pass regulations to prevent schools selling unhealthy foods and drinks	38	41

Source: Worsley (2007).

BOX 9.12 THE CASE FOR REFORM OF SCHOOL FOOD SERVICES

There are a number of problems associated with the *laissez-faire* approach taken in Australia and New Zealand:

- Children and parents perceive many of the foods commonly sold by school food services as 'unhealthy' (e.g. chocolate, hot chips and meat pies) (Cleland et al. 2004).
- The school canteen should provide children with the opportunity to apply theoretical learning to practise healthy food habits. The canteen should reinforce, not undermine, the health messages taught in the classroom. Many adults want more control of the types of foods provided at school (see Table 9.3).
- Many schools find it difficult to provide healthy food to children (Maddock et al. 2005). School canteens are often used solely as revenue-earners to subsidise other school activities such as sports teams.
- The number of volunteers who run school food services is diminishing. Only 37 per cent of primary and secondary schools involve adult volunteers in Victorian school canteens (Maddock et al. 2005).
- Many parents work long hours in paid employment outside the home, which makes food provision for children more difficult than in earlier generations. Over half of secondary schools provide food for children before school hours (Maddock et al. 2005), and the number of out-of-school-hours child-care centres is rising rapidly (www.facs.gov.au/internet/facsinternet.nsf/childcare/familiesoutside_school_hours_care.htm).
- The legal responsibility to provide safe food may extend to the adverse nutritional effects of foods (such as those associated with high saturated fat foods) many years after sale. Education systems may be liable for any long-term damage they inflict on their students' health.

Other OECD countries have not adopted such a *laissez-faire* approach. In France, Italy, the United Kingdom and the United States, for example, governments subsidise school food services (www.localfoodworks.org). This is done in a variety of ways, and with varying degrees of success. In these countries, the feeding of children is perceived as a shared responsibility of the community (governmaent) and parents.

The development of school canteen policies

Strong advocacy is needed to raise public and political awareness about the need for high-quality food provision in schools. The New South Wales School Canteen Association is a useful model that attempts to promote healthy eating among children through its networking among schools, food identity and New South Wales Health.

Schools pay an annual subscription to the association, which provides them with canteen guidelines, training and lesson plans and access to healthy food products provided by companies that aameet the association's nutritional criteria (www.healthy kids.com.au). Recently, the Go For Your Life Program in Victoria, Australia has provided many resources for healthy eating and physical activity promotion in schools (www.education.vic.gov.au, search for Go For Your Life).

BOX 9.13 WHAT IS A HEALTHY SCHOOL FOOD SERVICE?

A healthy school food service:

- makes it easy for students to choose healthy snacks and meals;
- offers a variety of nutritious foods;
- promotes foods that are consistent with the *Dietary Guidelines for Children and Adolescents in Australia*;
- can be an avenue for consistent and continual health education;
- complements the diverse elements of the school curriculum;
- involves students, parents and the wider school community;
- is an integral part of the entire healthy school environment.

A whole-school approach to healthy eating

The school's food services and curriculum programs on healthy eating should be complementary. Positive peer pressure within the education setting can create a culture in which nutritious foods and a healthy lifestyle are actively chosen. This culture should permeate the entire school environment and can have an impact on choices made by students about food consumption when they are not at school.

A positive attitude towards a school canteen that supports healthy eating should be promoted and endorsed. This is facilitated through a whole-school approach to nutrition.

The World Health Organization encourages schools to take a health-promoting schools approach to support healthy eating.

The Health Promoting Schools framework provides a useful model for schools to promote and protect the health of students. The framework highlights the importance of the relationships between:

- curriculum, teaching and learning;
- school organisation, ethos and environment;
- community links and partnerships.

A health-promoting school is a place where all members of the school community work together to provide students with integrated and positive experiences and structures that promote and protect their health.

Further information can be accessed from the *Health Promoting Schools* (http://www.ahpsa.org.au/) website.

Source: http://www.education.vic.gov.au/management/schooloperations/healthycanteen/pol_wholeschool.htm.

Although considerable policy development is required at state level, school communities can do much to improve the quality of food that they supply by adopting school food policies (see Boxes 9.13, 9.14). Only through 'bottom-up' approaches can the school community make its food policies sustainable.

School food services can be run on a financially sound basis while providing healthier, appetising food choices. To do so, staff (and students) require training in sound business, marketing and nutrition practices. A stimulating example of the linking of schools with the local community is Collingwood College, which has school garden and cooking programs that involve children and their families with food markets, food retailers and regional personnel of the Department of Human Services (Alexander 2003).

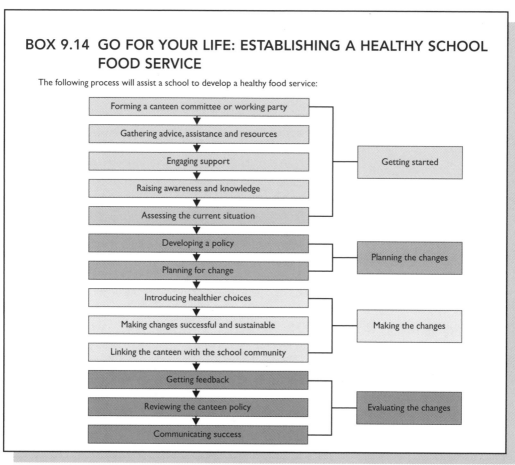

BOX 9.14 GO FOR YOUR LIFE: ESTABLISHING A HEALTHY SCHOOL FOOD SERVICE

The following process will assist a school to develop a healthy food service:

- Forming a canteen committee or working party
- Gathering advice, assistance and resources
- Engaging support
- Raising awareness and knowledge
- Assessing the current situation

Getting started

- Developing a policy
- Planning for change

Planning the changes

- Introducing healthier choices
- Making changes successful and sustainable
- Linking the canteen with the school community

Making the changes

- Getting feedback
- Reviewing the canteen policy
- Communicating success

Evaluating the changes

Source: http://www.eduweb.vic.gov.au/edulibrary/public/schadmin/schops/healthycanteen/gfyl_getstarted.pdf.

Nutrition promotion in secondary schools

Almost all secondary schools provide food services and teach curricula that provide opportunities to promote healthy eating (Maddock et al. 2005). They provide a setting in which the teaching of life skills is particularly relevant to students, who begin (at various ages) to lead independent adult lives.

Secondary schools share many of the issues facing primary schools—for example, the need for consistency between what is taught in the classroom and the quality of foods provided by the school food service. However, they tend to be larger and their students tend to be less teacher-centred and more peer-oriented. Although fewer healthy eating interventions have been conducted in high schools, more of them appear to have been effective (Worsley and Crawford 2005a).

Secondary schools have distinct advantages which can facilitate nutrition promotion. Their large size may bring economies of scale that allow professional food service managers to be employed; more of them involve students in managing school food services (Maddock et al. 2005); and secondary students are often intensely interested in food and in experimentation with food (French et al. 2001; Worsley and Skrzypiec 1997) and so may be amenable to change.

In some school systems, these advantages are further underscored through the provision of health and home economics classes in which the principles of healthy eating are taught. Many students study courses in health and human development and food technology to Year 12, which provide opportunities to teach the skills of healthy eating. Unfortunately, the practical effects of these courses on students' food habits have never been evaluated.

The secondary curriculum provides many opportunities for teachers and others to promote healthy eating among secondary students—for example, through exposure to novel foods (such as tropical fruits and vegetables), and by developing students' budget planning, shopping and food-preparation skills. This can be done in partnership with diverse organisations such as horticultural groups, food retailers and sports clubs. The curriculum can help students meet their ethical and moral objectives through food (such as the adoption of vegetarian diets by those interested in animal welfare) and by helping them use foods to enhance social acceptance without harmful dieting practices. Secondary education can also enable students to anticipate their future needs by acquiring adult life skills, such as baby care and other independent living skills.

The rigid timetabling requirements of secondary schools, combined with their large size, make them profitable sites for vending machine companies. In 2004, approximately one-third of Victorian secondary schools had at least three of these machines (Maddock et al. 2005). In the United States they are even more common, and food services are often dominated by the marketing activities of confectionery and fast food companies. However, even in such health-hostile contexts, students' food choices can be influenced by reductions in the prices of low-energy foods in the school cafeteria (French et al. 2001), rises in the price of high-energy foods in school canteens, and the positioning of low-energy foods close to student's vision (at the front of the counter) in school canteens (D. Wilson, personal communication, 2003). Useful British examples of healthy school food service are provided by Wheelock (2007). Box 9.15 describes the CHIPS study, an instructive American example of a food service intervention.

BOX 9.15 THE CHIPS STUDY

This study examined the effects of adding low-fat snacks to 55 vending machines in 12 secondary schools and 12 worksites in Minneapolis, USA. The products had four prices—equal price to 'normal', and 10 per cent, 25 per cent, and 50 per cent reductions.

In addition, there were three promotional conditions (none—the usual status quo, vending machines with low-fat labels, and low-fat labels plus promotional signs which were used singly and in combination).

Sales of low-fat vending snacks were measured continuously during a 12-month period, in both the schools and the worksites.

The price reductions on low-fat snacks were associated with significant increases in low-fat sales (see diagram below). The promotional signs only had weak effects.

FIGURE 9.3
NUMBER OF LOW-FAT PRODUCTS SOLD PER MACHINE PER TREATMENT PERIOD, BY PRICE REDUCTION CONDITION, CHIPS STUDY, 1997–99

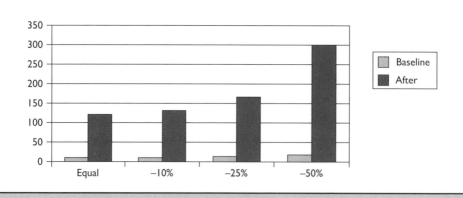

Source: French et al. (2001).

Conclusions

School-based nutrition and health promotion can be effective in influencing children's and young people's eating behaviours, attitudes and beliefs. However, the acquisition of healthy eating habits requires *a supportive school food environment* which is consistent with the nutrition curriculum taught in the classroom. School canteens should actively promote healthy foods and beverages and discourage unhealthy products. *School food policies* which involve students, teachers, canteen staff and parents are essential for the shaping of such a supportive environment. Teachers, canteen staff, parents and other adults should act as healthy *role models*. They should provide *feedback and encouragement (positive reinforcement)* for healthy eating.

Timetables and classroom rules should enable, rather than inhibit, students' consumption of healthy foods and beverages. *The food and nutrition curriculum* should have an experiential and practical skills acquisition emphasis. It can be taught in a variety of disciplinary areas (e.g. health studies, home economics, maths, English and social studies).

Discussion questions

9.1 Outline some of the food consumption and nutrition problems facing babies and infants and their families.

9.2 How might toddlers' diets be shaped in healthy ways by parents and carers?

9.3 'Nutrition promotion at school is largely a waste of time.' Discuss the arguments for and against this view.

9.4 How should a school go about setting up a school food policy?

9.5 What sorts of knowledge and skills should adolescents possess when they leave school?

9.6 Describe what you think would be the ideal food service in a secondary school.

10

Nutrition promotion at worksites

Introduction

Many people spend much of their waking lives at work. Work often provides a sense of meaning in their lives, and strong social support networks. Worksites are places full of intense social interactions; they are where people maintain strong friendships and where they derive much of their incomes. Worksites vary from huge organisations which employ hundreds or thousands of people at one site down to one-person businesses which may employ one or a few individuals. In aggregate, however, small businesses employ huge numbers of people; perhaps up to 80 per cent of the working population is employed by small business.

The average time spent at work has increased in some countries, like Australia, the United Kingdom and the United States during the past decade. In other countries, like France, and in some companies the number of hours worked is restricted, either by legislation or by company policy. The longer people are at work, the greater is the need for the provision of nourishment and, correspondingly, the opportunities for nutrition promotion.

As well as varying in size, worksite settings vary by industry sector and according to their organisational values and practices. Typically, worksites are found in manufacturing (factory sites), in offices (e.g. clerical and IT workers), in service industries like hospitals and universities, in sales areas like shops and department stores, and 'on the road' (e.g. sales personnel). Food may be provided at or near these sites by cafeteria, vending machines or food outlets near the worksites. Even where food may not be available, with personnel having to bring their own, usually times are allocated for the consumption of foods and beverages. These off-work times may be very short (e.g. a single 30-minute lunch break) or they may be more extensive (e.g. morning break, long lunch, afternoon break). Recently, in English-speaking countries, there has been a trend to reduce the time allowed for food consumption — indeed, the practice of voluntary overtime has become prevalent. But however long they may be and however they are organised, these 'eating episodes' offer opportunities for nutrition promotion.

Organisations also vary according to their values and practices. All companies and many government departments operating under 'free market' conditions are under intense pressure to maximise their profits and to minimise labour costs — this tends to reduce spending on nutrition promotion. The organisation's values may oppose or reinforce these market trends. For example, the organisation may place strong emphasis on worker health, internal cooperation and sharing, which may result in the provision of pleasant and healthy food services, child care and flexible working. Some companies operate to three bottom lines: profit, social responsibility and environmental sustainability. However, most do not, working only to maximise profits.

More generally, the culture in which the company operates is likely to affect the working relationships between management and workers, and between the workers. In his pioneering work during the 1970s, Hofstede (2005) showed that cultural values such as 'masculinity' affected the role of women in the workplace (e.g. they may be forbidden from taking management roles) or the degree of egalitarianism tolerated. These powerful cultural values can be expected to affect the provision of services like child care and the standard of health and nutrition promotion in the workplace. It is wise for nutrition and health promoters to familiarise themselves with an organisation's values and working practices before deciding whether it is feasible to work with them. Oldenburg et al. (2002) have provided a checklist to assess the suitability of the organisational climate for health promotion.

Historically speaking, the idea of the worksite is fairly recent, dating from the first Industrial Revolution. It is certainly true that people have always worked to keep themselves alive, but the modern workplace which separates people (usually men) from family and home life is recent. It generated a separation and a tension between work life and home life. Work life was often very arduous and exhausting, and workers were often unable to take part in domestic work. This separation has been maintained to the present day in the form of family versus work obligations. In a sense, it has been compounded through the major shift in work life during the past half-century: the entry of women into full-time paid employment outside the home. Many women resume paid work soon after childbirth. In the mid-twentieth century, most women stayed at home to care for children and other dependants. Today, children, sick and aged people still require care, but women's availability to do this work has been reduced. The result is a plethora of arrangements which include the re-entry of some men into domestic life and work, professional child care, out-of-hours and after-school care, aged residential care, and much more. The net effect has been to generate great stress for workers, particularly for women, to which employers may or may not respond. Three consequences of this tension between family and work responsibilities are the rising demand for breastfeeding facilities at work and elsewhere in public life, the rising demand for child-care facilities and lower birth rates.

The 'old' separation of work and family is becoming blurred. This places more demands on employers for health and nutrition promotion, and in so doing raises the importance of the workplace. This may be a blessing for nutrition promoters since the worksite is often a highly controlled environment. In many cases, workers are exposed to a narrow range of foods at work—that is all they have to eat. If the available food choices are altered to become more nutritious, then the nutrition status of the workers is likely to be improved. Unlike retail food shops in which customers have very wide choices, worksite venues are more restricted and provide less choice. This can present nutrition promoters with opportunities.

Opportunities for health and nutrition promotion at worksites

Worksites are prime foci for health promotion activities, and most of this activity is supported financially and in other ways by employers. Health promotion is fairly common in companies, and nutrition-related activities are quite popular in the USA, EU and elsewhere (e.g. Chapman 2004). Most of these activities remain unevaluated and unreported so we know little about their effectiveness in promoting employees' health, but clearly a number of benefits result from them for employers to continue to support them.

Advantages and disadvantages of worksites for health and nutrition promotion

Worksites have many advantages and some disadvantages for health promotion (see Box 10.1). The key things to note are that people spend many hours at work, and employers often have vested interests in maintaining and promoting the health of their workforces. The main disadvantages have to do with the fact that the primary aim of most organisations is not the promotion of the health of their employees—health promotion activities come after the profitability or other prime purposes of the company or organisation. Change is a feature of most modern organisations, so health promotion has to be systematically organised over the long term if it is to have lasting value for employees.

Worksite nutrition promotion has unique opportunities and challenges which will be discussed below. Perhaps its main advantage is that employees usually have to eat at some time during the working day—this provides opportunities to supply healthy foods and to provide information about them which is of relevance to employees and their families.

BOX 10.1 THE ADVANTAGES AND DISADVANTAGES OF WORKSITE HEALTH PROMOTION

Advantages

- Working adults spend a significant amount of time in worksite settings and usually can be reached effectively through these kinds of settings.
- Work organisations have a clear economic and enhanced performance rationale for conducting a health promotion program.
- The compensation and benefits aspects of employment provide a strong potential platform for formal incentives for health promotion.
- Programmatic economies of scale are possible in medium and larger worksite settings or in serving large numbers of small worksites.
- Peer and social support are potentially available in worksite populations.
- Work cultures are amenable to intentional influence in support of health promotion.
- Work organisations have the capability and resources to conduct fairly complex programming.

> *Box 10.1 continued*
>
> **Disadvantages**
>
> - Work organisations have a natural volatility that makes continuity, follow-up and consistency of effort a challenge.
> - The economic pressures associated with the business cycle often prevent adequate funding of programs.
> - Traditional health promotion programming is difficult to rigorously evaluate, particularly the determination of the economic return associated with programming.
> - The demands of modern work limit the amount of work-invasive programming that is feasible in worksite settings.
> - Some of the potential distrust between employees and employers can limit participation and effectiveness of programming.

Source: Chapman (2004).

Why should companies promote health and nutrition?

The provision of food services, child care and health services can cost employers considerable sums. So why would they want to provide them? There are several reasons. First, unhealthy workers cost employers large sums in the form of absenteeism and lowered performance. Similarly, if there are no flexible schemes to enable parents to care for children, then absenteeism, poor performance and high staff turnover (and associated retraining costs) are likely to occur. Good health and nutrition services are often in the employer's best financial interests. Second, these services may be negotiated with unions and staff associations as part of enterprise bargaining agreements. They may be offered in lieu of pay rises. There may be considerable tax advantages for employers in doing this (especially in the United States) and they can contribute to staff morale.

Third, these services may be mandated by law. That is, there may be government health policies which ensure that adequate health and nutrition services are provided. Most examples of these are found in the school sector—for example, the US Federal government requirement that school areas must employ a nutritionist to oversee food service provision. In theory, there is no reason why such requirements could not be mandated for worksites.

Fourth, the current American trend of litigation against food companies by individuals who claim their ill-health has in some way been caused by their voluntary use of the company's products presents implications for employers. If retail fast food chains are found to be liable for the purchases of voluntary customers, the legal liability of employers who supply foods under restricted circumstances may also be in question. This trend is likely to cause employers to question whether they wish to provide food on their premises, and if they do to ensure that it complies with minimum safety and nutrition standards. This may be a useful 'trigger' for the provision of higher quality food at worksites. A final, important factor is the influence of the 'corporate responsibility' movement which encourages ethical behaviour by companies, a key part of which is respectful treatment of employees.

What can employers offer in terms of nutrition promotion?

It is clear that the research literature reflects only a very small part of the nutrition services offered at worksites. Companies frequently provide services for their employees, such as restaurants and cafeterias, health education and training and child-care and breastfeeding facilities, among others. Usually, the effectiveness of most of these services is not evaluated or reported publicly. Much of the service provision at worksites is contracted out ('outsourced') to specialised catering companies.

Approaches to worksite nutrition promotion

Many companies provide some forms of worksite nutrition promotion, either as a 'normal' activity of the company or occasionally in the form of formal intervention projects. Some examples of research interventions are discussed below. Current best practice guidelines for worksite promotion in general are listed in Table 10.1. The health promotion process can be divided into four principal stages: program planning, design, operation, or implementation and evaluation. In practice, these stages often occur in a cyclical manner—evaluation, for example, preceding changes in plans and redesign.

Interventions have many drawbacks (see Chapter 7), namely their lack of long-term sustainability, their narrow end-points (such as measurement of changes of intakes of a few 'target' foods or nutrients), their inflexibility and inability to deal with real-life 'open systems' in which new factors can enter at any time (changes are

TABLE 10.1 WORKSITE PROGRAM HEALTH PROMOTION PRINCIPLES

Planning	Design	Operations	Evaluation
Strong management support	Implement annual health risk assessments	Provide appropriate incentives for program participation and annual assessments	Formally evaluate program goals and objectives e.g. annually
Align program aims with business aims	A mix of approaches and methods	Use multiple channels for effective communication	Evaluate changes in risk and health behaviours
Address perceived needs of target population and stakeholders	Active recruitment	Create supportive cultures to change social norms and health behaviour expectations	Evaluate organisational gains periodically to assess programs economic and non-economic benefits for the organisation
Involve employees in planning and implementation	A range of prevention goals primary, secondary and tertiary	Create supportive environments to make healthy behaviours easier to do	
Have clear goals—make them familiar to all	A mix of on-site and virtual interventions, e.g. email, sms, mail		
Interdisciplinary team to design and implement	Target high risk individuals effectively		
	Make program as accessible as possible		
	Integrate program with other health activities inside and outside the organisation		

Source: Chapman (2004).

difficult to incorporate without compromising the intervention design), and their high cost. However, they do provide information about the approaches which are most likely to bring about change, and if they are designed to represent typical worksite situations they can serve as useful 'demonstration projects' on which nutrition promoters and companies can base their own practice.

The columns in Table 10.1 list some of the most important activities in worksite health promotion. For example, under 'program planning', the first box emphasises the need to gain and maintain strong, 'real' support from senior management (and from employee unions, if they are present). The absence of such support might cause health promoters to refrain from further dealings with that company. Obviously, they would try hard to get it, but if it is not forthcoming then any attempts at health promotion may be futile or at least ineffective in terms of the effort involved. A second key aim in program planning is to formally assess the needs (and wants and views) of the key stakeholders in the organisation: the employees, their families, union representatives and various levels of management—indeed, anyone who has an interest in the success of the health promotion program. An example of needs assessment (relating to the California 5 a Day Program) is described below.

Program design is undertaken according to the needs assessment and to the other program planning findings. Health risk appraisal (HRA) usually includes the measurement of individuals' biomedical indices, such as body weight, blood pressure, serum cholesterol and so on. One of the problems in nutritional risk assessment is that the reliability and validity of short, easy-to-use instruments (e.g. for assessing fruit and vegetable intake, or saturated fat intake) are sub-optimal, despite recent progress.

Many issues are raised during the operation or implementation of any program. Chapman (2004) lists five, three of which have been themes in recent nutrition promotion interventions. These are the maximisation of participation—especially by individuals in most need—together with the creation of supportive environments such as the use of employee advisory boards to adapt and deliver the program to employees (e.g. Treatwell) and the use of effective communications. This means that promoters need to communicate in ways which are appropriate to groups of employees (e.g. orally via games and parties rather than via written materials, for example). Consistent follow-up is essential to ensure the program is actually 'delivered' as planned.

Evaluation is essential to any program, whether it is a formal intervention or a systemic program. Again, Chapman (2004) describes the various aspects which are common to health promotion. However, two are unique to worksite promotion:

the evaluation of organisational gains—which is important to maintain the support of senior management; and the evaluation of participant satisfaction—which is necessary to ensure the program is meeting the perceived needs of employees.

Reviews of the nutrition intervention literature

There have been several useful reviews of the literature in this area (Glanz and Mullis 1988; Janer et al. 2002; Wilson et al. 1996; Seymour et al. 2004). Seymour et al. (2004) is particularly useful, and will be referred to below. They reviewed over 90 interventions at point of sale at worksites, universities and in supermarkets. They retained 30+ of these studies but rejected the others, which they decided were of poor quality either because proper comparison groups were not included or because the studies were inadequately described. The review identified a number of shortcomings. Practically all of the interventions were short term only. The assessment of dietary changes requires further refinement: the evaluations have not considered likely changes in the consumption of a variety of foods. For example, an intervention might show that intakes of broccoli increased, which sounds good but they may have failed to have assessed the intakes of 'companion' foods such as cheese—increases of broccoli *and* cheese may not be as nutritionally desirable as increases of broccoli by itself. More sophisticated evaluations of change are required.

After *needs assessment* (see below) has been undertaken, one of three approaches may be taken, usually separately but sometimes in combination. These three approaches vary according to the amount of freedom individuals have to choose their foods.

Information and education

Information may be provided about foods that are on sale, typically through labelling the content (e.g. fat and energy content) either on the product package, on vending machines or on menus in cafeterias. The aim of this approach is to enable employees (or customers in supermarkets) to 'decide for themselves'. This approach attempts to influence consumers' purchasing decisions—it assumes that people have a fair degree of choice and ability to choose what they purchase and consume. Seymour et al. (2004) call this the 'behavioural approach'. Nutrition promotion (and food marketing) in retailing or among schoolchildren tends to use this approach because the consumer in these settings is essentially free to choose from a wide variety of products. Information can be presented in various ways (e.g. framed in terms of risk reduction; tailored to suit individuals' concerns and interests; or formatted in various fonts, layouts, colours and other eye-catching devices). Seymour et al. (2004) argue that in worksites this is probably not the best approach because employees do not necessarily have wide choice over their food purchasing at work.

The environmental change approach

This takes account of the reduced freedom of employees at worksites to choose a variety of food products. An *environmental nutrition intervention* is defined as one that affects availability, access, incentives or information about foods at point of purchase (Seymour et al. 2004; see also Glanz and Yaroch 2004). Rather than try to persuade people to make healthier food choices, the choices they can make are reduced in various ways. To take an extreme example, only healthy foods like fresh fruits, vegetable meals and reduced fat milks may be provided in the cafeteria and in vending machines. Most environmental nutrition interventions have not been so crude; some have merely positioned healthier foods closer to consumers' gaze, while others have provided loyalty schemes—customers receiving cash back after buying a set number of serves of the healthier food choices—and several have reduced the prices of healthier foods (French 2003). The provision of access to special rooms or facilities for breastfeeding employees is another example of environmental nutrition promotion. Seymour et al. (2004) suggest that this approach tends to be far more effective in changing food choices in worksites than the 'behavioural approach'.

Changes in the social environment, such as the tapping of social networks to gain employee or management support, are often called *social ecological* approaches. These may include a wide variety of activities such as employee advisory boards, and the organisation of food supplies for staff meetings, parties and get-togethers, outings, leisure activities, family involvement and more. *Communication approaches* are very much part of this social ecological approach. They focus on the communication of persuasive messages. While the classical behavioural approach is concerned with individuals' motivations, attitudes and beliefs, communication approaches usually recognise the social nature of people so they recruit opinion leaders and attempt to build up or utilise social support networks. Peer-led nutrition education is one common form of communication.

Policy approaches

These have rarely been attempted in worksite nutrition promotion. Some years ago in Australia, Goodman Fielder set up its own corporate nutrition policy, part of which aimed to provide healthy food options for its staff. More recently, Unilever has devised a global policy to reduce the amounts of salt and saturated fats in its products. However, working food and nutrition policies are more commonly found in local community groups, in pre-school centres and primary and high schools (see Chapter 7). Seymour et al. (2004) recommend consideration of national food and nutrition policies which would specify a range of actions at point-of-sale outlets. They point to the high efficacy of American legislation which forced the fortification of foods with a number of micronutrients in the 1930s and the compulsory fortification of certain foods with folate in the late 1990s, both policies apparently being highly successful in reducing the prevalence of nutrient deficiencies. They note that several environmental nutrition strategies have been suggested which could form part of national policy, such as: fruits and vegetable available free or at reduced costs at locations including worksites; limiting the sodium content of canned foods; taxing of high-fat, high sugar snack foods; regulating the temperature and freshness of cooking oil for French fries; reduced portion sizes; the nutrition labelling of restaurant foods; and partnerships between weight loss companies and restaurants.

Such changes may appear radical and unacceptable to many companies. If they voluntarily adopted these strategies, it might put them at a competitive disadvantage with respect to other companies. This is where a legislated nutrition policy would have advantages because it would prevent health-conscious companies being penalised compared with less responsible companies. The present nutrition situation has many similarities with the state of farm tractor safety a generation ago. Most tractor manufacturers realised that roll bars would prevent injuries, but their high costs prevented them from fitting them as standard equipment—companies that did so were put at a price disadvantage in the marketplace. However, when ethical manufacturers requested government to legislate for compulsory fitting of roll bars this impediment was removed, and all tractors now have them. Environmental nutrition and food and nutrition policy could have major effects on nutrition promotion in worksites (and in other settings) in the coming years.

Food and nutrition policies are perhaps the most coercive forms of nutrition promotion; they trade off the individual's freedom to choose in order to make major health gains. This is likely to be controversial but at the worksite the empirical evidence suggests that most people are eager to choose 'healthy foods' over less healthy alternatives so long as the foods meet their taste and price expectations.

Selected examples of approaches to worksite nutrition

Needs assessment

The needs of employees for health and nutrition services, healthy food choices and nutrition information are often assessed prior to mounting a program. Several eating pattern and dietary assessment instruments have been developed, along with food service evaluation instruments—for example, the Checklist of Health Promotion Environments (Oldenburg et al. 2002) is a useful instrument to help assess general health promotion needs. This is a 112-item checklist of workplace environmental features which are believed to be related to physical activity, healthy eating, alcohol consumption and smoking. The assessment is centred on the physical characteristics of the worksite (e.g. the presence of showers, stairs), the features of the information environment (e.g. the presence of nutrition leaflets), and the properties of the immediate neighborhood around the workplace (e.g. the presence of bike tracks).

Needs assessment of fruits and vegetables and physical activity at the worksite

This was a major needs assessment exercise (Buckman et al. 2004). The California 5 a Day Worksite Program conducted telephone interviews and focus groups with California business leaders from small, medium and large-sized companies between August and October 2002. Focus groups were also held with low- and middle-income working women throughout California. The aim of the research was to determine the best ways to increase fruit and vegetable consumption and physical activity in the workplace. The interviews with businesspeople were conducted to find out:

- the types of unhealthy behaviours that affect employees;
- what is being done at worksites to improve the health of employees;
- the reasons businesses have for implementing programs to improve the health of employees;
- why other business do not have health programs for employees; and
- the best ways to increase fruit and vegetable consumption and physical activity at work.

The focus groups with working women aimed to understand:

- the barriers to healthy eating and physical activity in the workplace;
- the factors that would encourage them to eat more fruits and vegetables and do physical activity at work.

The business leaders' views

Unhealthy behaviours

The business leaders reported several unhealthy behaviours among their workers: smoking (mentioned by 58 per cent), lack of physical activity (38 per cent), poor diet (28 per cent), overweight (10 per cent), alcohol consumption (8 per cent), over-worked (5 per cent) and stress (3 per cent), though 10 per cent did not mention any unhealthy behaviours. While these percentages do not reflect the actual prevalence of these behaviours or their importance, they do show that business leaders were conscious of the extent of the burden of unhealthy behaviours.

Business responses

Almost all the leaders (88 per cent) felt there were benefits associated with worksite programs, such as improvements in employee productivity, reduced illness-related absenteeism and lower employee health care costs. However, only 50 per cent of the leaders indicated that their worksites had health promotion programs. Those leaders who did not have any saw them as too costly, or as unwanted by employees.

Desirable policies and actions

Fifty-five per cent of the leaders thought increased fruit and vegetable consumption and daily physical activity were important as they contributed to good health. Three-quarters supported the provision of healthy foods in worksite criteria and vending machines and elsewhere; 47 per cent were in favour of the provision of incentives for walking, making fitness facilities and equipment available, and pro-viding flexible work schedules to facilitate physical activity among employees.

During focus groups, the leaders noted that they required convincing evidence that promotion of healthy eating and physical activity at worksites improves business profits and productivity. They suggested that there were opportunities to form partnerships with health insurance plans to offer these services. Tax incentives to do so would be motivational for them. They thought that healthy eating might be promoted through the offering of fruits and vegetable in vending machines and in cafeterias, subsidising employees to purchase healthy food items at work, offering coupons or discounts for healthy food purchased at work or at nearby food outlets, locating farmers' markets near worksites, and providing personal information about nutrition.

Despite concerns about liabilities for injuries during physical activity promotion, the business leaders were generally in favour of nutrition and physical activity promotion. Suitable policies and government actions (such as tax incentives and insurance cover) might help convert their interest into practical programs.

The working women's views

The working women perceived several barriers to healthy eating and physical activity at work, including: vending machines which sell mainly junk food; provision of unhealthy foods at work functions (e.g. doughnuts offered during work meetings); and the fact that the easiest foods to access outside work were at fast food outlets, which were seen to be convenient and inexpensive. Lack of time during the work day was the key barrier to physical activity at work; even when room was given over to physical activity, most women felt their employers would not like them to spend time exercising.

They thought the best way to encourage nutrition and physical activity was to surround employees with healthy choices and provide them with opportunities to engage in healthy behaviours. The working women's views overlapped considerably with those of the business leaders. The three most promising strategies based on this needs assessment are summarised in Box 10.2.

BOX 10.2 STRATEGIES FOR IMPROVING THE NUTRITIONAL QUALITY AT THE WORKSITE, DERIVED FROM THE CALIFORNIA 5 A DAY WORKSITE PROGRAM

I **Improve access to healthy foods and physical activity in the workplace by:**

- offering healthy food at meetings;
- offering healthy snacks;
- encouraging healthy food choices at work celebrations;
- delivering healthy foods;
- making arrangements with restaurants for discounts and incentives for healthy foods;
- working with catering trucks to encourage them to offer low-cost healthy food choices with an emphasis on fruit and vegetables; and encouraging them to provide signs and other advertising to inform workers about their healthy food selections;
- establishing farmers' markets at or near workplaces;

Box 10.2 *continued*

- providing healthy, appealing food in cafeterias at reasonable prices;
- providing vending machines with healthy foods such as fruit and vegetables.

It was also felt that:

- (US) employers should take advantage of existing tax laws to provide healthy foods for their employees on a pre-tax basis. This could enable the provision of tasty healthy meals at very low cost;
- companies which relocated should try to do so near recreational facilities such as parks and walking trails, mass transit, restaurants and stores which offered healthy food choices;
- physical activity should be made easier—for example, by providing time for exercise such as scheduled activity breaks, encouraging 'walking meetings', setting dress codes that allowed more comfortable clothing and shoes, and making space available for physical activity;
- employers should provide physical activity facilities such as on-site gyms, or reimbursement of off-site gym memberships;
- active commuting should be promoted (e.g. offering financial incentives for employees who walk, ride a bike, take public transport or carpool to work; locating workplaces within safe walking distance of dining, shopping and public transport; providing lockers and showers; offering safe, secure and free bike storage);
- facility design such as open accessible attractive stairways; free facilities for storage of bicycles; showers and changing facilities; space for physical activity and workplace location near public transport and safe, well-lit walking and biking facilities.

2 **Foster supportive work environments that encourage healthy lifestyle choices, through:**

- worksite assessment and information to make healthy eating and physical activity core values of the organisation; fostering teamwork and social support such as team-based low- or no-cost nutrition and physical activity programs; linking with existing programs such as the 5 a Day Worksite Program, or the Take Action! employee physical activity promotion program.

3 **Establish public policies that bolster health promotion efforts at worksites, such as:**

- employer liability for physical activity promotion;
- nutrition standards in workplaces (e.g. food-service policies in all public buildings which mandate that at least 50 per cent of food served meets guidelines for healthy food choices);
- facility design policies which set standards for the construction of workplaces so they encourage physical activity, together with laws that support healthy workplaces;

> *Box 10.2 continued*
> - insurance premium breaks—working with insurance companies to provide premium breaks for employers on a sliding scale based on their preventive health and wellness initiatives. This would include providing incentives for health maintenance organisations to take more active roles in disease prevention at worksites;
> - offering pre-tax flexible options to help employers to pay for eligible health and wellness-related expenses on a pre-tax basis.

Source: Buckman et al. (2004); many resources are listed in this report.

Approaches to program implementation

Recent trends in worksite nutrition have included the development of environmental and policy-oriented programs which have been features of interventions such as Treatwell and Working Well (Sorensen et al. 2004). These interventions have emphasised changes in the physical and information environment such as the provision of wider variety of food choices in cafeteria and better labelling of the health attributes of foods. They have also focused on the social environment, particularly the acquisition of organisational support and commitment from management and from neighborhood institutions (such as fast food outlets near the worksite). This is consistent with the general social ecological model (see Chapter 6). This model stresses the importance of participation strategies and change-friendly social contexts (such as work groups, families and leisure groups). The older emphasis on individual psychological factors has been maintained but now uses computing technology to provide tailored interventions to suit particular constellations of attitudes, beliefs, knowledge (or lack of it), cognitive stage of change and other individual factors. Finally, recent nutrition programs have often had multiple end-points—they deal with multiple risk factors such as physical inactivity, smoking and so on, as well as changes in food choices, especially fruit and vegetable choices.

Most published examples of nutrition promotion at worksites are American. This reflects the importance attached to the area by companies and funding agencies in the United States. Other countries are interested too, most notably the Netherlands (Steenhuis et al. 2001). Most studies have been integrated health promotion programs, of which nutrition promotion has been only one part. Three major sets of end-points distinguish the various examples: disease reduction programs such as various heart health and cancer risk reduction; fruit and vegetable and other healthy food promotions; and the promotion of women's health and breastfeeding.

The Working Well trial

The Working Well trial (Biener et al. 1999), conducted between 1985 and 1992, was the largest randomised worksite health promotion trial in the twentieth century. It was one of the first formal environmental nutrition interventions. It focused on the smoking and nutrition behaviours of 21 801 employees (68 per cent men, 92 per cent white) at 57 matched pairs of worksites in sixteen states of the United States. The worksites included manufacturing, communication, public service and utilities. They were matched for size, company type, sex distribution, blue- or white-collar occupations, presence of cafeterias, type of smoking policy and baseline survey response. Wage structures, social and marital status were not examined, nor were the effects of the intervention on productivity. The intervention lasted for two years at each site.

Working Well emphasised employee participation in the planning and implementation of the intervention. It aimed to go beyond the prevailing focus on the modification of individuals' behaviours to enhance the physical and social environments of the worksites. It incorporated several novel features in addition to its main aim of reducing cancer risk by changing employees' consumption of fat and dietary fibre and smoking cessation:

- *Promotion of worksite environmental change* — the smoking intervention increased restrictions on smoking in the workplaces and increased reminders of the importance of a smoke-free workplace. Environmental nutrition interventions helped to implement polices which mandated health food choices at work functions, increased the availability of point-of-purchase labelling of nutrition information, and increased healthy food choices in cafeterias and vending machines.
- *Changing the social environment* — Working Well aimed to change social norms relating to food choice and smoking, and to utilise social support networks and worker participation through measures such as the creation of employee advisory boards.

In the design of the trial, 57 worksites were randomly allocated to the nutrition and smoking cessation intervention and 57 matched sites were allocated to the control group. At the intervention sites, the Employee Advisory Boards worked with the study interventionists to increase accessibility to healthy foods, and developed and implemented company policies which supported healthy eating and smoking cessation.

The changes that occurred were assessed by surveys of employees and key informants in the companies. For example, changes in the social environment affecting eating were assessed in an employee survey by items which inquired about the employees' perceptions of the extent of coworker encouragement to eat a low-fat diet and the perceived level of managerial concern about the workers. Food service managers were interviewed about efforts to offer more low-fat and high-fibre foods in the cafeteria and about the implementation of point-of-purchase nutrition information. A summary of the stakeholders involved and the kinds of information collected in the trials is shown in Box 10.3. A Healthy Food Environment Index summed the number of healthful changes made at each site divided by the total number possible of environmental or policy changes that could have been made.

BOX 10.3 THE STAKEHOLDERS AND VARIABLES MEASURED IN THE WORKING WELL TRIAL: SOURCES AND TYPES OF ORGANISATION DATA

Baseline organisational informants	Data supplied by informants
Personnel director	Workforce demographics
Chief executive officer	Rates of absenteeism, turnover, accidents
Health promotion director	Labor–management cooperation
Benefits manager	Smoking policy content, enforcement, compliance
Health and safety professional	Catering policy
Food service manager	Experience with participatory strategies
Vending company representative(s)	History of health promotion programs
Union representative	Organisational profitability
	CEO support for health promotion
	Type of food and vending service
	Average utilisation of food service
	Availability of healthy food
	Food preparation facilities

Source: Patterson et al. (1998).

After two years, the Healthy Food Environment Index score was higher in the intervention sites than in the control sites (36.5 per cent vs 23.5 per cent, p<0.012). That is, employees in the intervention sites perceived that they had more access to fruits and vegetables and to nutrition information at work than those at the control sites. In addition, a wide range of players had increased their support for the trial.

Key weaknesses of the trial included poor measurement of the many aspects of the study—often only one questionnaire item assessed particular changes in the physical or social environments—and the fact that social desirability biases may have influenced employees and other informants to report desirable changes in the food environment. Some of the nutrition environment changes were easier to implement than others—for example, point-of-purchase labelling of foods in vending machines increased but the provision of low-fat, high-fibre foods in the cafeteria did not. Labelling of food requires less organisational change, with the product purchaser seeking out readily accessible information and passing it on to other employees. The provision of low-fat, high-fibre foods, on the other hand, requires changes by the vendor (usually an outside company) and by the worksite which may not be attractive. High-fat, low-fibre foods can be highly profitable and so there may be less motivation to either change the products on offer or to find vendor companies which will offer them. The investigators concluded that:

> These results might suggest a program strategy in which environmental changes that are simple and under local control are tackled first, and those that involve more complex relationships among organisations are approached later. The synergistic effects of environmental change programs and individually focussed health promotion hold great promise for reducing the burden of disease and disability among employed people. (Biener et al. 1999, p. 492)

The Next Step trial

This program (Tilley et al. 1999) targeted automotive industry workers who were at increased risk of colorectal cancer at 28 worksites. It encouraged prevention and early detection practices and promoted low-fat, high dietary fibre eating patterns among 5042 employees. The worksites were randomised to a two-year nutrition intervention or to a control, no-intervention group. The nutrition intervention consisted of classes, mailed self-help materials and personalised dietary feedback. The nutrition outcomes were assessed by mailed Food Frequency Questionnaires.

One year after the commencement of the intervention, fat intake was reduced by 0.9 per cent energy, dietary fibre intake increased by 0.5 g/1000 kcal, and fruits and vegetables increased by 0.2 servings per day (all p<0.007). Two years after commencement, the control worksites had also changed positively so the intervention effects were relatively smaller. Generally, the intervention effects were greater among people under 50 years of age, among active people and among people who attended the worksite classes.

The study illustrates that individually oriented nutrition promotion can bring about modest effects which are greatest among people who appear to be more motivated. Subsequent analysis of the trial (Kristal et al. 2000) showed that the intervention increased predisposing factors for dietary change (skills, knowledge and beliefs as defined in the precede–proceed model) and the likelihood of shifting to or remaining in the action and maintenance stages of changes (as defined by the transtheoretical model). Changes in these mediating variables accounted for between 34 and 55 per cent of the dietary intervention effects. The authors concluded that greater emphasis should be given to environmental changes and the building of social support, along with more detailed analysis and reporting of larger studies.

While this program provides evidence for the utility of the transtheoretical model it did not utilise environmental supports, and the project's cost benefits and sustainability were unclear.

Treatwell 5 a Day

This program (Sorensen, Hunt et al. 1998; Sorensen, Stoddard et al. 1999. Hunt et al. 2000) was one of nine intervention studies supported by the US National Cancer Institute which were designed to promote fruit and vegetable consumption (the 'Five-a-Day for Better Health' program). Treatwell was conducted in 22 community health centres which ranged in size from 27 to 640 employees (nine had fewer than 100 employees). Since a key aim of the study was to examine the effect of family involvement on worksite nutrition promotion, the health centres were randomly assigned to one of three conditions: a minimal control intervention; a worksite-only intervention (WO); and an intervention which focused on both the employees' families and on the worksite (worksite plus family, or WPF).

The main findings showed that employees in the WO increased their daily servings of fruit and vegetables by 7 per cent (0.2 servings), but those in the combined WPF treatment increased their intakes by 19 per cent (0.5 servings approximately), no change being observed in the minimal control intervention (p<0.05). Analyses of the data showed that the WPF family intervention was associated with greater number of hours spent by Environmental Advisory Board (EAB) members on project activities, more events offered per site and per employee and greater consumption of fruit and vegetables, even though there were more EAB members and more participation per event in the worksite only condition. Across both conditions, the more EAB hours spent in program activities, the greater the number of events. Small worksites had greater employee awareness of the program and higher participation in program activities.

The study shows that the worksite can be used to involve employees' families and, in so doing, better compliance with program goals can be achieved. That is, if the employees' social support network is involved, the greater the outcome. This social network theme was further supported by another feature of this 'community-organising' study. At its commencement, EAB were set up at each worksite. These included worksite coordinators who represented a number of departments and occupational categories. The EABs used consensus-building and task-orientation techniques to plan and implement the interventions at each worksite—in other words, to tailor them to the peculiarities of individual worksites.

The investigators felt that the amount of time spent by EAB members in project activities was probably the most important employee involvement factor. The findings support the importance of using community or worksite members in program planning and implementation.

Future directions

Key issues in worksite nutrition promotion are summarised in Box 10.4.

BOX 10.4 KEY CRITERIA FOR WORKSITE NUTRITION PROGRAMS

Several studies have described the utility of Employee Advisory Boards which should be set up prior to any promotion program.

The aims and definition of healthy eating

These vary—for example, reductions in consumption of saturated fats but not all fats; increased consumption of fruit and vegetables, dietary variety, unprocessed grains. Other aims include increased knowledge of food content, cooking and shopping skills (intermediary goals). There are a number of effects:

Box 10.4 continued

- *direct effects* on the target population—increases in single foods or combined (e.g. broccoli only or cheese and broccoli?);
- *secondary effects* on the rest of a person's diet at home; indirect food effects on other people's diets (e.g. family, children, other workers)—positive or negative—are rarely considered;
- *side-effects*—positive changes in eating may lead to changes in physical activity, and vice versa;
- *unwanted negative side-effects*—changes in target food (e.g. low fat intake increases) may be accompanied by increases in other sources of energy (e.g. sugars, dietary changes with increased smoking or no change in physical activity).

Theoretical approaches

Do interventionists know what are the most likely influences on behaviours (e.g. precede–proceed)? Is the theory chosen feasible in the context of the setting?

Related design issues: The problem of healthy worker effect

Healthy workers tend to be those who come to work and who participate in most forms of programs, including health promotion programs.

Methods and techniques

- *Physical environment*—What is being changed to facilitate changes in food consumption or other target behaviours—incentives, space, labelling, prices, etc.?
- *Social environment*—Are there any Employee Advisory Boards? Are families involved? What is being done to create diffusion, participation and social support among employees?
- *Learning and personal factors*—Are there courses, information sources (e.g. leaflets, programs, posters etc.) which inform people about the aims of the program and how to meet them? Are demonstrations available? Are senior staff modelling the behaviours?
- *Has a needs assessment been carried out* to ascertain staff views and needs?

Equity and access issues

Are high-risk groups reached? Are those in socially disadvantaged groups reached? What is done to ensure the program reaches those who need it most, irrespective of their social or other disadvantaged backgrounds (e.g. low literacy, disability, special needs of women of reproductive age)?

Evaluation

Is there any form of evaluation? Types of evaluation include the following:

- *Formative evaluation*—Are all relevant stakeholders involved?
- *Process evaluation*—Are all relevant stakeholders involved? Are process evaluation findings acted on?
- *Outcome (impact) evaluation*—Are the measures used as undemanding as possible? Are valid and reliable measures being used? Are unobstrusive food consumption indices used to

---

Box 10.4 continued

complement self-report indices? How are evaluation data converted to meaningful communications for employees? Are unwanted effects being considered and checked? How are evaluation findings used to further the aims of the program? Are there plans for long-term follow up or monitoring of effectiveness of the program? Are unobstrusive indices of progress towards goals in the three key areas of physical environment, social support and learning needs being implemented?

- *Cost benefits*—What are the benefits and costs of the program in financial and human terms (e.g. use of volunteer labour)?

Reporting

In what form are project reports presented? Have stakeholders been kept up to date through scheduled meetings and communications? Are the findings recorded in durable form for access by the health community? How can other worksites and health promoters access the findings about the program?

Policy

Are policies being developed to ensure the smooth running of the program?

- *Sustainability*—What is being done to ensure the long-term sustainability of the program?
- *System institutionalisation*—What steps have been taken to build the program into the worksite's normal system functioning?

Sorensen et al. (2004) suggest that environmental and policy approaches are probably the most likely ways to improve the consumption of health foods such as fruit and vegetables. The main lessons to be drawn from recent interventions include the following:

1. *The identification of barriers to, and facilitators of, organisational and environmental changes within worksites*—Strategies are required to gain management commitment to comprehensive programs. This requires understanding of the complex ways in which different levels of management can be recruited to support healthy food choices at work. A key part of the recruitment of management support is the establishment of a 'business case' for the promotion of healthy food choices at work.

2. *Creation of effective partnerships*—Partnerships, such as those between government agencies (e.g. Departments of Health), local community groups (e.g. environmental groups, schools) and private businesses (e.g. local food outlets, gymnasiums), can enhance worksite health promotion programs.

3. *Make healthy foods more available*—There is no reason why healthier foods cannot be tasty, convenient and reasonably priced. This usually requires the

development of partnerships between food suppliers, employee groups and unions, marketing organisations, public health agencies and employers.

4. *The combined influence of price, perceived health benefits, taste and convenience* on employee demand for healthy foods requires more research and should be addressed in worksite programs. Foods provide more than a single benefit for most people, and convenience, price and taste are important influences which can moderate the influence of perceptions of healthiness. Food Cent is one of the few examples of nutrition promotion which actively dealt with consumers' concerns about price, acceptability and convenience in addition to healthiness (see Chapters 3 and 7).

5. *The rapidly changing nature of the workforce* needs to be taken into account in the design of nutrition promotion programs. These include worksite downsizing, technological innovations, and part-time and casual employment. Examples of 'newer' targets for nutrition promotion include construction workers, sales staff, transport workers and temporary and contract workers. In addition, blue-collar employees in small businesses require more attention from promoters. In particular, the special needs of women of reproductive age, such as facilities for the breastfeeding of babies and child care, need to be addressed with some urgency (Dodgson et al. 2004; McIntyre 2002). Similarly, the needs of shift workers and employees with disabilities require much more consideration.

Generally, there needs to be more recognition of the external forces which impinge on the worksite, such as the 'family–work' balance. Sorensen et al.'s (2004) suggestions for further advancement of the area are summarised in Box 10.5.

BOX 10.5 FACTORS WHICH PROMOTE DISSEMINATION AND ADOPTION OF PROGRAMS

Interventions should be:

- straightforward and require little modification;
- compatible with and be seen to add to existing organisational practices;
- flexibly implemented, preferably after a pre-trial;
- enlist a program champion;
- health promotion should be in the organisation's mission statement;
- health promotion should be included in job descriptions and/or funded in employee health promotion programs;

Box 10.5 continued

- good communication systems among managers and employees help diffuse programs;
- maintain key features of the intervention but adapt them to new elements;
- to facilitate transfer, train personnel in the use of the intervention;
- use an appropriate ecological framework;
- plan for sustainability and dissemination of positive outcomes;
- develop and maintain effective partnerships between public and private groups across health and non-health sectors such as health agencies, manufacturers and food service suppliers and community groups.

Source: Sorensen et al. (2004).

Conclusions

Worksite programs can bring about changes in food consumption. A variety of different methods and approaches can be effective. Perhaps it is too early to decide which methods and approaches are most effective, because there are problems with the assessment of change—especially dietary change. For example, people may eat in a healthy manner at work but compensate in an unhealthy manner at home, thus yielding no major nutrition gains.

Generally, a combination of information provision, changes in the food supply and changes in the physical and social environments appear to be helpful in changing employees' worksite food consumption behaviours.

Discussion questions

10.1 How would you conduct a needs assessment at a worksite? What stages would you work through?

10.2 Describe some of the factors which make worksite nutrition promotion attractive.

10.3 What are the advantages and disadvantages of worksite nutrition promotion for employers?

10.4 Outline the design, methods and outcomes of the Treatwell program.

10.5 Compare and contrast the various approaches to worksite nutrition promotion.

10.6 Describe some of the difficulties in measuring the effects of worksite nutrition programs.

10.7 List the best practice principles of worksite health promotion as described by Chapman, giving nutrition examples.

11

Hospital and health service-based programs

Introduction

In affluent countries, most of the health budget is spent on curative services; only a small fraction (2–3 per cent) is spent on preventative services. The sums of money involved are huge—about 8 per cent of the Australian Gross National Product (GNP) ($60 billion per annum on government health services), 12 per cent of the American GNP and 6 per cent of the British GNP. Most of this is spent on the employment of huge numbers of health professionals. There is raging debate about the efficacy and coverage of the various national health service systems—for example, the Wanless Report (2002) describes the huge health and economic benefits for Britain that would flow from major investment in public health programs. This report confirms that there is great potential for health and nutrition promotion in the health services.

The health services mainly have curative or caring roles. That is, they help people to overcome existing illnesses or at least attempt to reduce the suffering caused by illness. Most of this work is done in clinical, one-on-one conditions. For example, the doctor meets with the patient to examine, treat and counsel them. It is perhaps only in the past 30 years that it has been realised that these clinical consultations can have powerful preventative effects. For example, simple advice to quit smoking can be effective among many people (e.g. 7.5 per cent in the Sick of Smoking study—Wilson et al. 1990). Because of the immense amount of research on disease risk factors, clinicians are in a good position to screen patients who are at high risk of disease and then to treat them.

The identification of high-risk individuals has begun to have effects at the population level (Holliday 1999). While Rose's (1985) model of disease causation rightly suggests that it is those at moderate risk who provide the most cases of any disease, the treatment of high-risk individuals can prevent substantial numbers of people from becoming 'cases'. Serum cholesterol screening and associated treatment with statins and other drugs provide an important example of this approach. Despite the debatable cost effectiveness, we can regard the screening of at-risk individuals as an important part of disease minimisation and thus of health promotion

Two additional factors are cause for optimism about the role of clinicians from a variety of professions in health and nutrition promotion. First, clinicians can give useful advice and support to many people in their attempts to change their food and lifestyle habits. Their advice is often highly influential since health professionals are generally highly respected. Not only do numerous clinicians consult directly with substantial numbers of people, it is also likely that they can indirectly affect their patients' families and friends. While these social flow-on effects have rarely been studied, evidence from marketing suggests that 'good' and 'bad' news diffuses through social networks. Second, there is increasing evidence that small but specific changes in diet and lifestyle can slow down or reverse disease processes (see Chapters 1 and 9).

In order to promote nutrition in the health sector, we need to ask two questions:

- Do health professionals have the requisite knowledge, skills and resources to assist in behavioural change?
- How do we find ways to ensure that as many of them as possible provide this kind of support?

While the general answers to these questions are mainly negative, many health service practitioners provide excellent examples of nutrition promotion, as we will see below. However, the potential for massive improvement is great because of the large numbers of highly skilled professionals employed in the health services.

Change is big time

Health services are changing rapidly in their structure and in the ways they operate. For example, new technologies are changing the relationships between specialists and technicians (e.g. in the detection of bowel cancers), new treatments are being evaluated all the time, and advanced procedures are being conducted in family practices and community health centres rather than in hospitals. Underlying all this change are several factors which are conducive to nutrition promotion.

Innovation and new technologies

It is now possible to treat people successfully who could not be treated a decade or even five years ago. This often means tailored nutrition and lifestyle management programs are required to maintain their health. This may contribute to people's rising expectations about their health and treatments.

Increasing longevity

More people are living longer with illnesses. Four out of five 80-year-olds have one or more degenerative diseases. Health and nutrition promotion can do much to control diseases like heart disease and type 2 diabetes relatively cheaply.

Rising costs

Many new technologies and drugs are very expensive. Combined with increasing longevity and higher quality of life expectations, this contributes to the ever-increasing costs of the health services. It has been estimated, for example, that about half of the adult population could be treated with expensive statin drugs to reduce the incidence of heart attacks. This scenario wouldn't leave much money for all the other activities of the health services! Fortunately, nutrition and other forms of health promotion are probably as effective and far cheaper. This simple example illustrates the problems facing governments around the world—they need to find effective, less-expensive ways to treat and prevent illness. This situation provides opportunities for nutrition promotion.

People and settings

Nutrition promotion is conducted principally in several health service sectors: hospitals, maternal and child health services, family physicians' practices and community health centres, and in government departments and non-government agencies. Within these settings, doctors, nurses, dietitians and various allied health professionals and alternative therapists have major but differing roles in nutrition promotion. The nutrition promotion issues encountered by these groups in these settings are outlined below with some examples.

Hospitals

Hospitals are shrinking and people are spending less time in them. It is likely that hospitals will morph into high-intensity treatment centres for very seriously ill acute patients. Their traditional curative services are steadily being taken over by large general practices, community health centres and specialist 'day' hospitals. Although they are unlikely to disappear, they are likely to become regional hubs with major outreach networks into the community. Despite this 'shrinkage', they deal with large numbers of people who are often at points in their lives when they become conscious of their susceptibility to life-threatening diseases, and most of them feel motivated to adopt new lifestyle habits ('teachable moments').

There are two pressing nutrition issues for today's hospitals in addition to those associated with the 'normal care' of seriously ill individuals. They concern malnutrition and the prevention or amelioration of diseases. It has been estimated that between 20 and 30 per cent of hospital patients in the European Union (Beck et al. 2001) are under-nourished in some way. The situation is similar in other affluent societies such as Australia and the United States. Patients, especially lone elderly patients, often come to hospital in an under-nourished state following years of poor eating habits. Others either become under-nourished in hospital or become more malnourished the longer they stay. The forms of under-nutrition include protein energy malnutrition and micronutrient deficiencies (e.g. calcium and vitamin D— Nowson and Margerison 2002). Paradoxically, obesity may also be a serious problem. The coexistence of obesity with micronutrient deficiency, observed in societies undergoing rapid nutrition transition, may also be seen in hospital patients. These forms of malnutrition threaten patients' lives, frequently prevent the use of advanced treatments, and cause complications and prolonged stays in hospital.

Council of Europe report on malnutrition in hospitals

This landmark report notes that under-nutrition has occurred among patients in European hospitals for over a century (Beck et al. 2001). The report suggests that there are five common barriers which prevent many patients from being fed properly:

1. *A lack of clearly defined responsibilities in planning and managing nutritional care* — Routine nutritional risk screening and assessment are usually not performed, though sometimes weight, weight loss and BMI may be assessed. Nutritional counselling and nutritional support for under-nourished patients and high-risk patients are rarely provided. Decreased lengths of stay and lack of time, staff, nutrition education and interest are related factors.

2. *A lack of sufficient education about nutrition among all staff groups* — The nutrition education of physicians and nurses is poor and has not kept up with the latest research. Best practice nutritional support is often not provided. There is a

particular problem in the inability of many nurses to identify at-risk patients. There is widespread poor communication and coordination between staff, leading to poor nutritional care. While dietitians are usually the most up to date in terms of nutrition education, their educational levels and responsibilities are highly varied. Food service staff may be unaware of the importance of nutritional care and so food service departments may not have much influence over departmental budgets and resources, compared with clinical departments.

3. *Lack of influence of the patients* — Many patients have a poor view of hospital food — even before they enter hospital. Although menu choices are frequently provided, many patients may select low-fat foods, being ignorant that weight loss in hospital may increase the risk of complications. More assistance in menu choice and greater availability of between-meal foods and beverages is required. In reality, patients have little influence over the foods they are served while in hospital.

4. *Lack of cooperation between different staff groups* — Lack of appetite appears to be a major cause of hospital under-nutrition, but most staff do not know this and fail to work together to motivate patients to eat. Motivation to eat more requires close cooperation between all staff members. The report underscores the importance of establishing a positive food culture to which all staff subscribe.

5. *Lack of involvement of the hospital management* — Food service departments tend to be regarded by hospital managements as a general facility, not as a patient treatment service. This lack of support is often compounded when services are outsourced, with nutritional quality standards often being omitted from contracts.

The development of under-nutrition in health care facilities is a form of iatrogenesis. As the Council of Europe report suggests its causes are complex, ranging from the non-provision of nutritious foods to lack of assistance with eating (e.g. for patients who are unable to eat unassisted and have to be fed—which is time-consuming and often hard work), loss of appetite and/or refusal to eat. The management of patients with these difficulties is a responsibility of the nursing, dietetic, catering and medical staff. Ignorance of patients' nutritional needs among these staff and failure to communicate about patients' feeding problems (e.g. between nurses and dietitians or between nurses' shifts) is an important set of causes. This is a ripe area for nutrition promotion. The Council of Europe report makes a series of recommendations to overcome the problems it identified (see Box 11.1).

BOX 11.1 SUMMARY OF PROPOSED COUNCIL OF EUROPE GUIDELINES FOR THE PREVENTION OF UNDER-NUTRITION IN HOSPITALS

1. **Lack of clearly defined responsibilities in planning and managing nutritional care**
   - (i)   Clearly assign management and staff categories' responsibilities regarding nutritional care.
   - (ii) Standards of practice for assessment and monitoring of patients' nutritional risk should be developed at the national level and task responsibilities clearly defined.
   - (iii) The hospital's responsibility should *not* be limited to the hospital stay.

2. **Lack of sufficient education about nutrition among all staff groups**
   - (i)   Improvement in the educational level of all staff groups is required in the form of continuing education in general nutrition and nutrition-support techniques.
   - (ii) There should be particular focus on the training of non-clinical staff members, along with definitions of their areas of responsibility.

3. **Lack of influence of the patients**
   - (i)   The provision of meals should be individualised and flexible.
   - (ii) All patients should be able to order food and extra food and be informed of this ability.
   - (iii) All patients should participate in planning their meals and have some control over food selection.
   - (iv) There should be immediate feedback from patients about their likes and dislikes of food that is served. This feedback should be used to develop appropriate, target group-specific menus.

4. **Lack of cooperation between different staff groups**
   - (i)   Hospital managers, physicians, nurses, dietitians and food service staff should work together to provide optimal nutritional care.
   - (ii) Hospital management should give priority to cooperation (e.g. through organisational research to optimise cooperation).
   - (iii) Organised contact between the hospital and primary health care should be established.

5. **Lack of involvement from the hospital management**
   - (i)   The provision of meals should be regarded as an essential part of the treatment of patients, not just as a catering service.
   - (ii) The hospital management should acknowledge responsibility for food service and the nutritional care of patients, and should give priority to food policy and the administration of food services.
   - (iii) The hospital management should take account of the costs of complications and prolonged hospital stay due to under-nutrition when assessing the cost of food service.

Source: Adapted from Beck et al. (2001).

The health-promoting hospital

Reorientation of hospitals and health services is one of the five strategies advanced by the Ottawa Charter (1986). In various parts of the world, there have been many attempts to shift hospitals from purely curative roles towards participation in health promotion. In 1997, the Vienna Recommendations were adopted by the World Health Organization in an attempt to provide countries, regions and hospitals with guidance to bring this change about (see Box 11.2). They are based on an earlier framework which was outlined in the Budapest Declaration on Health-promoting Hospitals (1991) and on the Ljubljana Charter on Reforming Health Care (1996). Key types of interventions are described in Box 11.2.

BOX 11.2 PRINCIPLES OF THE VIENNA RECOMMENDATIONS

A health-promoting hospital should:

1. promote human dignity, equity and solidarity and professional ethics, acknowledging differences in the needs, values and cultures of different population groups;
2. be oriented towards quality improvement, the well-being of patients, relatives and staff, protection of the environment and realisation of the potential to become learning organisations;
3. focus on health with a holistic approach and not only on curative services;
4. be centred on people providing health services in the best way possible to patients and their relatives, to facilitate the healing process and contribute to the empowerment of patients;
5. use resources efficiently and cost effectively, and allocate resources on the basis of contribution to health improvement; and
6. form as close links as possible with other levels of the health care system and the community.

Source: WHO (1997).

The Vienna Recommendations note that:

- hospitals play a central influential role in the health system;
- large numbers of people pass through or work in them (up to 20 per cent of the population in some countries are hospital patients each year; 30 000 hospitals in Europe employ 3 per cent of the total workforce);
- they can be dangerous places in which to work, in terms of infections, chemical, physical and stressor hazards;

- they produce a lot of waste, and so can help to reduce environmental pollution; and
- they have great potential for health promotion and are increasingly concerned with patients' lives before and after their stay in hospital.

While the Vienna Recommendations deal with health promotion in general, it is easy to see that they apply equally to nutrition promotion—for example, the prevention of malnutrition requires participation by and communication between patients and several groups of staff, as well as high levels of organisational management.

The health-promoting hospital in practice

The idea of the health-promoting hospital is now quite mature and many attempts have been made to put it into practice, with mixed success. Several commentators have noted that different countries, regions and hospitals have adopted the model with varying degrees of enthusiasm. For example, there has been considerable activity in the United Kingdom, Ireland, Australia, Germany and Poland, and relatively little in Spain, Romania, the Ukraine and Turkey (Whitehead 2004). The types of interventions involved are shown in Table 11.1.

TABLE 11.1 DEFINITIONS OF THE INTERVENTION CLASSIFICATIONS

Intervention	Details
Primordial prevention	The complete absence or removal of a disease or risk factor e.g. small pox has been removed from the world population; tobacco smoking is becoming rarer in Northwest America (below 10 per cent prevalence).
Health development	Aims to create social, economic and environmental change that support population health. Examples include policies for taxation, education, food fortification and labelling and transport.
Primary prevention	Aims to avoid disease or injury from recurring by identifying and reducing communities' and individuals' risk of exposure of behaviours. Examples include sun smart and smoke-free workplace campaigns.
Secondary prevention	Aims to screen individuals for signs of disease in early stages and provide advice or treatment to cure it or prevent it from progressing (also called 'preventive medicine'). Examples include blood pressure or cholesterol screening, breast and colon cancer screening.
Tertiary prevention	Aims to prevent the reoccurrence of disease and limit disability or complications arising from an irreversible condition such as cardiac rehabilitation.

Source: Adapted from Bensberg (1998).

However, the switch from purely curative aims to broader health promotion agendas is often inhibited by a lack of clear aims and strategies, few funding resources, poor training facilities and a lack of national policy commitment or priority for healthy promotion in hospitals (Whitehead 2004). It is difficult for hospital health promotion to progress from specific localised projects to comprehensive health-promoting hospital organisations involving staff and patient participation and total quality management. Hospital-based professionals need to become more focused on health promotion, but their training and funding encourages them to remain treatment-focused (Whitehead 2004). This is compounded in times of financial restraint when hospitals 'set aside their mission to promote and protect the health of their surrounding communities' (Weil and Harmata 2002; Wright et al. 2002; Whitehead 2004).

Despite these difficulties, there is a growing expectation that hospitals will lead local public health reform. Wright et al. (2002) describe the new Public Health hospital as 'one that develops its staff to move away from increasing medical sub-specialization to an increasing understanding of the wider health agenda'. Johnson and Baum (2001) provide a useful typology of hospital health promotion which outlines the main ways in which hospitals can promote health. From their observations of an active health-promoting hospital in Adelaide, together with observations from the literature, they found that the key elements of successful hospital health promotion were:

- strong leadership at different levels of the hospital, especially from senior management;
- incorporation of health promotion in the hospital's vision and strategic role statements;
- strategic, operational and evaluation plans for health promotion;
- staff development and education; and
- resources such as staff, physical facilities and finance.

They suggest that there are four main types of health-promoting activities undertaken by hospitals.

Type 1: 'Doing a health promotion project'

Usually these are isolated projects undertaken by individual staff members. They aren't part of the hospital's strategic approach to reorient its services towards community health promotion. They may be aimed at patients and their families, the staff, the organisation as a whole, the physical environment or the community served by the hospital. Examples might include the provision of a water fountain for visitors and staff, a healthy eating menu guide for the staff canteen or distribution of a healthy eating leaflet for diabetic patients. These sorts of projects often peter out as the initiator loses interest or leaves the hospital. However, they may also represent the start of an interest in health promotion which may be supported by senior management to grow into longer term commitments.

Type 2: Delegating health promotion to the role of a specific division, department or staff

The management of the hospital may decide that certain staff or departments will 'do health promotion'. For example, a special health promotion unit may be set up which will advise patients and visitors, or a diabetes educator may be appointed to provide counselling and support for people with diabetes. This sort of specialisation may be very effective, but in a narrow, contained way. Certain categories of patients and staff may benefits, but the mission of the hospital as a whole remains essentially curative.

Type 3: Being a health promotion setting

The hospital decides to become a setting in which health promotion activities are aimed at patients and their families, staff, the organisation and the physical environment of the hospital. For example, concerted nutrition programs might be included such as the provision of healthy meals and snacks for hospital shift workers; 5 a Day or similar programs may be promoted to patients, their families and visitors via leaflets, advisory programs and hospital TV; the food service may use nutritional HACCP procedures to ensure that patients and staff receive nutritious meals; and so on. However, the approach stops at the hospital doors, with little attempt being made to promote health in the local community.

Type 4: Being a health promotion setting and improving the health of the community

In this approach, the hospital actually attempts to reorient its services to support its community's health. It has to 'systematically develop effective and collaborative working relationships with patients and their families, other service providers and the broader community to achieve the best outcomes'. The hospital that Johnson and Baum (2001) studied was the Adelaide Women's and Children's Hospital, which supplies specialist services for the whole of South Australia. Examples of its outreach included the activities of its Children's Health Development Foundation, which has provided curriculum materials and cooking lessons in schools and established networks of food producers, nutritionists and community groups around South Australia as part of the Eat Well South Australia initiative.

Most staff working in hospitals of types 1 and 2 would remain in traditional treatment-oriented roles, while those in types 3 and 4 would undergo considerable reorientation of their work activities to encompass the broader aims of health promotion.

Prevention and amelioration via nutrition promotion

Hospitals present many opportunities to help patients (and their families) to change their eating and lifestyle habits. People who have often undergone much suffering and experienced the threat of death are often in a contemplative frame of mind and are motivated to do anything to prevent a recurrence of the episode which brought them to hospital. Physicians and surgeons, together with dietetic and allied health staff, are well placed to use the illness episode as a starting point for lifestyle change. Several types of actions have been adopted, such as:

1. *Introduction of healthy menu choice* and encouragement of patients to taste healthier foods such as low-saturated fat foods, fruits and vegetables. These menu choices may be accompanied by dietetic counselling or information (e.g. in the form of leaflets) which encourages the adoption of healthier eating and lifestyle habits. In the United Kingdom in the 1980s, Don Nutbeam, an eminent health promoter, visited UK hospitals, took samples of the foods served to patients (and staff) and called press conferences to alert the public to the (usually) low quality of the foods on offer—often the quantity of saturated fat in the foods was above national health service guidelines. Dr Nutbeam always made sure the business phone number of the hospital manager was available during these conferences! This sort of tactic tended to encourage the following of government guidelines, at least for a few months! A big problem for the hospitals was that high-saturated fat foods such as butter and processed meats were less expensive than 'healthier' alternatives, due in part to EU subsidies to agriculture. So there was the ironic situation of patients hospitalised for heart disease, caused in part by high intakes of saturated fats, being fed subsidised saturated fat products while in hospital.

2. *Rehabilitation services.* Many patients who are hospitalised with serious diseases such as heart disease, stroke or type 2 diabetes are encouraged to join

rehabilitation programs which are usually offered by their outpatient services. Here they will be counselled about their diets, lifestyles and medications. This may be done individually by a dietitian or nurse educator, or in groups with other patients. In some hospitals, the diagnosis of one family member with a condition like diabetes is a signal for the patient and their immediate family to be invited to receive dietary and lifestyle counselling. The importance of rehabilitation is illustrated by several trials:

- The *Oslo Heart study* was conducted in the Oslo General Hospital, starting over a quarter of a century ago (Hjermann et al. 1981). Patients at high risk of heart disease were either given the 'normal' clinical advice (controls) or they were given specific advice to quit smoking and shown how to reduce the saturated fats in their diets at regular six-monthly cholesterol checkups. Over time, more people in this treatment group survived compared with the control group. The trial was among the first to demonstrate the utility of simple advice in clinical settings to persons at high risk of heart disease.

- The *Lyons Heart study* (de Lorgeril et al. 1999), named after the French city in which it was conducted, compared two groups of patients who had suffered myocardial infractions. The control group received the standard treatment and advice prevalent in France during the early 1990s; the other group was asked to consume a margarine containing several n3 fatty acids. During the subsequent five-year period, 50 per cent more patients in the treatment group survived. This study demonstrates the effectiveness of hospital-based specific dietary change programs.

- *COACH*—this emerging program in Australia aims to empower cardiac patients to monitor their own progress towards rehabilitation goals and to encourage their doctors to provide them with appropriate medication. As part of the program, they are given dietary advice by specially trained dietitians which is consistent with their medication regimen.

Maternity hospitals

Maternity hospitals and birthing centres are extremely important settings in which new mothers can bond with their babies and initiate breastfeeding. This is not an easy process for many women, and requires the development of a set of skills which can be aided by maternity nurses and other health workers as well as the mother's immediate family. The evidence is that if breastfeeding is started soon after birth, then it is easier for the mother to carry on feeding the child in this way. There is sound evidence, from systematic reviews, that both professional and lay support, such as hospital-based counselling of mothers-to-be and assistance with the initiation of breastfeeding soon after birth raises the prevalence and duration of breastfeeding (Sikorski et al. 2003).

In contrast to this healthy approach, for many years hospitals offered new mothers infant formulae and other inducements from formula companies soon after birth. This helped to reduce breastfeeding rates, bringing about less than optimal infant health. The activities of some companies in the Third World have been notorious.

This sort of promotion is banned under the provisions of the WHO International Code of Marketing of Breast Milk Substitutes (WHO 1981). Almost 120 nations have ratified the code, which prohibits aggressive infant formula marketing promotions. The code is quite explicit and provides clear guidance to health service staff (Box 11.3). Several of the infant formula manufacturers now have policies which claim to abide by the code, and important professional organisations such as the American College of Pediatrics have developed supportive polices—for example, the college will terminate any support it has received from any formula company which markets its products directly to the public. Unfortunately, unless individual nations have passed legislation to enforce the code, its provisions are frequently ignored. Inspection of women's magazines and visits to maternity hospitals demonstrate how widely this important code is flouted.

BOX 11.3 THE WHO INTERNATIONAL CODE OF MARKETING OF BREASTMILK SUBSTITUTES

NO advertising of breastmilk substitutes

NO free samples to mothers

NO promotion of products in health-care facilities

NO company 'mothercraft' nurses to advise mothers

NO gifts or personal samples to health workers

NO words or pictures idealising artificial feeding, including pictures of infants on the products

Information to health workers should be scientific and factual

All information on artificial feeding, including the labels, should explain the benefits of breast-feeding and the costs and hazards associated with artificial feeding

Unsuitable products, such as condensed milk, should not be promoted for babies

All products should be of a high quality, and take into account the climatic and storage conditions of the country where they are used

---

Source: World Health Organization publication WHO/MCH/NUT/90.1, 1991 cited at Baby Friendly USA, www.social.com/health/nhic/datahr2200/hr2282.html.

The Baby Friendly Hospital Initiative

The Baby Friendly Hospital Initiative (BFHI) is a joint project of UNICEF and WHO. It aims to increase the prevalence and duration of breastfeeding and attempts to establish an international standard for maternity services. The aim of this initiative is for babies to be exclusively breastfed until six months of age (Kramer et al. 2001).

The benefits of breastfeeding for the mother as well as the child are extensive. They include the following:

- Breastfed infants are protected from gastro-enteritis and atopic eczema (Promotion of Breastfeeding trial PROBIT; Kramer et al. 2001).
- It reduces infant mortality into the second year of life in less developed countries (WHO Collaborative Study Team 2000).
- Relative to formula-fed babies, breastfed babies have higher scores on cognitive tests (Anderson et al. 1999).
- There are long-term cardio-protective benefits—for example, systolic blood pressure was found to be higher in formula-fed seven-year-olds compared with breastfed peers (Wilson et al. 1998).

The BFHI endeavours to give every baby the best start in life by creating a health care environment where breastfeeding is the norm, thus helping to reduce the levels of infant mortality. (Los Angeles Breastfeeding Taskforce)

The initiative has been adopted in many countries, including Australia, New Zealand, the United States, Canada and the United Kingdom. It is centred on the Ten Steps to Successful Breastfeeding policy (Box 11.4). Hospitals which comply with this policy can apply for recognition as a 'Baby Friendly Hospital'.

Since the BFHI began, over 15 000 facilities in 134 countries have been awarded Baby Friendly status. Reports of the success of the initiative are given regularly in UNICEF's BFHI News. Generally it appears that in areas where hospitals have become Baby Friendly more mothers are breastfeeding their infants, with resulting improvements in child health.

To help with the adoption of this policy, the national BFHI organisations provide implementation guides, training course questionnaires for external teams to assess facilities for the award of Baby Friendly Hospital status, as well as manuals for the transformation of hospital practices and the verification of the ending of free and low-cost supplies of infant formula. It is worth noting that relatively few

---

BOX 11.4 BABY FRIENDLY HOSPITAL INITIATIVE: TEN STEPS TO SUCCESSFUL BREASTFEEDING

1. Have a written breastfeeding policy that is routinely communicated to all health care staff.
2. Train all health care staff in skills necessary to implement this policy.
3. Inform all pregnant women about the benefits and management of breastfeeding.
4. Help mothers initiate breastfeeding within half an hour of birth.
5. Show mothers how to breastfeed, and how to maintain lactation even if they should be separated from their infants.
6. Give newborn infants no food or drink other than breast milk, unless medically indicated.
7. Practice rooming-in: allow mothers and infants to remain together 24 hours a day.
8. Encourage breastfeeding on demand.
9. Give no artificial teats or pacifiers to breastfeeding infants.
10. Foster the establishment of breastfeeding support groups and refer mothers to them on discharge from the hospital or clinic.

Source: www.bfhi.org.au.

maternity hospitals have fully adopted this convention, so there is a long way to go. Globally, breastfeeding rates, even at three to four months of age, are low (e.g. in Australia only 40 per cent of infants at this age are exclusively breastfed). In affluent countries, breastfeeding rates tend to be lowest among young mothers and those in low-income and poorly educated groups. Paradoxically, in less affluent countries, breastfeeding rates tend to be lowest among the more affluent groups in the population (Rogers et al. 1997).

Physicians, family physicians, dietitians and other health professionals

Medical practitioners are the managers of patients' care. They consult with substantial proportions of the population each year, and often gain unique insight into patients' lives. Watt (1996) notes that physicians as a group are perhaps the last remaining professionals who have access to the private lives and living conditions of all sections of late modern society. They have major reach into the population and are highly influential; they are key opinion leaders. Patients listen to what they say and often act on their counsel. Unfortunately for nutrition promotion, despite this superb vantage point many physicians have little education about nutrition or training in the nutritional management of patients though the situation appears to be changing, at least in North America (St Jeor et al. 2006).

While specialist physicians such as geriatricians, endocrinologists and paediatricians have played major roles in nutrition promotion, it is family physicians (general practitioners) who are far more numerous and who have contact with large numbers of people. For example, Australia, with a population of 20 million, has 28 000 general practitioners who see about 80 per cent of the population at least once per year. The US Centers for Disease Control has nominated the doctor's office as a key setting for nutrition and physical activity promotion.

There has been much discussion of the roles of family physicians in general and in nutrition in particular (see supplements: *European Journal of Clinical Nutrition* 1999, vol. 53, Supplement 2, S112–S114; *American Journal of Clinical Nutrition* 2003, vol. 77, no. 4, Suppl, 999S–1092S; *European Journal of Clinical Nutrition* 2005, vol. 59, S–S196). Primarily, they aim to be able to distinguish life-threatening illnesses from less serious conditions and to manage patient care. Surveys indicate that about a third of general practitioners have an interest in nutrition and actively promote healthy eating in their practices. Pomeroy (2007) suggests that these 'nutritionally oriented' general practitioners play reinforcing, supportive and explanatory roles. They reiterate the dietary proposals made by dietitians and specialist physicians, explain the need for dietary change to patients, and support them in their attempts to eat healthier foods. Despite this great promise, family physicians' role in nutrition promotion is problematic, for several reasons:

1. *Poor knowledge* — Their knowledge and behavioural change skills are generally poor. Many GPs simply do not know the value of nutrition and healthy eating; they are unaware of much recent research showing the importance of dietary and lifestyle change for disease control. Doctors also report that, while they know they are in the business of health promotion, they do not have sufficient grasp of behavioural changes theories (Pomeroy 2007).

2. *Lack of motivation* — This is associated with lack of knowledge. Many GPs think the task of attempting dietary change is too difficult, so they do not try to promote healthy eating.

3. *Drug culture* — There is strong pressure on medical practitioners to prescribe drugs for many conditions. The pharmaceutical industry spends huge amounts of money to encourage drug prescribing, frequently resorting to free gifts and financial inducements. While there are many pharmaceutical

representatives who teach doctors how to use drugs, there are few equivalents to teach doctors how to manage dietary and lifestyle change in their practices. Doctors complain of a lack of dietitians and nutritionists who could play such supportive roles, and often poor communication with these professionals (Pomeroy 2007).

4. *Lack of time and resources*—These are often cited by doctors as factors which prevent them from promoting nutrition. This is undoubtedly a factor in fee-for-service health systems, but it also reflects the low priority placed on health and nutrition promotion, which is not perceived as important enough to include in their practices. Some physicians do make conscious decisions not to include nutrition counselling in their practices because they know it takes time and will lose them some income (Worsley and Worsley 1991). The advent of general practice nurses and of health trainers (in the United Kingdom) and coaches may assist general practitioners to implement nutrition and lifestyle programs.

Despite these difficulties, there have been some major initiatives in doctors' offices during the past decade which suggest that nutrition promotion in these settings is effective. Here are some examples.

Margetts et al. (1999) have shown that general practitioners' dietary and physical activity interventions can reduce hypertensives' systolic blood pressure by 2–4 mm Hg (systolic). While these studies included anti-hypertensive drug treatment, they also included simple nutrition counselling and were quite successful in lowering patients' blood pressure.

SNAP: A population health guide to behavioural risk factors in general practice

SNAP (RACGP 2004) stands for: **S** quit smoking, **N** better nutrition, **A** moderate alcohol, **P** more physical activity. The SNAP guide was designed by the Royal Australian College of General Practitioners to enable GPs to reduce chronic disease behavioural risk factors. It presents them with the evidence of harm associated with smoking, poor nutrition, excess alcohol consumption and a sedentary lifestyle, and uses the transtheoretical model and the '5A' principles (Box 11.5) to help them offer treatment appropriate to their patients' medical and life needs.

The Dutch Royal College of General Practitioners Nutrition advisory system

Many patients attending general practices have long-standing chronic conditions such as heart disease and type 2 diabetes. The doctor's task is to help them maintain or improve their health and to prevent any worsening of the condition. This can be difficult to do because patients visit the doctor's office at irregular intervals and for a variety of reasons, so monitoring their health status is often haphazard and frustrating. For example, Mrs X comes in several times a year but no matter what advice is given she remains obese. The Dutch RCGPs (Van Binsbergen and Drenthen 2003) realised that the relative frequency of patients' visits to doctors is a major advantage in promoting their nutritional health, if only the doctor could remember from visit to visit the advice he or she has given them. The members of the College were well aware that even small changes in dietary and lifestyle habits and in body mass index could have major benefits for patients.

To overcome the ad hoc nature of GPs' approach to nutrition promotion, the College developed a computer information system that stores vital information about dietary and health habits and BMI. This information is upgraded at every visit, no matter how irregular patients' visits may be. The information system selects relevant nutrition information from its files and prints a letter to the patient and doctor highlighting salient points about the patient's condition and habits, and suggesting actions that the patient might take in the coming weeks (e.g. eat a piece of fruit at lunch). So typically, when a patient comes to see a member of the College, any information elicited by the doctor is immediately entered into the information system (body weight is always measured) and change strategies are indicated by the system and discussed by the doctor with the patient.

This sophisticated system is a form of *tailoring*—advice is provided which is appropriate to the patient's particular background and circumstances (e.g. to their particular disease states and age group). A major benefit is that it helps to motivate doctors, especially with regard to seemingly intractable patients. For example, some patients appear to remain obese no matter what is tried. Recording their body weight at each visit enables time trends in the patient's BMI to be displayed. Often such graphs demonstrate small but steady weight loss over a period of months or years. In the absence of the information system, these small changes would be missed and thus patient and doctor would be 'demotivated'. However, the detection of small trends is motivating and enables both doctor and patient to persist with their attempts at dietary and lifestyle change.

One limitation of this approach is that it depends on a centralised health service, where patients are enrolled in a particular practice, as in the United Kingdom. In other countries like Australia and the United States, patients may change doctors frequently and there are no centralised records. However, similar information systems may still be feasible in these decentralised systems. For example, in Minlaton in South Australia, patients have been issued with Smart Cards which any doctor can use to access their recent medical history. The patients carry their records with them rather than having them stored only in one particular practice.

Active Scripts and Life Scripts

The New Zealand Heart Foundation initiated Active Scripts during the late 1990s. Just as doctors prescribe drugs, it was reasoned that they could also prescribe dietary and physical activities. Special prescription pads were provided for this purpose. Evaluation of the pilot project suggests that many patients find these simple directives useful: the prescriptions can be put in the kitchen—for example, on refrigerator doors—to remind patients to eat or move in particular ways at various times of the day. Healthy shopping lists are a development of this idea, as are Green Scripts, which focus on healthy environmentally friendly activities. The basic notion is that people find it easier to attempt to make a few simple changes at a time. Active Script has now been adopted and developed by the Australian Department of Health and Ageing as the Lifescripts Program (see Box 11.5).

It is clear that doctors can play major roles in nutrition promotion, and they need to do so if the costs of caring for sick people are to be controlled. More education and professional nutritionist support for doctors is required.

Nurses, midwives and health visitors

A key role for nurses is to support patients (and their families) so that they can optimise their quality of life. So it is not surprising that they can be found in many health service and community settings such as critical care centres, maternal and child health services, birthing services and residential care facilities for aged persons and persons with disabilities, as well as in general practices and community and public health centres. Increasingly nurses are specialising according to the settings in which they work, such as community nurses, public health nurses, general practice nurses and so on.

BOX 11.5 LIFESTYLE PRESCRIPTIONS

Lifescripts are tools for GPs to use when providing lifestyle advice to patients. Advice may be about quitting smoking, increasing physical activity, eating a healthier diet, maintaining healthy weight, reducing alcohol consumption or a combination of these.

Lifescripts is a national initiative being implemented through local divisions of general practice, promoting risk factor management in general practice and primary health care services.

The Lifescripts Resources aim to make it easier for GPs and their practices to manage lifestyle-related risk factors by providing a framework for:

- raising and discussing lifestyle risk factors with patients;
- advice in the form of a written script and associated patient education; and
- referral to other providers to support a healthy lifestyle.

Lifescripts Resource Kit

The Lifescripts Resource Kit focuses on the five lifestyle risk factors for chronic disease: smoking; nutrition; physical activity; alcohol; and weight management. The Kit contains evidence-based tools that enable general practice to assist patients to modify their lifestyles. The Kit components are:

- Divisions of General Practice Resource Kit: a manual to assist divisions incorporate Lifescripts within their activities and to provide support and guidance to general practices;
- Practice Kit: posters, flyers and checklists for practice waiting rooms and a manual to assist general practices integrate Lifescripts into their activities; and
- Risk Factor Resource Kit: assessment tools, lifestyle prescriptions and 5As guideline cards for each risk factor.

The consumer resources include posters for practice waiting rooms, flyers and checklists to inform patients about the initiative and provide them with an opportunity to identify to their GP if they wish to discuss lifestyle risk issues. Separate Indigenous specific posters and flyers provide a more culturally targeted message for interested practices and Aboriginal medical services.

The key practitioner resources are the 'Lifescripts script pads', which resemble normal prescription pads. General practices can obtain five different script pads addressing smoking, alcohol consumption, nutrition, physical activity and weight management on which they can write lifestyle advice tailored to each patient's needs. They can also use the pads to refer patients to other services, where available, to further support healthy lifestyle choices. In conjunction with the script pads, assessment tools and guidelines for each risk factor are provided along with key articles and evidence supporting the links between risk factors and disease.

An educational CD-Rom is included in the Lifescripts Practice Kit to assist and enable GPs and other practice staff to train in motivational interviewing techniques, to improve their ability to encourage behaviour change in their patients.

Source: www.health.gov.au/internet/main/publishing.nsf/content/health-pubhlth-strateg-life scripts-index.htm.

Some groups of nurses such as maternal and child health nurses and school nurses spend large proportions of their professional lives dealing with food and health issues. The evidence is that in many countries their training in nutrition management and promotion is poor. Howie (1987) showed long ago that many New Zealand and Australian nurses had little education in nutrition and tended to rely on popular books for their information. This is regrettable because it is clear that they have major roles to play in the promotion of health. Increasingly, nurses are being employed to educate patients and other members of the public—for example, hospitals employ them to educate patients with diabetes and nursing mothers, and some family medicine practices employ them to run group counselling sessions and other health promotion activities.

In the United Kingdom, midwives and health visitors provide continuity of care for mothers of young infants, with health visitors taking over the mother's and baby's care from the midwife after the immediate post-natal period.

Dietitians and nutritionists

Originally, dietitians emerged in hospital settings as applied nutrition scientists who helped to care for seriously ill patients. This work has become highly technical and demands high levels of nutrition knowledge and skill. It is very much a 'one-on-one' activity. Dietitians became broadly distinguished from nutritionists, who often were laboratory-based scientists who undertook research and consulting activities. Of course, not all dietitians work in clinical settings; many work in private practice providing one-on-one counselling, while others work as community dietitians and still others work as nutrition advisers in the government and food industry sectors.

Another established group of 'applied nutritionists' are *home economists*. They emerged in the United States and New Zealand in the late nineteenth century as part of the 'cult of true womanhood' which reified the domestic role of women and the application of 'science' to mothering roles. The American notion of the 'home-maker' is part of the home economics movement. The women's movement of the 1960s and 1970s highlighted the restrictive right-wing world-view of women that was associated with home economics which went into decline in educational and academic circles in English-speaking countries (though not in Japan, Korea, Taiwan and other Asian countries).

However, the essential role of home economists in teaching people (of both sexes) how to manage aspects of their domestic lives remains. It is evident that in today's complex free market economies, skills in food purchasing, preparation, parenting and financial management are essential for men as well as women. Accordingly, there is a renaissance of home economics (also known as consumer science and human ecology) in the form of life skills education. Home economics is where health, community and family settings overlap. These educationists are key nutrition promotion professionals. They illustrate that nutrition promotion goes way beyond health and the health services (www.vhetta.com.au).

Public health nutritionists, community nutritionists and dietitians

Many dietitians and nutritionists have realised for a long time that community and environmental influences shape people's food habits, and so affect their nutrition status. Gussow and Contento (1984) articulated these sympathies in their review of world nutrition education. In recent times, nutritionists and dietitians have been joined by educators, social and behavioural science graduates, ecologists, economists and others in the general belief that nutrition promotion requires a mix of disciplinary knowledge and skills. Until about a decade ago, this was called 'community nutrition' but in recent years it has become known and defined as 'public health nutrition'. It incorporates the older view that nutrition knowledge should be applied to the improvement of the public's health, but in addition it acknowledges that social, environmental and behavioural factors influence a population's food intake.

In order to help populations to eat more healthily, a range of disciplines and skills is required. Traditionally, most dietitians and nutritionists have had little education in health and nutrition promotion, food psychology and sociology, community development and education principles and techniques. Thus the term 'public health nutrition', while in practice being synonymous with 'community nutrition', conveys this nuance of the inclusion of a broad variety of biological and non-biological disciplines. An alternative, the more value-neutral term 'population nutrition', is often used. It avoids the notion of 'public good' and 'public health'. Because the values of equity and good governance (in the classical liberal meaning of the term) are central to health promotion and community well-being, most practitioners prefer the term 'public health nutrition' (see Chapter 1). The World Association for Public Health Nutrition was set up in November 2006.

The main competencies required by public health nutritionists are outlined in Box 11.6. Examples of community nutrition and public health nutrition can be seen in all the settings chapters, but especially in Chapter 7.

BOX 11.6 PUBLIC HEALTH NUTRITION COMPETENCIES

Public health nutritionists are expected to show leadership in the following broad competency areas:

• Surveillance and assessment of the population's health and well-being
• Promoting and protecting the population's health and well-being
• Developing quality and risk management within an evaluative culture
• Collaborative working for health and well-being
• Developing health programs and services and reducing inequalities
• Policy and strategy development and implementation to improve health and well-being
• Working with and for communities to improve health and well-being

---

Source: Adapted from Nutrition Society of Australia (2007).

Other allied health professionals, such as *pharmacists, physiotherapists, speech therapists* and *alternative health practitioners*, have major opportunities to promote nutrition in their practices. They require training to maximise these opportunities. Nutrition is for all professions and people, and should not be restricted to one or two professions no matter how well qualified they may be.

Roles of professional nutrition organisations

Professionals tend to organise themselves into professional organisations, usually to protect or extend their material interests (such as their status and salary conditions). They also take positions and advocate on a variety of issues including nutrition issues. That is, they propose, advocate and sometimes implement policies which promote the population's nutritional status. Note that just because an organisation represents health professions, what it proposes is not always supportive of the public good!

Opportunities for professional organisations in neo-liberal political regimes

If government believes in small or limited government, then who will govern? If, like the Bush Republican government of the early twenty-first century in the United States, governments see obesity primarily as a matter of individual choice, who will enable institutions and organisations to take action in various settings to prevent obesity? Part of the answer is professional organisations. Box 11.7 lists some of the more common nutritionally related professional bodies. Professional organisations have, for example, called for changes in food regulations and developed codes of practice and policy outlines which have later been modified and implemented by governments. They have also entered into partnerships with private companies (e.g. the Coles-Dietitians Association of Australia fruit and vegetable in-store promotion). These partnerships can be of major benefit for the public. Sometimes, however, turf wars develop between rival non-government organisations and professional bodies, which can lead to public confusion and loss of confidence—for example, when two professional organisations give conflicting advice about heart health or cancer prevention. Most agencies now work hard to avoid such conflict. Partnerships with private industry can be lucrative and professional associations may be tempted to 'soften' their approach to controversial issues—risking their public credibility.

BOX 11.7 EXAMPLES OF PROFESSIONAL ORGANISATIONS WHICH PROMOTE NUTRITION

In the United States and Canada

- The American Dietetic Association
- The American College of Pediatrics
- The American College of Sports Medicine
- The American Heart Foundation
- Canadian Heart Foundation—transprovincial heart disease prevention programs
- The American Medical Foundation
- The American Public Health Association
- The Society for Nutrition Education and Behavior
- The International Society for Behavioural Nutrition and Physical Activity

In Australia and New Zealand

- The Public Health Association—has a nutrition policy group
- The Dietitians Association of Australia—policies on public health nutrition
- Nutrition Australia—the Healthy Eating Pyramid
- The Home Economics Association of Australia and VHETTA
- The National Heart Foundation—risk factor prevalence surveys, Pick the Tick
- New Zealand Heart Foundation—Pick the Tick, Heartbeat NZ, many community nutrition initiatives, Active Scripts
- State Anti-Cancer Foundations e.g. the 12345#+ Guide to Healthy Eating
- Nutrition Society of Australia, Nutrition Society of New Zealand

Box 11.7 continued

- International Diabetes Institute
- Osteoporosis Australia
- Australian Kidney Foundation

In the United Kingdom
- The British Medical Association—strongly promotes public health and public health nutrition through the *British Medical Journal*
- The Nutrition Society
- The British Heart Foundation
- The British Nutrition Foundation
- The Cancer Council of the UK (Cancerhelp)
- The Public Health Association

Note: A more comprehensive list is provided on the Nutrition Australia website.

Conclusions

There is much to be done to promote nutrition within the health services. Many patients and clients of these services suffer from acute or chronic conditions associated with various forms of malnutrition which can threaten their lives and delay their recovery. Perhaps because so many patients suffer from disease or are at high risk of disease, the notion of promotion and prevention is rather strange to many who work in the health services. They tend to be a sole curative focus. Fortunately, examples abound which show that nutrition promotion in such high-risk communities can have substantial health benefits. The key need for the future is the provision of thorough nutrition promotion education for health practitioners so that they can identify the opportunities for advocacy, and system reform which will make the health service environments truly health promoting.

Discussion questions

11.1 Describe some of the nutrition problems commonly found among hospital patients.

11.2 Describe some of the main influences that are causing change in health services.

11.3 Describe the types of factors that prevent general practitioners from providing dietary counselling services for their patients.

11.4 Outline the ten steps towards successful breastfeeding. Which factors might impede such progress?

11.5 Describe the main types of health-promoting hospitals as suggested by Johnson and Baum.

11.6 Outline the potential roles of allied health professionals in nutrition promotion.

11.7 Compare the potential roles of dietitians and public health nutritionists in nutrition promotion.

12

Nutrition promotion in the retail sector

Introduction

The retail sector offers many opportunities for nutrition promotion because it is here that food is transferred from the production and distribution systems into households, and subsequently into our bodies. The main advantage of the retail system is that it can provide nutrition promotion with extensive 'reach' into the population; millions of shoppers make several visits to retail establishments every week and spend time in them making decisions about food purchasing. The main disadvantage is that much of the retail system is under the control of large, powerful, complex private companies for which population health and nutrition are not major priorities. Nevertheless, the sector is full of opportunities.

The context: Food retailing and food services

Broadly speaking, the retail food industry is divided into three main camps: catering and food service outlets; small retail businesses such as butchers, greengrocers and individually owned supermarkets; and the large supermarket chains. The retail sector has attracted nutrition promoters for many decades. This is because it provides contact with substantial numbers of people at point of purchase (POP) when they are deciding which food products to buy. Small changes in consumers' purchasing habits may help them improve the nutritional quality of their diets.

Supermarkets and supermarket chains dominate the sector, with perhaps as many as 90 per cent of consumers' food purchasing decisions being made in them (Caraher et al. 1998; Euromonitor 2000). In addition, small independent companies such as greengrocers, butchers and bakeries compete for the shopper's money. Some of these may come together in the form of markets. Restaurants in all their forms, including fast food outlets and cafes, comprise an increasing retail outlet for ready made foods. Perhaps as many as one-quarter of meals in the United States, Canada, Australia and New Zealand are now prepared and consumed outside the home. Vending machines are yet another ubiquitous type of food retailing. Finally, many retailers like pharmacies and health food stores sell nutrients (usually in the form of dietary supplements) and nutritional advice directly to the public, and weight loss companies often sell meals direct to the public (e.g. Lite 'n' Easy, Jenny Craig). All these different types of food retailers offer opportunities for nutrition promotion.

The stakeholders in the retail sector are many and varied. First and foremost are the owners and managers of retail companies such as the proprietors of restaurants and the managers of supermarket chains and stores. Most retail food outlets are privately owned companies, and although nutrition promoters may be invited to work in these private premises they are limited by the objectives and rules of the owners and managers. The main aim of retailing is to make profits for the companies—not to promote health, although this may be a good way to maintain or increase company profits, and increasingly companies are concerned with their social and environmental responsibilities.

Of central importance are the shoppers. Retailers are in the business of meeting customers' needs and wants. This is an advantage for nutrition promotion since retailers will seek to provide any legal goods and services customers demand. So if sufficient numbers of customers are willing to purchase healthy foods and nutrition services, many retailers will provide them. This customer orientation provides the basis for successful partnerships between nutrition promoters and retailers.

There are many different types of customers, primarily women but also men, children (and their parents), older people, people with disabilities and disease (e.g. those suffering from diabetes), members of ethnic minorities, people with different sets of psychological aspirations and so on—all of whom may have specific wants and needs.

Other key stakeholders include manufacturers and fresh food suppliers who wish to sell their products in supermarkets and other stores. Manufacturers often have more limited usefulness in nutrition promotion than retailers since they have to sell a relatively limited number of products which may have nutritional 'problems' (e.g. they may contain high levels of energy, saturated fat or salt), and so they may not be as willing to join in nutrition promotion programs (though many have modified their products to do so and fresh food providers have much to gain from nutrition promotion programs). Local, state and national governments are also key players since they regulate fair trading and food safety issues. Local community groups may form special relationships with stores (e.g. arrange for charity donations in store) and many other groups and agencies often attempt to use retail stores to promote their views, products or services. This multitude of stakeholders provides fertile ground for the development of partnerships in nutrition promotion, but it makes program implementation complex and difficult at times. The aims and goals of the stakeholders need to be carefully negotiated and clearly expounded while the program is being designed.

Supermarkets

The private nature and profit orientation of supermarkets have a number of consequences. First, all space and labour resources are devoted to doing business, and have been for decades. Retailers often retain their core managerial employees for long periods of time, frequently for 20 years or more. Any space or labour diverted to nutrition promotion could be spent on profit-making activities. Therefore, health and nutrition promotion in themselves cost the company money. More than that, they are likely to cost individual category managers money—since many of them are paid bonuses in proportion to their sales turnover—as well as time and effort. At first glance, non-essential activities like nutrition promotion represent a threat to income. This can impede the cooperation offered by employees, as does the widespread image of nutritionists as 'diet police'. Nutrition promotion is not perceived to be central to retailing.

It follows, then, that nutrition promoters may be expected by retailers to pay for the space and labour that they use in the supermarket, or at least offer the company some alternative reward such as more positive images among customers which themselves are realisable in the form of increased market share and profits. This raises the question of whether health authorities are willing and able to pay for the costs involved (and these can be substantial) or whether third parties such as manufacturing associations and companies are willing to do so. This third eventuality is also fraught with problems as the manufacturer may not be acceptable to the retailer or may wish to use nutrition promotion for its own narrow pecuniary ends (e.g. by making unsubstantiated nutrition claims about its products). A workable compromise needs to be agreed upon before any program is commenced.

Feargal Quinn, proprietor of SuperQuinn in Ireland, has noted that retail chains succeed in business only by keeping close control of three factors: price, quality and service (Quinn 1990). Surveys have shown that many consumers are relatively dissatisfied with these aspects of retailing so the potential exists for innovators to capitalise on this discontent. Retailers are often concerned that if they spend more on services (such as nutrition promotion) their prices will rise and they will lose market share to low-service price discounters.

Trends in the industry

Several trends are becoming apparent in the industry. First, there is a trend to consolidation of ownership. During the past 20 years there has been a reduction in the number of supermarket chains. In Australia, for example, two major chains — Woolworths and Coles-Myer — dominate retail food spending. This may reduce competition and places extra pressures on manufacturers and suppliers; if a retail company refuses to stock a company's products, then the manufacturer's profitability is threatened. Similarly, the fewer the number of chains, the fewer the opportunities for nutrition promotion. This concentration of ownership has been accompanied by increasing internationalisation of retail chains as they spread around the world.

Second, the trend towards 'vertical integration' is also associated with concentration of ownership. Retailers have either taken over manufacturers and suppliers or, more commonly, they have been able to impose strong contracts on suppliers to force them to meet their demands for particular product standards such as the absence of 'additives' in manufactured products, as well as financial demands for shelf rentals and product promotion. Since retailers have major influence on consumers' nutrition status (Cheadle, Psaty, Curry et al. 1993; Cheadle, Psaty, Diehr et al. 1995), this means that this important part of the food supply is increasingly under the control of fewer companies. They may assist or impede nutrition promotion as they wish.

Third, supermarkets have become larger and stock many more products: 30 000 items per supermarket is not unusual and even 250 000 items can be found in some hypermarkets. This provides a buzzing confusion of very similar products for consumers to choose from — so much so that their ability to choose in a thoughtful manner is probably reduced substantially. This increase in size has often been accomplished by building stores on cheaper land on the outskirts of cities — stores which are in practice unavailable to poorer people who do not own cars.

Fourth, supermarkets provide foods from practically everywhere on earth and in all seasons. For example, strawberries are usually available at different prices throughout the year. This has implications for the economic sustainability of local agriculture and horticulture (since they may be displaced by cheaper imports) and for ecological sustainability, since products are transported without regard for the environmental impact of their transportation. That is, many products consume many 'food miles' in getting to supermarkets—retail prices do not reflect the fossil fuel energy used in growth and distribution (see Chapter 4). All of these trends provide challenges and opportunities for nutrition promotion.

Supermarket shoppers are increasingly interested in health, ethical and environmental issues. For example, the market for 'organic' produce has increased substantially in most Western countries, and surveys of consumers continue to show that they have strong interests in nutrients and nutritional issues (Chapter 3). Now that most Western consumers have lots of food available, they tend to worry more about its quality, its health properties, its ecological impact and so on ('the full stomach syndrome'—Spriegal, personal communication 2001). However, Rothschild's observation that there are three main groups of shoppers is apposite: some don't have the knowledge to choose healthy foods, others don't have enough money, and still others aren't motivated to do so (Rothschild 1999). This encapsulates the main aims of nutrition promotion: transfer knowledge, increase availability and motivate consumers to eat healthily—all of which can be achieved in the retail sector.

In Chapter 3 we saw that food consumers vary in many ways in terms of their life stages, and in their attitudes, interests and opinions. Supermarket managers have to provide an overall service which meets their varied needs and demands. They do so by placing categories of products in different positions in the supermarket (e.g. the dairy section, the greengrocery section, etc.) and by informing customers of their prices and 'special offers' through the use of signs, labels, charts and other devices.

Supermarket nutrition promotion programs: What do they try to change?

Much nutrition promotion in supermarkets has been about the provision of information or about changing shoppers' food purchases (and thus hopefully their families' food consumption). Historically, the emphasis was on reductions in the use of high- (saturated) fat foods, but more recently it has been on the promotion of fruits and vegetables, especially the 5 a Day Program in the United States. While behavioural change is an ultimate goal, retail nutrition promotion can serve several intermediate goals, each of which may have intrinsic value.

Awareness-raising

Customers can be made aware of key nutrition issues—for example, that Brand X products have less salt than other brands, or that information booklets are available for new parents, or that a certain percentage of turnover is given to ecological causes.

Communication of declarative knowledge

Customers can be informed about the nutrient content of products and the value of those nutrients—for example, that some vegetables are good sources of folate, which is valuable in helping to reduce the risk of heart disease and neural tube defects in babies.

Communication of procedural knowledge and skill acquisition

Shoppers can be shown how to select or prepare healthy food via pictures, leaflets, in-store radio or demonstrations. For example, they can be shown how to trim the fat off meat, how to prepare vegetable casseroles or how to reduce the risk of food-borne illness.

Changes in purchasing behaviour

Promotion of particular food products can result in short-term changes in purchasing of those items. For example, price reductions, demonstrations, incentives and promotions can cause the sales of products to increase.

Long-term purchasing

The Giant Foods studies (Levy et al. 1985; Rodgers et al. 1994; Schucker et al. 1992) showed clearly that long-term changes (over two years) can be brought about in the sales of promoted foods and therefore in the likely consumption of those foods.

How are point-of-purchase programs implemented?

Implementation is a crucial issue. It is one thing to decide on project aims and outcomes, but quite another to implement the project. For example, during one statewide nutrition campaign which supplied books to all supermarkets, only about one in four stores actually had the books on display. Information stands may be shifted away from their optimal positions by uninformed staff (and even put into storerooms!). This might be expected since such nutrition promotion trials—though agreed to by management—may impede the day-to-day running of the store—and the wages of the staff are paid for by the selling of merchandise, not through nutrition promotion.

If implementation is to occur in a proper manner, it must be planned well in advance and monitored frequently. Someone has to be responsible for the considerable amount of activity involved. For example, keeping a leaflets stand full and neat takes several minutes of work each day. If this is to be done by the retail company, a considerable amount of authority has to be delegated to someone in the store—it has to be made part of their official duty. To acquire such authority in a large chain requires senior management decisions and constant monitoring of performance. Clearly no outside agency has such authority. (In a small company, however, such problems are less since the owner of the store is usually present and can supervise implementation activities.) The alternative is to do what manufacturers do: hire a company to maintain the stands and displays (a 'detailer'). This costs money. The bigger the company and the greater the number of stores, the more complex the management of a nutrition program becomes, necessitating specialist coordination. In the United States, specialised dietetic companies coordinate and implement retail nutrition programs, producing and maintaining everything from leaflets and shelf stickers to tours and store 'newspapers'.

There are many ways to promote nutrition and healthy eating in supermarkets (see also Box 12.1) Here are some examples:

BOX 12.1 WHAT DOES A GOOD POINT-OF-PURCHASE NUTRITION PROGRAM LOOK LIKE?

- *The aims and targets need to be clear*—Is the aim to provide detailed information about one topic (e.g. cholesterol) or is it to raise awareness about a series of important issues, or is the aim to get people to change their behaviours? We would argue that we should try to give shoppers some perspectives about a few nutrition issues which are relevant to important consumer segments. Thus the need for a balanced varied diet, the roles of fats and fibre and factors involved in body weight control—especially exercise—are perhaps central to the dietary guidelines. Packaging information for particular groups of consumers (e.g. parents of young children) is difficult.

- *Both 'effort reduction' and 'benefit enhancement' approaches are important*—For example, we can help people to read food label information quickly and enjoyably, and we can explain the likely benefits of products. Most importantly, the consumer has to see some advantage in possessing nutrition information.

Box 12.1 continued

- *Consumer market segmentation and consumer (and staff) needs assessment are important* for the viability of any nutrition promotion program. Social and psychological needs (e.g. to be attractive and accepted by peers) are probably more important for message acceptance than nutrition needs. The 'nutrition product' must meet the needs of the consumer. Not all consumers are the same, nor do they all exist in isolation from families and community organisations (like health centres and schools) which can support the communication of messages through retail nutrition. The program should be designed to meet the needs and wants of major segments of shoppers.

- *A focus on negative nutrients like salt, fat and sugar is important,* but so is a focus on *positive nutrients and food groups.* Most programs so far have paid the majority of attention to negative nutrients because these are the items people check out first (will it harm us?), but there is plenty of evidence that many people (particularly women) are very interested in 'positives' like the virtues of iron, calcium and dietary fibre, as well as less tangible phenomena like 'eating in harmony with nature'.

- *A wide range of media is ideal* so that the attention and learning preferences of consumers can be accommodated (e.g. shelf labels, signs, shelf 'talkers', logos, wall charts and posters, trolley posters, recipe cards and leaflets, in-store videos and radio). There is a danger of over-reliance on print media—which can't be read or understood by sizeable proportions of people even in literate countries; pictures, demonstrations and audio media are useful alternatives. Strategies which involve more active participation by shoppers include nutrition information leaflets and brochures, in-store cooking demonstrations, nutrition bingo games and competitions, store guides and supermarket tours conducted by dietitians. Whatever the medium, the messages must be prominent and targeted.

- *From an educational point of view, communications requiring the active participation of shoppers are better than passive communications*—To date, most programs have failed to actively engage the customer in knowledge acquisition or behavioural change. The general reliance on passive brochures and leaflets contrasts markedly with the minimal use of active techniques such as questioning, games, competitions and supermarket tours. This is understandable—the latter are more expensive than the former!

- *Novelty, change and longevity* appear to be key features of program implementation. People soon tire of the same poster or set of leaflets.

- *Long-term commitment by retail companies, food manufacturers and suppliers and health departments is essential*—Most technologies take about ten years to become part of the mainstream, and there is no obvious reason for retail nutrition to differ. Financial resources, staff training and involvement are likely to be substantial.

> *Box 12.1 continued*
>
> - *Evaluation in terms of the processes involved and the outcomes achieved over time is essential to ensure value for money*—This requires the use of sound marketing and educational models in the pursuit of clear, feasible goals. Monitoring of changes in shopping behaviour, consumer awareness, attitudes, knowledge and changes in family consumption behaviours may be required by the retail, manufacturing and government health authorities.
>
> If these principles are followed, the evidence suggests that:
>
> - *Credible information about nutrition will interest customers*—if it is seen as relevant and new and is easy to find, understand and use.
> - *Presentation of information about food products will improve consumer confidence in the retailer and the products*—and it gives the information source a better image. This is likely even if the information is not used by many customers.
> - *POP interventions may even change buying behaviours!* (see Glanz et al. 1992; Glanz and Yaroch 2004; Seymour et al. 2004).

Source: Worsley and Adams (1992).

Shelf labels

Most stores put labels on the edges of shelves, usually underneath products. Usually these provide basic details of the product, such as its price and weight. American supermarkets provide the 'price per ounce' or other unit weight. This makes comparisons between similar brands easier to make. Some promoters have highlighted products' nutritional properties through simple shelf signs or logos indicating 'low fat', 'low salt', 'folate enriched' and so on. This draws attention to particular products.

Charts and signs

Large and colourful charts may be put near products to draw attention to their nutritional properties or to illustrate how to prepare them. For example, in fruit and vegetable sections charts may emphasise their high vitamin content, or in meat sections simple messages may be given about the best ways to cook meats. Very practical information can be given, such as signs which say 'Sugar-free checkout', which help parents avoid last-minute confrontations with their young children.

Leaflets and brochures

Many stores provide information about nutrition content or preparation procedures (often in the form of recipes) (see Figure 12.1). These are typically placed in green-grocery, meat or dairy sections, though they may be found in aisles of the stores in which baby foods and baby products are stocked—such as booklets on weaning foods and procedures. Usually they are provided by manufacturers to promote their branded products; nevertheless, the practice illustrates the major opportunities to disseminate information to people who have specific interests in different parts of the store.

In-store radio and announcements

Most stores play specially adapted music to slow down customers in order to encourage purchasing. These 'programs' are often interrupted with announcements of special offers and other business announcements. Nutrition promotion announce-ments—for example, about the excellent nutrition quality of fruit and vegetables (including the frozen and canned varieties)—can easily be made subject to commer-cial considerations (the airtime is often sold to manufacturers). Many customers do take notice of these announcements.

FIGURE 12.1 EXAMPLE OF A SUPERMARKET NUTRITION COMMUNICATION PROGRAM (A FREE MAGAZINE CIRCULATING TO HALF A MILLION SHOPPERS)

Information kiosks

Computerised information stands enable customers to find the locations and content information of all the foods in the store. This is particularly valuable for customers who suffer from diabetes, coeliac disease and other conditions, since it enables them to find foods which suit them quickly. A more expensive but probably more effective alternative is to have an information kiosk which is staffed by a health worker. Some years ago several of Tesco's larger stores in the United Kingdom employed home economists to answer customer inquiries, to provide in-store cooking demonstrations and to liaise with local schools.

In-store demonstrations

These are commonly used by food suppliers to promote their products in store. In the retailing industry, these are regarded as the most effective ways to launch new products and to communicate with customers. For example, there may be cooking demonstrations about the best ways to cook 'Trim Lamb' or other low-fat meat dishes. Baby food manufacturers like Heinz and Gerber used to conduct cooking demonstrations which were usually well attended. (They have now ceased because of accusations of undue influence over young mothers, though internal company evidence suggests that much of the communication was about basic cooking and baby care procedures with minimal product promotion.)

Incentives, coupons, etc.

Price reductions and coupons have been tried with some success to encourage shoppers to buy healthier foods. For example, the American WIC Program was revised to enable people from low-income backgrounds to purchase fruit and vegetables with coupons. Some stores have combined with health agencies to provide additional services on presentation of coupons earned by buying products from the store. In New Zealand, Cliona Ni Murchu is conducting a randomised control trial in a supermarket chain which services Pacific Island and Maori communities to estimate the relative effects of price reductions and nutrition communication on food purchasing.

Feedback on shopping receipts

Nutritional commentaries printed on shopping receipts can influence shoppers' purchasing patterns in the long term. For example, the fat content of goods purchased may be highlighted along with messages about the health properties of different types of fats (see Ransley et al. 2001 for details of methodology). The main problem is to decide what feedback is to be given. Should it be about saturated fat, total energy, salt or vitamin content, or something else such as 'food miles'? Clearly this form of 'nutrient profiling' feedback has great potential, especially if the shopping is done online, when there is greater opportunity to frame the feedback in attractive ways.

Shopping tours and demonstrations

These are fairly popular. Typically, they are led by a dietitian or a nutritionist. Groups of people who may suffer from a particular disease or condition such as diabetes or obesity will be taken around the store and the benefits of particular food products will be explained. Food Cent$ is an example of an integrated supermarket tour program (Foley and Pollard 1998). Low-income respondents were taken through a three-stage process based on the dietary pyramid. The first stage involved a visit to a supermarket where the members of the groups indicated their usual purchases (e.g. muffins, crisps, chocolate and other products). They were then asked to calculate the cost per kilogram of these products — which is usually very high. The subsequent stages involved the respondents learning how to prepare alternative foods which were tastier and more acceptable to their children than commercial products. The evaluation of Food Cent$ shows that the women quickly learned how to substitute cheaper, healthier products. Long-term follow up over several years shows that the participants continued to build their skills, becoming less reliant on high-priced, high-fat commercial products.

Staff training in nutrition and other lifestyle matters

Such training may be undertaken by companies. The UK Institute for Grocery Distribution, for example, routinely educates store managers in health and ecological issues so they can meet the demands expressed by customers. These range from inquiries about the nutritional content of products, the safety of can openers for older people and the ecological impact of packaging to the development of women's cooperatives in developing countries.

Food product labels on the supermarket's own brands

Because retail chains produce their own food products they are able to put nutrition information on their own labels. For example, Tesco has introduced energy density information on the front of its packs. Food label information is discussed in Chapter 13.

Computer alert and reporting systems

Electronic information systems are being used routinely in some supermarket chains to provide additional services for customers and to maintain the quality and safety of the foods that are offered for sale. For example, in the United Kingdom, Sainsburys uses its loyalty card system to collect information on the food preferences and food sensitivities of its customers. This enables the company to query customers at the checkout about their purchases of foods to which they have previously indicated they are sensitive, such as hazelnuts. It also allows the company to market specific groups of products like wines to customers who have indicated their preference for them — for example, via quarterly newsletters. The potential to provide feedback about the nutritional quality of purchases over a period of time is very great. In addition, particular groups of people like parents of young children could be sent special information about child feeding and rearing, as happens in the Coles Baby Club. The key issue here is the impartiality of the information given to customers; these schemes can degenerate into nothing more than highly targeted product marketing campaigns.

Combined approaches

Of course, nutrition promotion does not have to rely only on one of these approaches; combinations of methods can be used. For example, shelf signs and charts can advertise in-store demonstrations, shopping receipts can feed back information relevant to nutritional themes portrayed in the fruit and vegetable section or in the bread aisle, and informational approaches can be combined with incentive approaches (as Food Cent$ shows). Furthermore, new technologies such as hand-held barcode scanners, 'talking' shopping carts and shopping online are all likely to enhance the opportunities for in-store nutrition promotion.

Novelty and temporal aspects

If the in-store promotion remains static, customers will soon tire and lose interest. In our experience, the uptake of nutrition leaflets and signage lasts for about six weeks. However, customers will notice and attend to new promotions. The larger and more striking the signage, the faster customers will notice. Health-conscious customers (who may be up to a third of supermarket shoppers) will eventually spot relevant information, even if it is small or fairly inconspicuous. This temporal aspect is important, because if novelty and interest can be maintained there seems little reason why nutrition information programs should not last for months and years. The dietary guidelines offer a long-term 'curriculum' for supermarkets, with new themes, facts and suggestions being capable of being introduced at different times (e.g. monthly) throughout the year.

Advantages of point-of-purchase nutrition programs in supermarkets

Supermarket nutrition promotion programs have several major advantages (Pennington et al. 1988; Light et al. 1989), including:

- *Reach* — POP programs have excellent reach into most population segments. Surveys show that most shoppers visit supermarkets over twice a week.
- They are a *site of food purchasing* — Around 80 per cent of all food purchasing decisions are estimated to be made in supermarkets so they are key sites for influencing food purchases.
- They are *representative of the population* — Supermarket shoppers represent households from virtually all segments of society so they can act as major disseminators of nutrition and health messages.
- *Many shoppers are pro-nutrition* — As many as three-quarters of consumers rank nutrition and health information as very important relative to price and convenience in deciding which food products to buy. Nutrition and health considerations are very important within price categories.
- *Consumers like POP nutrition* — Many consumers believe that point of purchase is a suitable place for the provision of nutrition and health information.
- *In-store marketing techniques are effective ways to promote the sale of particular food products*, including healthy food products. There is evidence (see below) that POP programs may change people's purchasing habits, and thus their food consumption.
- *The presence of several enthusiastic retail sectors* (e.g. greengrocers, butchers, pharmacies, some supermarkets) — A number of companies, business organisations and government agencies have vested interests in the success of nutrition promotion — for example, government nutrition policies, national Heart and Anti-Cancer Foundations' programs (e.g. 5 a Day and similar fruit and vegetable promotions, the Australian and New Zealand 'Pick the Tick' programs), and rural industry marketing programs (e.g. Dairy Australia's Three Serves a Day program, the Australian Fruit and Vegetable Coalition).

Problems associated with point-of-purchase nutrition programs in supermarkets

Despite their advantages, supermarkets have several drawbacks as venues for nutrition promotion.

Dislike of shopping

Although 'going shopping' is the most popular recreational activity, 'doing the shopping' is among the least popular of tasks! For most people, shopping is an unpleasant, anxiety-provoking activity. Today's shoppers are more time stressed than their predecessors. More of them have part- or full-time jobs coupled with major family responsibilities and high expectations about their daily lives. Of course, it may be that POP nutrition programs are part of the answer to this problem — they may help make shopping more pleasant.

Distrust

There is widespread distrust of (the information provided by) the food industry, including supermarkets (see Chapter 3). Most consumers believe that the supermarket is set up only to sell as many products to them as possible, irrespective of their quality. Point-of-purchase nutrition programs can provide credibility to super-markets to the extent that they are supported by credible health agencies like National Heart Foundations or state Departments of Health.

Cost

Retail nutrition programs cost money; they have to be financed. Many supermarket managements do not see the provision of nutrition information as being part of their responsibility, particularly as profit margins tend to be low overall. Although profits may be high in some categories such as fruit and vegetables, these are offset by lesser profits in other areas. Sound financial arguments need to be advanced to gain the cooperation of supermarkets. Large retailers do not have unlimited budgets for new innovations, especially those that are unlikely to make additional profits — such as nutrition promotion programs. Large chains tend to be conservative and stay with retailing strategies which they know will work. However, some large British and American chains such as Tesco and Giant Foods have undertaken nutrition promotion as part of their general service provision. These high levels of service distinguish them from their competitors.

Company culture

Large supermarket chains tend to be highly bureaucratic and they can be difficult to work with, especially if the activities consume labour and financial resources. Supermarket chains have their own corporate cultures which may impede the activities of nutrition promoters who are likely to be reminded often that the company's business is making money, not health promotion. Small independently owned retail outlets are often easier for nutrition promoters to work with. Their management often has clearer and tighter control over its staff and events—if things are not proceeding according to plan, it is soon seen and can usually be rectified. In contrast, in the large chains there are 'house rules' about how to do things and the management is multi-faceted and often distant from the staff on the shop floor. Thus agreements made with senior management (say, at head office in the state or nationally) do not necessarily reach down to the managements of individual supermarkets. Independent outlets such as butchers, greengrocers and independently owned supermarkets face considerable competition from the large chains and often are well motivated to participate in any scheme which is likely to entice customers. They also have considerable population 'reach'—for example, in Australia approximately half of shoppers buy their meat and greengroceries from independent outlets.

Positive attitudes towards nutrition promotion on the part of senior management are essential (they must 'believe' in nutrition promotion). Such positive attitudes are usually related to the company's attitude to the provision of 'service'.

Isolation

Retailing is quite separate from other aspects of community life. In particular, health promotion and retailing are quite separate things in many people's minds. Thus health workers know little or nothing about the organisation and working of stores, and retailers—like most people—know little of the specialised health disciplines such as nutrition. This is despite the fact that they sell products which are loaded with nutrients! Similarly, the education and community welfare systems barely overlap with food retailing despite sharing knowledge, skills and needs which could be of use to each other. For example, a major complaint of adult customers is that they do not understand food labels. It is likely that in local schools children are being taught knowledge and skills about consumer mathematics which could be of great benefit to the supermarkets' customers—if only there was some communication between the two sets of institutions.

At present, people working in different local institutions tend to have blinkered vision about the purposes of their own and other workplaces. If nutrition promotion programs are to be successful, the barriers between the health, education, welfare and retail sectors will have to be broken down because consumers' needs span all these areas. It is especially important to integrate supermarket nutrition promotion with these sectors at the *local level*, if only because successful programs require a lot of resources and labour which supermarkets cannot provide. This, of course, is a major argument for local community nutrition promotion.

Independence and ownership of nutrition promotion

Is the nutrition promotion program a part of a health promotion network or an arm of the company? Generally, most retail nutrition projects are conducted for and often by companies. However, there are several potential advantages if a local health network conducts a nutrition promotion program. First, local resources such as the services of dietitians and health centres are more easily available. Second, the program is more easily perceived as being for the good of the community — and not as a company or industry promotion. Third, a wider array of material resources and volunteer labour may be available. The downside is that local community health networks are likely to be considerably over-stretched before any such program is initiated and the resources (such as they are) may be difficult to manage — particularly within the regimen of a large chain. However, community involvement may be more feasible if national organisations such as government Health Departments, community and industry organisations (e.g. Heart and Cancer Foundations, dietitians' associations, fruit and vegetable coalitions, rural industry and retailer associations) could support the preparation of materials and campaigns. An interesting example of a cross-sectoral collaboration is the Heart Foundation of Australia's food service Tick program, conducted with McDonald's (see Box 12.2).

BOX 12.2 MCDONALD'S GETS THE TICK

The Australian Heart Foundation decided in 2007 to award its Tick logo to McDonald's foods which are low in energy, salt and saturated fats. This has been controversial in the nutrition community because McDonald's is seen to epitomise all that is unhealthy about fast food. The Heart Foundation included this food service program because one in three food dollars is spent on eating out of home in Australia. Some 4.8 billion meals and snacks are eaten out each year, and one-third of Australians eat out every day. Unlike foods purchased in the supermarket, food served in restaurants has no nutrition information panel or ingredients list, so there is no way of knowing exactly what you are eating.

It took McDonald's more than twelve months of reformulation and taste testing to meet the Heart Foundation's nutrition standards. Only when the company met standards covering nutrition, promotion and advertising, food safety, staff training and production were they able to earn the Tick. Twice a week, every week of the year, a McDonald's store somewhere in Australia is randomly audited.

Initial effects of the program included the following:

- In the six months after the launch of the Tick in McDonald's, 12.9 tonnes of sodium were removed from the food supply. This is a result of McDonald's Australia reducing sodium in

---

Box 12.2 continued

all its bread rolls by 43 per cent to meet strict Tick standards and making the changes across the entire Deli Choice Roll range, so even those selecting non-Tick rolls benefit.

- The introduction of the choice of multi-grain Deli Choice Rolls added 25 tonnes of fibre to Australian tables in just three months.
- Just by cooking its McChicken patty and Fillet-o-Fish portions in the virtually trans-fat free oil required by Tick-approved meals, McDonald's Australia removed 18.7 tonnes of trans-fat from the food supply.
- Almost one in five McDonald's customers (19 per cent) reported that they switched to a Tick meal—48 per cent of customers switched from a non-Tick meal to a Tick meal at point of sale, while 29 per cent intending to buy a Tick meal bought a non-Tick meal (mostly 15–24-year-olds).

Source: Heart Foundation of Australia (2008).

Diffusion of 'expert' knowledge

It is common sense to suppose that if consumers have sound information about food and health issues, they should be able to make better decisions about them. However, there is a major problem in the dissemination of information which can be assimilated by people with little or no prior knowledge of nutrition who usually have plenty of more pressing things to do. The process of translating information and knowledge held by experts so that it can help consumers in their daily lives is often difficult. For example, many people are interested in ways of avoiding cancer, and nutritionists know that high intakes of fruit and vegetables can have marked preventative effects. What should nutritionists tell people about this 'story'? The problem is that nutritionists assume a lot of prior knowledge (about vegetable content and carcinogenesis) which most lay people simply do not know. The result is that draft messages have to be written in very simple terms which nutritionists find miss out 'crucial' bits of information. Indeed nutritionists may differ substantially in their versions of the 'truth'. The production of information materials takes a lot of time and requires checking and rechecking among both experts and lay people.

A related problem concerns the narrow focus of many nutritionists. For example, the dietary guidelines are summaries of nutritionists' views of crucial information about food which is important for the general population. They take little account of the views of consumers. In one study we found that almost half of shoppers were thinking of the possible dangers of pesticide sprays when they bought vegetables. This is not mentioned in the dietary guidelines, nor are quality issues (like concerns about adulteration of foods). This illustrates the partisan nature of expert knowledge—it is organised in discrete pieces which are not intended to fit together. In contrast, the concerned parent, for example, has to try to fit all the bits into a jigsaw in order to safeguard the health of their child. Nutrition promotion needs to relate to broad issues which meet the daily concerns of consumers about food and health as well as those of 'experts'.

The effectiveness of supermarket point-of-purchase interventions

Glanz and Yaroch (2004) found that four types of intervention strategies—the provision of point-of-purchase information; increased availability, variety and convenience; reduced prices and coupons; and promotion and advertising—were effective ways to influence eating behaviour. Similarly, Seymour et al. (2004) also found a variety of positive effects. Ten intervention studies met their stringent design and evaluative criteria. All of the interventions promoted specific food products and all of them reported sales figures, though two also reported dietary consumption data. The durations of the interventions ranged from one week to two years in the case of the large multi-store Giant Foods study.

Five of the eight information-only studies found increased sales of the target foods but it was unclear whether any particular informational method was superior to others. Kristal et al. (1997) and Curhan (1974), in contrast, used financial incentives. Kristal et al. (1987) provided produce coupons in information leaflets (about fruit and vegetables) at weekly intervals. They found that 36 per cent of customers used them, but no changes in fruit and vegetable sales or in dietary intake were observed. Curhan's study was more complex, consisting of several interventions. When prices of produce were reduced by a minimum of 10 per cent only, the sales of soft fruit increased (banana sales increased by 18 per cent). When produce display space was increased to at least 200 per cent of the original allocation, sales of produce increased substantially (between 28 and 59 per cent, depending on type of produce). In the third intervention, produce was placed in high-traffic areas of the supermarket and sales of hard fruit like apples appear to have increased.

Curhan's interventions lasted for only one week each, during a seven-month period (Curhan 1974). However, the three most effective studies in terms of increases in sales figures were also the longest in duration (two years) conducted in

Giant Foods supermarkets (Levy et al. 1985; Rodgers et al. 1994; Schucker et al. 1992). They were conducted as part of a multi-component study in Giant Foods.

Critique of interventions

There is little doubt that POP interventions in supermarkets can work in that they can influence sales figures. Combined with the association of supermarket food availability and population food consumption (Cheadle, Psaty, Curry et al. 1993; Cheadle, Psaty, Diehr et al. 1995), it is clear that these micro-environments can be influential at least in the short term and probably in the long term. However, the evaluation of supermarket nutrition programs is not straightforward. For example, sales figures might show that sales of fresh broccoli may have increased, sales of frozen broccoli decreased and sales of cheese (containing saturated fats) may have increased. Would we regard such changes as nutritionally positive, neutral or negative? Comprehensive, wide-ranging evaluation methods are required in order to answer questions about the nature of changes resulting from programs, such as whether the overall quality of consumers' purchasing improved; whether the changes occurred mainly in the supermarket or at home afterwards; and whether vulnerable groups (like people with diabetes) were more affected than others. More fundamentally, the feasibility and choice of intended program outcomes, the definition of 'success' and the longevity of program effects require particular attention.

Supermarket nutrition promotion faces a number of key practical problems, including the following:

- What should be in the ideal supermarket program—what is the 'curriculum'?
- What sort of food and nutrition issues should be promoted in supermarkets and which stakeholders should influence the choice of issues?
- Most retail staff are untrained in food and health issues—who is responsible for training them?
- Supermarkets improve the availability of foods but are often absent in poor areas. Can government assist with the siting of supermarkets in these areas?
- Disabled and elderly people often experience difficulties in accessing supermarkets or shopping in them. Should governments and retail companies collaborate to make foods more accessible?

Supermarket chains are large, powerful organisations. The chequered history of small-scale interventions with individual supermarkets suggests that strong government policies are required if the potential of retail nutrition promotion is to be realised. Such national and regional policies would integrate promotion activities in the retail sector with those in other settings so that the normal systemic activities of supermarkets support population nutrition. Meanwhile, consumer organisations act as strong advocacy groups for change. Box 12.3 lists the health indicators which the British National Consumer Council uses to make annual public comparisons of supermarket chains.

Food product label information

One of the most important channels of information for the supermarket shopper is the product label. This contains an ingredients list, a nutrition information panel and increasingly, especially in the United States, a nutrition claim (e.g. salt-reduced) — possibly a health claim which suggests the health benefits associated with prolonged use of the product. In addition, other information on the label includes the pre-scribed name of the product, the use-by date, country of origin, the name and

---

BOX 12.3 THE NATIONAL CONSUMER COUNCIL'S HEALTH INDICATORS

Nutritional content
Salt content of ten everyday own-label processed foods

Labelling information
Use of front-of-pack, colour-coded signpost labelling in line with Food Standards Agency criteria
Use of guideline daily amounts (GDAs) on nutrition labelling on back of pack

In-store promotions
'Healthy' foods (fruit and vegetables) as a proportion of in-store price promotions
The presence of sweets and fewer 'healthy' snacks at the checkout

Customer information and advice
Information and advice on healthy eating available in store through leaflets and magazines, and via retailers' national telephone helplines and websites.

Source: National Consumer Council UK; see also Rayner et al. (2005).

address of the producer/supplier, storage instructions and details of its weight and/or volume. Regulatory authorities in most countries (e.g. the Food and Drug Administration in the United States, FSANZ in Australia and New Zealand) tightly control the information on the labels through a series of regulations.

Nutritionally relevant parts of the product label include the ingredients panel (see Figure 12.2), the nutrition information panel (Figures 12.3, 12.4 and 12.5), health claims (Box 12.4) and 'front-of-label information such as 'Traffic lights' (Figure 12.6) or daily intake guides (Box 12.5). It is a great deal of information which can lead to a lot of visual clutter.

FIGURE 12.2 THE INGREDIENTS PANEL FOR A WHITE BREAD PRODUCT

```
Ingredients list
Bread flour, baker's yeast
Gluten, salt
Vegetable oil, soya flour
Sugar, emulsifiers (481, 471, 472e)
Preservative (282)
Enzyme (amylase), water added
```

FIGURE 12.3 THE CURRENT AUSTRALIA NEW ZEALAND NUTRITION INFORMATION PANEL

NUTRITION INFORMATION

Servings per package: 3
Serving size: 150g

	Quantity per serving	Quantity per 100 g
Energy	608 kJ	405 kJ
Protein	4.2 g	2.8 g
Fat, total	7.4 g	4.9 g
– saturated	4.5 g	3.0 g
Carbohydrate, total	18.6 g	12.4 g
– sugars	18.6 g	12.4 g
Sodium	90 mg	60 mg
Calcium	300 mg (38%)*	200 mg

* Percentage of recommended dietary intake

Ingredients: Whole milk, concentrate skim milk, sugar, strawberries (9%), gelatine, culture, thickener (1442).

FIGURE 12.4 CURRENT AMERICAN NUTRITION INFORMATION PANEL

Sample label for
Macaroni and cheese

① Start here ➡

Nutrition facts

Serving size: 1 cup (228 g)
Servings per container: 2

② Check calories

Amount per serving

Calories 250   Calories from fat 110

③ Limit these nutrients

	% Daily value*
Total fat 12 g	18%
Saturated fat 3 g	15%
Trans fat 3 g	
Cholesterol 30 mg	10%
Sodium 470 mg	20%
Total carbohydrates 31 g	10%

⑥ Quick guide to % DV

Dietary fibre 0g	0%
Sugars 5 g	
Protein 5 g	
Vitamin A	4%
Vitamin C	2%
Calcium	20%
Iron	4%

• **5% or less is low**

• **20% or more is high**

④ Get enough of these nutrients

⑤ Footnote

* Per cent daily values are based on a 2000 calorie diet.
Your daily values may be higher or lower depending on
your calorie needs.

	Calories:	2000	2500
Total fat	Less than	65 g	80 g
Sat fat	Less than	20 g	25 g
Cholesterol	Less than	300 mg	300 mg
Sodium	Less than	2400 mg	2400 mg
Total carbohydrate		300 g	375 g
Dietary fibre		25 g	30 g

Source: www.fns.usda.gov/tn/parents/nutritionlabel.html.

FIGURE 12.5 EXAMPLE OF A EUROPEAN NUTRITION LABEL

INGREDIENTS
Wheat Flour, Rye Flour, Water, Wheat Bran, Vegetable Fat, Sugar (Sucrose), Salt, Bakers Yeast. pH-controlling agents: Vinegar, Sour Dough.

STORAGE INSTRUCTIONS
Store in a dry, dark place.
For 'Best Before End' see side of pack.

NUTRITION INFORMATION

Typical Analysis	Per 100 g	Per slice (8.4 g)
Energy	1527 kJ	128 kJ
	(365 kcal)	(30 kcal)
Protein	11.4 g	1.0 g
Carbohydrate	65.0g	5.4 g
(of which sugars)	2.6 g	0.2 g
Fat	6.0 g	0.6 g
(of which saturates)	1.0 g	trace
Dietary Fibre	7.5 g	0.6 g
Sodium	0.7 g	trace

18 slices per box

FIGURE 12.6 AN EXAMPLE OF THE UK FOOD TRAFFIC LIGHT SYSTEM

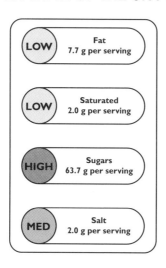

LOW Fat 7.7 g per serving

LOW Saturated 2.0 g per serving

HIGH Sugars 63.7 g per serving

MED Salt 2.0 g per serving

With traffic light colours, you can see at a glance if the food has high, medium or low amounts of fat, saturated fat, sugars and salt in 100 g of the food. In addition to the traffic light colours, you can also see the amount of these nutrients that are present in a portion or serving of the food.

Red = High
Amber = Medium
Green = low

Note: Traffic light labelling is a voluntary scheme developed by the UK Food Standards Agency. The lower and upper boundaries for the moderate (amber 'traffic light') category were set in 2006 by the UK Food Standards Agency. *Signposting*, http://www.food.gov.uk/foodlabelling/signposting/.

BOX 12.4 WHAT ARE HEALTH CLAIMS?

What is a claim?

A claim means any statement, representation, information, design words or reference in relation to a food that is not mandatory in the Food Standards Code.

Types of claims

A content claim is a statement regarding the amount of a nutrient, energy or a biologically active substance in the food. Health claims are claims that describe a relationship between the consumption of a food or constituent and particular benefits of the food in relation to health. Health claims are further divided into high-level health claims and general-level health claims. The categorisation of the claim is based on the extent to which the claim identifies particular benefit(s) for consumers in consuming that food.

General-level claims are those which:

- describe or indicate the presence or absence of a component in food (nutrient content claim)—e.g. 'This food is high in calcium';
- refer to the maintenance of good health—e.g. 'Helps keep you regular as part of a high-fibre diet';
- describe a component and its function in the body—e.g. 'Calcium is good for strong bones and teeth';
- refer to specific benefits for performance and well-being in relation to foods—e.g. 'gives you energy';
- describe how a diet, food or component can modify a function beyond its role in normal growth and development—e.g. 'exercise and a diet high in calcium help build stronger bones';
- refer to the potential for a food or component to assist in reducing the risk of, or helping to control, a non-serious disease or condition—e.g. 'Yoghurt high in X and Y as part of a healthy diet may reduce your risk of stomach upsets'.

High-level claims

High-level health claims reference a serious disease or condition, or a biomarker of a serious disease or condition. They include:

- claims that refer to the potential for a food or component to assist in controlling a serious disease or condition by either reducing risk factors or improving health—e.g. 'This food is high in calcium. Diets high in calcium may increase bone mineral density';
- claims that refer to the potential for a food or component to assist in reducing the risk of a serious disease or condition—e.g. 'This food is low in sodium. Diets low in sodium may reduce risk of elevated blood pressure.'

Regulation of claims

The level of a claim determines how the claim is regulated, including the evidence required for substantiation.

Source: Adapted from: Food Standards Australia and New Zealand, www.foodstandards.gov.au/ newsroom/factsheets/factsheets2006/nutritionhealthandre3396.cfm.

BOX 12.5 THE DAILY INTAKE GUIDE

This guide is intended to help consumers to understand information about the amount of energy and nutrients a product contains and how much a serve contributes towards their daily requirements. Food companies are already required by law to provide information to consumers about the nutrient content (protein, carbohydrate, sugars, fat, saturated fat and sodium) and energy of every serve of a product. The Daily Intake Guide provides additional information in a simple thumbnail presentation. Consumers can easily see information about the composition of the product and its relevance to their diet. The new labels work like this:

FIGURE 12.7 GUIDELINE OF DAILY AMOUNT ENERGY DENSITY SYSTEM, NUTRIGRAIN

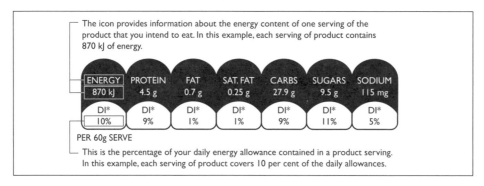

For example, what consumers know by reading the nutrition information panel is that by eating a 60 gram serve of a particular product they will be consuming 870 kilojoules. Daily intake labelling tells them that eating those same 870 kilojoules represents 10 per cent of their daily energy needs.

Source: Australian Food and Grocery Council, www.mydailyintake.net.

Purposes of food product labels

Food labels have several purposes: to inform, educate, advise or warn consumers (e.g. 'not to be consumed by people suffering from phenylketonuria'); to enable customers to compare products (e.g. product X contains less fat than product Y); to facilitate fair trading (e.g. that the tomato sauce really does contain lots of tomato); to trigger or reinforce behaviours (e.g. cook on high for 90 seconds); and perhaps to reassure (brand X is wholesome or is 97 per cent fat free—though the term is ambiguous). Two points are immediately obvious. First, a lot of prior knowledge is required to understand labels—for example, what exactly is an appropriate level of salt for a can of baked beans? Second, this is a big load for a little piece of paper! With the advent of functional foods and compulsory nutrition information panels, the role of the label is changing. Manufacturers and regulators are placing increasing onus on consumers to read labels carefully.

Consumers' use of labels

Study after study has shown that shoppers attend to nutrition and other information on the food product label (Synovate 2005; FSANZ 2003). Most use nutrient or ingredient information 'often' or 'sometimes'. For example, about one-third of women shoppers look at the nutrient panel during any shopping trip (Worsley 1996) and about two-thirds of shoppers report reading nutrient or ingredient information 'usually' or 'sometimes' before buying products (Synovate 2005; FSANZ 2003; Worsley 1996). About half of shoppers interviewed in one study claimed to have *calculated* the nutrient content of a product recently (Scott and Worsley 1994; Worsley 1996). Over 80 per cent claim that they compare products by comparing the nutrient content on labels.

However, consumers' understanding of label information is often vague and superficial (Raynor et al. 1992). Educated consumers tend to be more confident about their ability to use labels but they do not perform better than less-educated people (Raynor et al. 1992). Most consumers find numeric information difficult to use—for example, a survey of Sydney shoppers showed that one-third wanted a healthiness symbol only, instead of numeric information about nutrient contents. Most cannot judge 'servings' or technical jargon (Raynor et al. 1992); even percentage signs confuse many shoppers. In part, this is because of what the UK Food Standards Agency calls the 'gobbledegook' (information overload) on labels (www.foodstandards.gov.uk/foodlabelling).

Grunert and Wills (2007) reviewed the previous three years of nutrition label information research in Europe. They identified over 50 studies. Very few of them were based on any theory, few involved any academic experts and almost none of them convincingly demonstrated any effects on consumer purchasing or other consumer behaviours. They called for more studies which were theory based and which measured behavioural outcomes or observations of shopping choice behaviours. Clearly, many of today's consumers refer to label information but evidence of their utility for them is lacking.

Consumer wants

Consumers are often interested in details of 'negative' food components such as 'chemicals', salt, sugar, MSG and fat. Most think the use-by date, additive details, ingredients list, storage instructions and health claims are important. Most shoppers want information about the benefits or threats to health, such as indications of additive and preservative contents, cholesterol, total fats and salt calorie/energy content of foods and the percentage of ingredients by weight (Crawford and Baghurst 1990; Crowe et al. 1992; Lewis and Yetley 1992; Worsley 1996). Many want 'authorised' health claims similar to those provided by the Federal Drugs Administration in the United States (Crawford and Baghurst 1991). Above all, consumers want simple clear information about a range of topics such as safety, nutrition and health (Worsley 1996; Scott and Worsley 1997; Food Marketing Institute 1992), fairness (e.g. is the food from a 'cheap labour' country—Worsley and Scott 2000) and environmental aspects (Worsley 1996). This raises issues about the ideal format of the label—should it conform mainly to consumers' wishes or those of scientists and industry players? *Who should decide about label content and format?*

There is much scope for negotiation between consumers' 'felt needs' and 'expert-defined needs'. If food labels are to be truly effective in helping consumers choose healthier foods, then further research into consumers' perceptions of food and nutrition is required along with consumer education programs which explain how to use them.

Human information processing and the food product label

Human information processing ability is limited: our short-term memory lasts for only a few seconds and we organise information into 'chunks'—normally we can only handle about '7+/–2' chunks at any time (Miller 1957). To overcome these limitations, we use various psychological devices. For example, we may process information in a hierarchical manner, attending to the most important information first such as information about safety and 'negative' food components which may harm us (humans tend to be risk-averse). Or we can transform information to make it simpler, transforming numeric information such as the number of mg of vitamins or daily reference values into evaluative information like 'high', 'low' or, even better, 'good' or 'bad'. Such transformation is hard work for many people and most shoppers don't have the time or are unwilling to make the effort. Ideally, this work should be done for them, as is the case with the 'Traffic Lights' and logo systems. Alas, scientists such as nutritionists find it difficult to evaluate the amounts of nutrients as 'good' or 'bad' or even as 'some' or 'a lot'.

Logos

Human information processing limitations may explain why the use of logos like the Tick is so popular. If a product has a logo, then it is probably going to deliver some health benefit—indeed, when questioned, most consumers can explain the purpose of the Tick logo quite well. While nutrition purists do not like such logos, they may be preferable to the rough rules of thumb (or 'heuristics') that shoppers inevitably make up for themselves. An example of a consumer heuristic is reliance on the *length* of the nutrition information panel as a guide to its nutritional quality! In one study, the product with a longer nutritional information panel outsold its identical companion product which had a shorter nutritional information panel. Young and Swinburn (2002) have shown that the New Zealand Pick the Tick (PTT) logo program has been responsible for the removal of a substantial amount of salt from the food supply. The logo encouraged manufacturers to reduce the amount of salt in their products in order to gain the PTT logo.

Interesting variations of logos are the N number scheme proposed by Wheelock (1992) and the Food Traffic Light scheme proposed by the UK Food Standards Agency. Wheelock proposed that processed foods should be awarded a single number according to the number of nutrients they contain above trace amounts. Thus the higher the number, the more healthy the product is likely to be. The UK Food Traffic Light logo will award a red, amber or green-coloured circle to products according to their fat, salt and sugar contents (www.eatwell.gov.uk/foodlabels/trafficlights). This follows extensive consumer testing of this and related signposting schema ('guideline daily amounts'—Synovate 2005). An even simpler device used by some manufacturers is to highlight (and so draw attention to) the amount of a key nutrient (e.g. sodium) or energy content in the nutrition information panel.

The problem with logos and other qualitative devices is that someone has to decide what are 'good' or 'bad' or 'high' or 'low' amounts of nutrients. Governments and scientists are loathe to do this as such judgments can be quite arbitrary. However, consumers' health could well be fostered by illustrative devices which enable them to shop quickly and healthily, even if they have some degree of arbitrariness. *Nutrient profiling* is a set of techniques which can be used to arrive at these broad categorisations. Scarborough et al. (2007) define nutrient profiling as 'the science of categorising foods according to their nutritional composition'.

Information processing theory (Payne et al. 1992) provides some guidelines for point-of-purchase nutrition promotion. The main aim should be to reduce the information processing load put on the customer. This is contrary to the regulatory trend for the provision of ever more detailed information on the label. We should reduce the complexity of information provided, use evaluative formats as much as possible, list high-risk information (usually safety information) prominently, standardise formats and layouts, use large print, pictures, shapes and colours, and try to reduce visual and auditory 'pollution' from labels (and in retail shops). The extent to which this ideal situation is not achieved is the result of the 'politics' of food and the actions of industry and regulatory stakeholders whose goals often differ from those of shoppers.

Nutrition promotion in restaurants and other retail food service outlets

There are many types of retail food outlets which are increasingly being used by consumers—for example, cafes, restaurants, fast food chains, work cafeteria, transport outlets, street foods and vending machines in public places (see Box 12.6). For the purposes of brevity, all these outlets will be referred to as 'restaurants', following Glanz and Mullis (1988).

Generally, restaurant servings are larger and they tend to contain greater amounts of fat than meals prepared at home (Young and Nestle 2002). Partly in response to this situation, the National Heart Foundation of Australia announced the extension of its Pick the Tick program to food services in 2006 (www.national heartfoundation.com.au).

There are many opportunities to promote healthier choices in restaurants. The methods used are similar to those in other areas of nutrition promotion. People may be given information—for example, heart health information on menus and on posters over serving counters or on table 'tents'—or they may be given incentives in the form of redeemable coupons (if you collect healthy eaten tokens you may redeem them for healthy choices). In addition, the food service may be changed to provide more extensive ranges of fruit and vegetable dishes, or the prices of food may be altered. For example, vegetables and fruit may be served free with every meal, or vegetable and fruit dishes may be priced to be cheaper than less-healthy, high-energy dishes. There are lots of ways of influencing customers—information, incentives, food supply and the food service environment—and there are some good examples of their use.

Universities and health agencies provide some early examples—for example, the Stanford Heart Disease Prevention Program pioneered healthy menu information in the 1970s. In the 1980s, the Australian Hotels Association in conjunction with the Victorian (Australia) Food and Nutrition Policy set up an impressive program which aimed to prevent smoking in pubs and provide a healthier range of foods. The demise of the policy in the early 1990s, with the withdrawal of state government funding, saw the cessation of this program.

Heartbeat Wales in the mid-1980s developed a restaurant program which aimed to provide a non-smoking area in which lower fat foods could be prepared in hygienic conditions. Restaurants that complied with the terms of the program received Heartbeat awards which they displayed prominently to attract customers. Heartbeat was later extended to the rest of the United Kingdom and to New Zealand, where it was supported by the New Zealand Heart Foundation. Evidence from the United Kingdom and New Zealand (Price et al. 2000) suggests that the Heartbeat awards helped to improved caterers' knowledge of nutrition and encouraged the provision of healthier food choices.

In New Zealand, a development of the Heartbeat program resulted from examination of the quality of fats and cooking methods used by fish and chip shop

---

BOX 12.6 OTHER NUTRITION-RELEVANT RETAIL CHANNELS

- *Fast food chains*—Sell a limited range of predominantly high-energy products (chips, burgers and soft drinks). Some have extended their product mix to cater for the health conscious market (e.g. McDonald's Salads Plus).
- *Vending machines*—Found in transport interchanges, schools and at worksites. Predominantly stock high-energy foods and beverages. Several studies show that low-energy alternatives sell well if they are identified and lower priced.
- *Worksite and school cafeterias*—Often sell high-energy fast foods. School cafeterias were recently targeted by government-inspired bans on highest energy products—especially confectionery and soft drinks (United Kingdom, New South Wales, Victoria).
- *Pharmacies*—Sell nutrient and herbal supplements, not always conducive to healthy eating. Major point of contact for consumers worried about their health.
- *Book stores*—Sell books about nutrition, many of which are not evidence-based or down-right quackery. Ditto popular magazines and many internet sites.
- *Weight loss companies*—Most follow a self-regulatory code of good practice—some promote moderate diet change and healthy eating. Largely unevaluated.

proprietors. A subsequent program (Morley et al. 2002) provided training for these businesspeople in the selection and use of non-saturated fats and in the use of high-temperature cooking methods which reduce the absorption of fats by potatoes. Evaluation of the program suggests that use of saturated fat in this sector has declined (Morley et al. 2002). This program has recently been extended to the state of Victoria in Australia.

Several catering companies attempt to promote nutritious meals when financially feasible. A good example is Sodexho, a French-based company which is one of the largest food service companies in the world. The company endorses a variety of 'healthy eating' menus along with the view that eating should be a social, relaxing occasion. Examples of its operations show that healthy food choices can be popular and relatively inexpensive.

The effectiveness of restaurant nutrition promotion

Restaurant nutrition promotion can have several outcomes which can be difficult to evaluate. For example, restaurant patrons may like novel offerings, copy them at home but not order them again from the restaurant. Probably the main consideration here is whether there can be increases in the sales of fruit and vegetable dishes and other healthy foods with accompanying decreases in the sales of dishes which contain large amounts of energy, saturated fats, salt and sugars. To date, realisation of this ambitious aim has not been demonstrated, but there is little doubt that nutrition promotion within individual restaurants and chains can alter sales of products like fruit and vegetables (Seymour et al. 2004).

Most of the restaurant interventions reviewed by Seymour et al. (2004) showed statistically significant increases in the sales of target menu items. For example, in Albright et al.'s (1990) study, when a red heart logo was used to indicate that certain entrées were 'good for health' (defined by the investigators as low in fat and cholesterol), sales increased in two of the four restaurants in the study. Similar increases occurred when Colby et al. (1987) promoted low-salt, fat and cholesterol lunch specials using 'healthfulness', 'flavour' or 'daily special' messages. Again when low-fat sandwiches and dishes were marked as 'Good for you' in Target Discount Store cafeterias, sales of the items increased (Eldridge et al. 1997). The problem in comparing these types of studies is that they have different end-points (e.g. increase in sales of fruits and vegetables, or 'heart healthy' items) and they use combinations of different promotion methods. Overall, however, 'simply providing information in the restaurant setting appears to be associated with increased purchase of targeted items, suggesting that the specific information strategies were not as important as the act of intervening' (Seymour et al. 2004).

Conclusions

This chapter has shown that retail nutrition promotion is a promising area. However, it is also clear that simple, one-off behavioural interventions, while generally effective, have not been sustained or had demonstrated effects on population health. Multi-sectoral approaches involving the cooperation of local and national governments and promotion efforts in other settings, especially in the local community, are more likely to have prolonged, substantial effects. The role of government in setting and supporting broader macro-environmental strategies and nutrition policies which retail establishments can use to change their operations in sustainable ways is probably essential if real progress is to be made in this area.

Discussion questions

12.1 Describe the health-promoting supermarket of the future.

12.2 Design a supermarket tour for diabetics. Which parts of the supermarket would you stop at and examine?

12.3 What are the advantages and disadvantages of point-of-sale programs in supermarkets?

12.4 Design a retail food and nutrition information system for parents of young children.

12.5 How would you combine nutrition promotion in community settings with a retail nutrition promotion program? What would be its likely advantages and disadvantages?

12.6 Design the perfect food label for a breakfast cereal.

13

Nutrition communication in the media

Introduction

We live a media-rich society. Thanks to the wonders of the printing press, copper wire, optic fibres, computers and satellites we are completely surrounded by and bathed in communication networks. These include the mass media ('broad' media) such as radio, TV, newspapers and magazines as well as other ('narrow') media such as leaflets in doctors' surgeries or in supermarkets, pamphlets and books. The range of applications of the internet can be both 'narrow' (e.g. sites for bird-watchers) or broad (e.g. YouTube). All this exists in addition to the constant chatter and communication of our social acquaintances, who increasingly talk to us through mobile phones, video cameras, SMS messages and other channels. The mass media in particular are all-pervasive and help to shape our perceptions of the world and other humans. They reflect, in a somewhat distorted manner, prevailing views of food and nutrition. All these channels of communication present nutrition promotion with great opportunities, but also threats (e.g. the promotion of nutrition quackery).

Mass communication theories

Mass communication is a pervasive feature of global society. The mass media have become central to everyday life because they can communicate information across wide sections of an increasingly fragmented and specialised society. The lives of most individuals depend on the activities of many people around the world (e.g. the workers in China who make clothes for much of the world's population) and so they need communications about what is happening in the world (deFleur and Ball-Rokeach 1989). Communications technologies allow us access to more information about the world than ever before.

Mass communication plays several roles, including:

- the portrayal of reality (which is often a distorted picture—Pollay 1986, 1987);
- the provision of information and different points of view (e.g. in documentaries);
- entertainment; and
- persuasion to behave in certain ways—most notably, to purchase products and services.

The communication of information and the persuasive uses of the mass media are particularly relevant to nutrition promotion. The mass media are often blamed for encouraging poor health and obesity through their broadcasting of advertising for high-energy foods, fast cars and general hedonism and risk-taking. However, it is also clear that they can provide life-changing information and vivid examples of healthy living which can be utilised by health and nutrition promoters. The challenge for public health is to ensure that the media are part of the solution of health problems rather than one of the causes of ill-health. To this end, it is worth examining some basic communication models.

The basic model: Face-to-face communications

According to classical communications theory, the most basic communication process occurs between two actors, the communicator and the recipient (McCron and Budd 1979, 1981; deFleur and Ball-Rokeach 1989). The communicator encodes signs and symbols which they expect will elicit a desired response in the recipient (e.g. to educate, inform, persuade, entertain). This is the message. The attending recipient extracts the information that the communicator put in ('decodes' the message—see Figure 13.1). 'The communication act is about the conveying of meaning.' (McCron and Budd 1979, 1981) The signs and symbols used in communication have no meaning in themselves, except that which has become associated with them through education, socialisation and cultural learning. The study of signs and symbols is called *hermaneutics*.

FIGURE 13.1 SIMPLE MODEL OF THE BASIC COMMUNICATION PROCESS

Communicator →	**Message** →	**Recipient**
encodes		decodes
puts meaning into		draws meaning from
signs and symbols		signs and symbols

The model in Figure 13.1 looks simple, but problems in encoding and decoding do occur. Both communicators and recipients must share the same vocabulary and must attach similar meanings to the symbols being used. For example, nutritionists and lay people may imply different meanings to terms such as 'energy' (chemical energy or vitality?) and 'fat' (something which is always bad, or a mix of beneficial and harmful fatty acids, or the state of obesity?).

Often the recipient may not understand the communicator. In face-to-face communication this is usually signalled back to the communicator, perhaps by a look of puzzlement or by a question (e.g. Can you say that again?). As McCron noted long ago, 'the recipient is not a passive sponge, receiving exactly the message intended by the communicator—rather he [sic] extracts meaning by operating on the message' (McCron 1978). The recipient constructs meaning according to an interpretive framework. This construction depends on factors such as those aspects of the message selected for attention, the recipient's opinions, beliefs and attitudes about the subject-matter of the message, and often their beliefs and attitudes about the communicator (Hovland 1959).

Generally, face-to-face communication is more effective than mediated forms. Because of its interactive nature, various techniques can be used to ensure that the recipient understands the message, such as feedback, recapitulation and restatement. The recipient can query doubtful meanings and the communicator can test whether the message is understood in the manner intended (e.g. by checking with questions or by looking at the non-verbal expression of the recipient). Typically, this powerful method of communication is used in clinical and educational settings. The main drawback is that it is expensive, but it may be quite cost effective. The tailoring of nutrition messages through the use of surveys and computer programs is one attempt to replicate the advantages of face-to-face communication without its high costs.

Direct communication between individuals is still the main way in which humans communicate with each another. Mobile phones, texting and the internet are all embedded in one-on-one communication. Indeed, one of the ways mass communication campaigns work is by stimulation of conversation between members of the 'target audience' and their friends and relations.

The direct effects model

When communication is mediated, the major loss is that the communication process becomes much less interactive. Usually, the recipient is unable to send messages back to the communicator, except in the cases of online communications (chat rooms, emails), letter columns or talkback shows. In practice, the interactivity is low. In addition, the spontaneity of the communication is altered. With print media, the message can be examined repeatedly—but the author has little control over who reads the message and so writes in rather general terms, trying to strike a chord with everyone. Radio and television are faster moving. If the recipient misunderstands the message, then there is no going back (unless the program is recorded, but replays are very time consuming). As a result, television tends to deal superficially or generally with most topics—the messages must not be too detailed or much of the audience will become confused and will switch channels. The average 30-minute television documentary contains less information than can be written on a newspaper page (Patterson and McClure 1976). However, the distinctive features of television and the internet are that they convey instantaneous, emotional information very well. They can produce vivid memorable, vicarious experiences (such as the effects of smoking and poor diet on the insides of arteries!).

The mass media grew rapidly from around 1860 to 1914. Products such as Pears Soap were advertised widely with apparent success. During World War I, whole societies were organised for total war, mainly via the mass media. Propaganda was used to motivate populations to make more and more sacrifices. Perhaps as a result, the myth arose that the media have great power to alter the behaviour of most members of society—hence the maxim: 'Control the media and you control society'. This view is usually referred to as the 'direct effects' (or 'bullet' or 'hypodermic needle') model of mass communication. According to it, people are merely passive recipients of the media messages. Controllers of the media activate stimuli, which in turn activate a widespread and immediate response in society. This theory assumes that every human is really just like every other, all acted upon by external forces—the 'billiard ball' view of passive humanity.

The model remains popular today. We often hear health experts say something like: 'Give the public the correct information about nutrition and they will make better food choices.' In reality, not everyone in the population is the same or so passive. Communication researchers in the 1930s through to the 1960s became aware that audiences are quite mixed. They identified three important ways in which people differ:

1. *People differ in their psychological needs* such as the need for social companionship or for independence. Food beverage advertisers cater to these needs, especially among adolescents (e.g. Coca-Cola advertising is centred on images of happy groups of young people).

2. *People belong to different social categories* such as social status, ethnic and age groups. These categories are associated with different life situations and different needs. Food advertisers also take these categories into account when targeting their communications. Perhaps nutrition educators should do likewise, focusing on the shared needs and life experiences of broad categories of people. The Verb campaign (see Chapter 6) shows examples of different groups of Americans such as black American youth acting in healthy ways.

3. *People belong to different but overlapping social networks* which serve different functions. For example, there are networks for emergency help, others for gossip, for work, leisure and so on. Information can be transmitted around these networks because their members share common assumptions and language about the nature of reality. If we can tap into these networks, we may be able to influence the information present in them. For example, weight-loss companies rely on both mass media advertising and word-of-mouth communication between women.

Some people have very extensive social networks consisting of acquaintances from different walks of life like business, sports and theatre groups. These people are called 'connectors'. They are often involved in spreading new fads, fashions and social trends. When connectors combine with obsessive but sociable individuals who are steeped in knowledge about particular topics (e.g. where to get the best bargains) — so-called 'market mavens' — they can spark major changes in population behaviours. Good examples are the renewal of popularity of Hush Puppies and the prevailing desire for 'bottled water'.

The early researchers found that many people were either unaware of media campaigns or, if they were aware, they did not act on them. Studies of persuasive communications conducted around the time of World War II showed that even face-to-face communication was complex, with factors such as communicator credibility, order of presentation of information and recipient personality traits and fear arousal involved.

The two-step (and n-step) models

In studies of the communication of agricultural innovations (e.g. tractors, artificial fertilisers) conducted in Kansas in the 1940s (Rogers and Shoemaker 1971), it was found that certain people monitored media and relevant agricultural information sources closely. They were the first with the news and the first to try the innovations. Other people only took up the innovations when they saw their leader peers achieve success with them. These innovators became known as 'opinion leaders'. Market researchers try to identify opinion leaders and try to persuade them to adopt new products and services, knowing that they will influence others. In the Stanford Three City Heart Disease Prevention Program of the 1970s (see Chapter 7), many more people knew about the content of the program's TV advertisements than had seen them (Maccoby and Farquhar 1975). In these small towns, social networks were very effective in spreading information. The two-step model can be extended to n-step models as followers of opinion leaders communicate with others about the success or otherwise of innovations, and so on.

Active, social, partly mediated humans

Although n-step models go some way towards explaining the influence of the mass media, they tend to ignore the fact that most people are embedded in social networks in which their primary groups are strong influences—often stronger than any media influence. Humans are not blank slates on which messages for action can be written by the media. Instead, they have their own agendas. People select information from the media which is consistent with their own beliefs. Health information in the media is likely to be ignored or selectively interpreted: 'the question "What do the mass media do to people?" is inadequate . . . it must be replaced by the rather more difficult question: "What do people do with the mass media?"' (McCron and Budd 1979).

Properties of the mass media

This constructionist view led in the 1970s and 1980s to studies of what the media-makers put into mass communication (e.g. Mendelsohn 1968, 1973). The biases of the mass media, such as their preference for dramatic 'happenings' (e.g. bank robberies caught on video surveillance cameras) and emotional outbursts (teary interviewees are referred to as 'talent') sound trivial, but as the media are 'our eyes and ears on the world' the more we depend upon them, the more our decisions and views will be constrained by their biases. Gerbner et al. (1969) and others have suggested that the long-term implications of the mass media for most areas of life are quite profound, including:

- *They can set the climate of public opinion*, making people more aware of certain events and people than others (Noelle-Neuman 1973). The right-wing Fox network in the United States is infamous for its partisan presentations of events.
- *They can be used to set the agenda for debate*, particularly in areas where individuals have little other information. Perhaps the emphasis on slimming and obesity is a result of the visual and emotional biases of the media. Every single item in the media is put there by some group or individual. We have to ask 'Why?' and 'Why now?' and 'Whose interest is this item serving?' Why has the sudden interest in obesity occurred during the last few years? Whose interests does all the publicity serve?
- *They define normality*. Gerbner et al. (1969) observes that commercial advertising effects may be relatively weak compared with the effects of *programming*. When we watch programs, we tend to have our guard down and we can come to assume that what we see on TV shows is 'normal'—for example, that everyone drinks alcohol at celebrations or that most people eat fast foods regularly.

Use of media communications in nutrition promotion

This discussion of the mass media has several implications for nutrition promotion:

1. *Promoters need to be aware of the varying views of health and nutrition* held by the population and the situations and belief systems which influence them—for example, the importance of 'nutrition' relative to other goals in people's lives.
2. *Nutrition promoters should hold the media accountable* for what they broadcast about food and nutrition. They should examine the ways in which the media model, report and interpret nutrition matters and try to influence them. Are the implicit and explicit messages in popular TV shows pro-nutrition? If not, nutrition promoters need to advocate for change. The Parents' Jury is a good

example of an advocacy group holding the media accountable for their advertising content. The media production community is a valid community with
which nutrition promoters should interact. This requires close monitoring of
media content by nutritionists, as has been done by the German Federal
Institute of Food and Nutrition.

3. This implies that *nutrition promoters should strive to achieve consensus* among
   themselves, especially regarding complex issues. A broadly based food and
   nutrition policy is required. Sometimes, nutritionists appear to contradict one
   another in the mass media, which confuses the public. For example, in
   Australia in 2006, there was a heated attack by well-known nutritionists on
   the authors of a popular nutrition book (the CSIRO's *Total Healthy Eating
   Plan*). Whatever the merits of the book, the clear implicit message to the
   public was that nutritionists don't agree among themselves. This sort of public
   disagreement in the mass media threatens the creditability of all nutrition
   messages.

4. *Use of a range of pre-existing networks.* The mass media are only one set of information sources. They and other sources are open to influence by nutrition
   promoters—for example, adults can be reached through supermarkets, workplaces, social clubs, trade unions, churches, health services and schools
   (via their children). The influence of such networks may be much greater than
   the mass media (Wallack 1981, 1983). The identification of opinion leaders,
   'connectors', 'market mavens' and other communicators is an important
   priority for nutrition promoters because these people can spread nutrition
   messages throughout the population.

The status and role of nutrition knowledge

Communication processes transfer messages or information from one person to
another. These messages are often transformed into 'knowledge', on which people
base their decisions in daily life. The key question is: 'What is knowledge and does
it affect people's decisions, especially in the food and nutrition domain?'

One of the traditional aims of nutrition educators has been the transfer of nutritional knowledge to the general public. The assumption here is that if people
understand nutrients they will be more likely to adopt healthier food consumption
patterns. The recent history of mass primary, secondary and tertiary education
suggests that this is a reasonable assumption. Mass education (the imparting of
various forms of knowledge) has been associated with major changes in society, and

generally educated people behave differently from less educated people (Davies 2000; Ippolito 2003). Unfortunately, there is little evidence that nutrition education causes people to eat more healthily. People who are highly educated in nutrition, such as dietitians, often display 'nutritionally questionable' eating habits. However, many studies have shown that health and nutrition education among schoolchildren often leads to increases in knowledge which can be retained for months or years (e.g. Whitehead et al. 1973; Tones and Tilford 2001). So what might be the role of nutrition knowledge, and is nutrition education a worthwhile activity?

Knowledge and information

The concepts of 'information' and 'knowledge' lie at the heart of nutrition communication. In classical engineering definitions, information is *that which reduces uncertainty* (Shannon 1949). Information itself may or may not convey meaning, depending on the prior knowledge of the receiver. For example, the human genome project has produced a great deal of information which is quite meaningless for most people with the exception of interested molecular biologists. Most lay people cannot understand the significance of the project since they do not know much about the context in which the information is produced.

Knowledge differs from information in that it is contextual and is organised as a system of validated or testable beliefs. The associative model of human memory (Anderson and Bower 1973) compares a person's knowledge framework with a fishing net draped over a beached boat—there are knots (pieces of information) which are linked by strings (concepts or schemas). One of the aims of nutrition education is to help organise people's nutrition information into organised systems of knowledge which can assimilate new pieces of information ('facts') and reject any inconsistent information on the grounds that they are likely to be false (Gussow and Contento 1984).

Information and knowledge are stored in long-term memory as networks of associated concepts (Anderson and Bower 1973). Psychologists use the term *schema* to refer to interrelated sets of beliefs organised along an overarching theme (Skemp 1979), such as the green nature of many vegetables. Some types of knowledge matter more to some people than others so they may work hard to develop quite elaborate schemas. For example, people who have experienced misadventures with household bleaches often have well-developed tree-like schema linked to perceptions about the safety of various substances (Ley 1991). Knowledge is rarely passively absorbed by people; instead, it has to be actively created by the thinker through their personal experiences. This is why educationalists often employ 'discovery' or 'experiential' learning to help students acquire new knowledge. This often occurs most readily in small groups (Johnson and Johnson 1985a), perhaps because humans pay a lot of attention to the behaviours and opinions of other humans.

Properties of knowledge

Knowledge usually has benefits for the individual as well as for the population. Some of its individually relevant properties include its ability to make sense of the world and to predict the consequences of our actions (e.g. if we eat puffer fish we will probably become violently ill). Knowledge is also important for our emotional and material well-being. For example, most people know who cares for them and who they can rely on in times of trouble (e.g. when there is a medical emergency). Knowledge can have 'sleeper effects'—it may not be used for years after its acquisition. The beliefs we learn at school about infant feeding, for instance, may be of no use to us until we have our own children. It is difficult to predict just what humans will do with any given set of knowledge. For example, knowledge about the sources of dietary fibre could be used to prevent human constipation and bowel cancer or to choose food for the pet dog! Knowledge is very flexible stuff.

For nutrition promotion, knowledge has several very important characteristics. First, it defines 'common sense'. Think about the common sense of parents. They *should* know where food comes from and what sorts of food help infants to grow and thrive. They *should* know that girls put on fat around the hips as they approach puberty and they *should* know that they do not need to 'go on a diet'. Unfortunately, in today's post-modern society this 'common sense' may not be well transmitted between generations.

Second, knowledge generates behavioural possibilities. For example, the widespread belief that 'fat is bad' helps to generate slimming behaviours, dieting, low-fat food sales and anti-obesity gene treatment research.

Third, knowledge may not be *sufficient* to bring about changes in food consumption habits, but it may be a *necessary* factor in such change. People with sound nutrition knowledge are many times more likely to consume large amounts of fruit and vegetables than those without this knowledge (Wardle et al. 2000). Obviously motivational factors are also important, but without basic knowledge innovations in behaviour are unlikely. Finally, most people's knowledge is highly interrelated.

Unlike that of specialists in academic disciplines, lay people's knowledge tends to be fuzzy and overlapping—so knowledge of the fat content of foods may be closely associated with a person's knowledge of soap operas and fashion magazines. It is hard to cut up knowledge.

Nutrition communication and sources of nutrition information

Most people are in communication with others about food, health and nutrition for much of the time. For example, we know (from studies referred to in Chapter 12) that:

- members of most families talk about food and health several times a week;
- around one in three shoppers checks the nutrition panel on at least one product during any shopping trip;
- nutrition considerations play major roles in the purchasing of food products (e.g. the perceived fat content of foods).

Source credibility

Sometimes when we hear a statement we might say 'That's true, I agree with that!' It may not matter who said it. In the 1972 US presidential election, supporters of George McGovern often watched the campaign adverts of his opponent, Richard Nixon, just to confirm their opinion that Nixon was a bad guy. However, in nutrition and health matters, the source of the message often matters a great deal. So if a food company makes a claim about its product, we may say 'They would say that!' because we can see that they have a vested interest in claiming that their product conveys some benefit. If the source of a message is perceived to be credible, then the message is likely to be attended to; however, if it is not credible it will be rejected.

Nutrition promoters often have credibility with the public, perhaps because they are nutritionists and nutrition is valued. However, they can lose their credibility in several ways. First, if they are sponsored by a food company, many people will say they have been 'bought out'. Second, they will lose credibility if there is a contradiction between their behaviour and their message—for example, if they tell people to go easy on cakes and buns but are seen eating lots of them. Third, this will happen if they appear to share little in common with the target group—for example, street people may be less likely to accept advice from older 'middle-class' promoters (who may be seen as 'out of touch' or as well-off hypocrites).

Vividness and narrative are key properties of information which make it attractive to people. Some information is 'sticky': it attracts people's attention. Sticky information usually involves vivid descriptions of events (e.g. I ate food X and was immediately nauseous) and it often involves a human story with a beginning, middle and end—a narrative. Memorable information rarely consists solely of statistical or abstract information. Most people like messages which have relevance to their lives, and which are simple to understand. People are not particularly interested in information about the number of milligrams of sodium chloride in bread, but they may be more interested to know that high salt intakes can lead to high blood pressure, stroke and death.

Sources of nutrition information and their perceived trustworthiness

Many people rely on various sources of nutrition information. Some of these sources and their credibility are shown in Table 13.1. The mass media, food labels, books, family and friends, doctors, dietitians, the National Heart Foundation and health magazines are among the most popularly consulted sources. Generally, women use a variety of sources more than men, who rely most on their families.

The most trustworthy sources in lay people's eyes are orthodox health sources, followed by friends and family, food labels, cooking magazines, the mass media and alternative health sources. Men tend to distrust all sources more than women.

Although the orthodox health sources have most credibility, the mass media, family and friends and food labels are used more often. The challenge is to ensure that family and friends and the orthodox sources, in particular, have sound and relevant knowledge about food and health issues and that they are more accessible to members of the community. New internet developments such as the Healthy Eating Club and the Harvard medical newsletters may be part of the answer as the use of the internet becomes more widespread. It is also clear that food labels are an excellent form of communication (as discussed in Chapter 12).

TABLE 13.1 CONSUMER'S TRUST IN SOURCES OF NUTRITION INFORMATION

	Trust		Use often	
	% women	% men	% women	% men
1. Mass media				
Advertising (TV, radio, magazine ads, etc.)	22	15	19	12
Television programs	36	31	18	10
Radio programs	26	29	10	4
Articles in women's magazines		33	18	18
Newspaper articles	33	27	18	10
Articles in cooking magazines	54	49	22	10
2. Significant others				
Friends	43	36	16	5
Family	55	57	20	15
Workmates	23	14	8	2
Teachers/schools/higher education	29	21	2	2
3. Alternative sources				
Health food shops/staff	39	26	5	2
Alternative health practitioners	39	34	8	3
Articles in vegetarian magazines	29	29	5	3
Articles in health magazines	54	49	13	9
Slimming clubs	17	10	3	2
4. Specialist media				
Articles in science magazines	37	39	4	3
Articles in sports magazines	18	20	2	3
Books		49	42	24
Articles in health magazines	54	ns	13	9
Articles in vegetarian magazines	29	29	5	3
Internet	4	5*	0	0
5. Orthodox sources				
Dietitians/nutritionists	74	69	10	9
National Heart Foundation/Anti-Cancer Foundation		85	80	10
Doctors (medical)	83	81	12	7
6. Food labels				
Food labels	46	37	50	36

Source: Worsley and Lea (2003).

Opinion leaders and reference groups

Often when people do not have direct knowledge about the properties of products or other things, they rely on the opinions of other people who they admire—so-called 'opinion leaders'. They may also seek guidance from local 'market mavens' and 'social connectors'. Alternatively, people may judge the adequacy of their own behaviours—for example, the healthiness of the foods they eat—by comparing them with those of other groups of people—that is, the 'social norms' of their 'social reference groups'. For example, teenagers may compare their eating behaviours with those of their school friends, favourite sports teams or media stars. This often leads them to develop 'optimistic biases'—thinking, for example, that their diets are much healthier than those of others. Most people try to gain the approval of significant other people in their lives, such as their parents, friends, spouses and workmates. Therefore, communicators try to influence these opinion leaders and reference groups in order to make the social norms which influence people's behaviours more health-promoting.

Risk communication and management

Much nutrition communication concerns the transmission of information about risk. Consumers often attend to media sources of information in order to reduce their risk of harm or to increase the benefits which might follow from their actions. Advertisers often play upon consumers' perceptions of risk (e.g. the risk of social exclusion from being overweight).

In the 1980s, health psychologists and economists became interested in the ways lay people estimated the risks associated with exposure to various threats to health such as living close to nuclear power stations, using oral contraceptives and driving motor cars. They found that lay people's estimations of the risks associated with these threats varied a great deal, but rarely coincided with the real risks associated with these threats. We can estimate the real risks associated with a threat fairly directly by counting the numbers of deaths (or injuries) associated with it (sometimes called the *sideral risk*). The *subjective risk* perceived by lay people is often much higher or much lower than the actual risk.

This mismatch between subjective and actual risk estimations has been demonstrated in many studies. Early risk investigators such as Slovic (1987) found that subjective risks are influenced by several factors (Sandman 1989). Two of them—the degree of control people have over the threat, and ignorance of the way the threat affects personal well-being—are major influences. The more control people have over a threat to health, the less hazardous it is seen to be. Car driving is seen to be less risky than living near high-voltage power lines despite the fact that more deaths are associated with car driving. The less people understand how a threat works, the more dangerous it is perceived to be. So chemical additives over which people have almost no control (other than avoiding the product), and which have mysterious effects, present more of a subjective risk than fats which people can control and with which they are familiar—even though fat-rich foods do harm many more people.

Risk-management professionals have found that the subjective risks associated with health threats can be so high that people become angry and upset—in short, they become outraged. The term 'outrage' is used for these subjective reactions to threats. Generally, Risk=Hazard+Outrage (Sandman 1993). They have developed a set of 25 principles for the management of risk communication which can be useful for nutrition promoters (Sandman and Lanard 2003; 12 of the principles are listed in Box 13.1).

BOX 13.1 CONSENSUS RISK COMMUNICATION RECOMMENDATIONS*

1. **Don't over-reassure.** When people are unsure or ambivalent about how worried they should be, they often become (paradoxically) more alarmed when officials seem too reassuring. This can lead to anger and scepticism as well, and to loss of essential credibility if the truth turns out more serious than predicted. A potential crisis is a classic high-magnitude low-probability risk; if you keep assuring people how unlikely it is, they tend to focus all the more on how awful it would be.

2. **Put reassuring information in subordinate clauses.** When giving reassuring information to frightened or ambivalent people, it is helpful to de-emphasise the fact that it is reassuring. 'Even though we haven't seen a new case in 18 days, it is too soon to say we're out of the woods yet.' This is particularly important when the news is good so far, but there may be bad news coming. Practice converting one-sided reassurances into two-sided good-news bad-news combinations until the technique comes naturally.

3. **Acknowledge uncertainty.** Sounding more certain than you are rings false, sets you up to turn out wrong and provokes adversarial debate with those who disagree. Say what you know, what you don't know and what you are doing to learn more. Show you can bear your uncertainty and still take action.

Box 13.1 continued

4. Don't overdiagnose or overplan for panic. Panic is a relatively rare (though extremely damaging) response to crisis. Efforts to avoid panic—for example, by withholding bad news and making over-reassuring statements—can actually make panic likelier instead. Officials need to rethink their tendency to imagine that people are panicking or about to panic when they are merely worrying . . . or perhaps disobeying or distrusting you.

5. Don't ridicule the public's emotions. Expressions of contempt for people's fears and other emotions almost always backfire. Terms to avoid include 'panic', 'hysteria' and 'irrational'. Even when they are accurate, these labels do not help—and usually they are not accurate. Even when discouraging harmful behaviour such as stigmatisation, it is important to do so with sympathy rather than ridicule. If you are frustrated with the public express your frustration privately, so it doesn't leak out unless you want it to.

6. Establish your own humanity. Professionals are understandably preoccupied with looking professional. But especially in a crisis, the best leaders reveal their humanity. Express your feelings about the crisis, and show you can bear them; that will help the rest of us bear our own feelings, and help us build a stronger alliance with you. Express your wishes and hopes as well. Tell a few stories about your past, your family, or what you and your officemate said to each other this morning in the crisis.

7. Tell people what to expect. 'Anticipatory guidance'—telling people what to expect— does raise some anxiety, especially if you're predicting bad news. But being forewarned helps us cope, it keeps us from feeling blindsided or misled, and it reduces the dispiriting impact of sudden negative events. Warning people to expect uncertainty and possible error is especially useful. So is warning people about their own likely future reactions, particularly the ones they want to overrule: 'You'll probably feel like stopping the medicine before it's all gone.'

8. Offer people things to do. Self-protective action helps mitigate fear; victim-aid action helps mitigate misery. All action helps us bear our emotions, and thus helps prevent them from escalating into panic, flipping into denial or declining into hopeless apathy. Plan for this well in advance: mid-crisis is a harder time to start figuring out what to offer people to do—including the lesions of volunteers who will want to help.

9. Acknowledge errors, deficiencies and misbehaviours. People tend to be more critical of authorities who don't talk about the things that have gone wrong than they are of authorities who acknowledge those things. It takes something like saintliness to acknowledge negatives that the public will never know unless you tell. At least acknowledge those that the public does know or is likely to find out. Make these acknowledgements early, before the crisis is over and the recriminations begin.

Box 13.1 continued

10. **Be explicit about 'anchoring frames'.** People have trouble learning information that conflicts with their prior knowledge, experience or intuition. The pre-existing information provides an 'anchoring frame' that impedes acquisition of the new information. It helps to be explicit about the change—first justify their starting position (why it was right, or seemed right; why it is widespread), then explain your alternative (what changed; what was learned; why their starting position turns out, surprisingly, to be mistaken).

11. **Don't lie, and don't tell half-truths.** It doesn't require an out-and-out lie to devastate the credibility of crisis managers, and thus their ability to manage the crisis. A carefully crafted misleading half-truth can do the same harm, and so can a cover-up of information people later feel they should have been told. Such strategies may work for a while, at least for those who aren't paying close attention. But in a serious crisis many people are paying close attention. They may smell a less-than-candid official line long before they can specify the half-truths and omissions. And the price is high.

12. **Be careful with risk comparisons.** Why are some risks more upsetting than others? The statistical seriousness of the risk is certainly relevant, but so are 'outrage factors' like trust, dread, familiarity and control. In addition, a risk that threatens health care systems, economies and social stability is likely to be seen as threatening individual health as well, even when it does not. Efforts to reassure people by comparing improbable but upsetting risks to more probable but less upsetting ones feel patronising and tend to backfire.

*The 12 consensus risk communication recommendations are based on: Sandman and Lanard (2003).

Source: Sandman and Lanard (2003); also Crisis Communication I: How Bad Is It? How Sure Are You? Crisis Communication II: Coping with the Emotional Side of the Crisis. Crisis Communication III: Involving the Public. Crisis Communication IV: Errors, Misimpressions, and Half-Truths at http://www.psandman.com.

Risk management is crucial because nutrition promoters are in constant communication with the community. When community outrage is high but based on false premises, the health promoter's job is to calm the outrage. For example, many people incorrectly believe that traces of chemicals used in food production cause cancers and other diseases—the nutrition promoter's role is to reassure them. On the other hand, community interest in health-threatening practices may be low. For example, few people are aware of the deleterious role that salt added to foods during production has on health—in such situations, the nutrition promoter's role is to increase community interest ('outrage') through advocacy and communication until action is taken to reduce the hazard.

Food marketing and communication campaigns

'The field of marketing attempts to influence voluntary behaviour by offering or reinforcing incentives and/or consequences in an environment that invites voluntary exchange.' (Rothschild 1999) Marketing is centred on the customer and on the four Ps—product, price, promotion and place—which are four things that may be manipulated to induce the customer to purchase a product or service.

In order to communicate with groups of people and persuade them to consume healthier foods, nutrition promoters need to place themselves in consumers' shoes and meet the four Ps (and four Cs); The product or service therefore has to meet a 'want or conscious need of the consumer(s)'; the cost in terms of finance and effort must not be too high; the new behaviour has to be easy or 'convenient' to do; and it has to be communicated via appropriate channels which actually reach the person(s).

Social marketing, as noted in Chapter 5, is the application of basic marketing principles to the promotion of social goals such as health or nutrition promotion. Grier and Bryant (2005), in their useful review of social marketing in public health, define it as: 'The use of marketing to design and implement programs to promote socially beneficial behaviour change'. They cite two cases studies of social marketing applied to nutrition.

The Texas WIC program
The Women, Infants and Children (WIC) program (Bryant, Kent et al. 1998; Bryant, Lindenberger et al. 2001) aimed to change public perceptions of its functions from being perceived only as an agency for free food for poor people to being in addition a source of nutrition education, child immunisations, health check-ups and referrals. A social marketing plan, based on extensive market research, saw the number of clients rise from 582 819 in October 1993 to 778 558 five years later.

The Food Trust
This is a non-profit organisation in Philadelphia, in the United States. The trust's main aim is to increase access to nutritious foods at a reasonable price. Its Corner Store campaign was designed to decrease the incidence of obesity and diet-related disease by improving the snack foods available for young people in local corner

stores. Before the commencement of the campaign, the trust's staff interviewed key informants about the most appropriate social marketing strategies and conducted survey research about the foods in corner stores in five local communities. This showed that healthy food choices were largely unavailable. Assessment of the food environment enabled decisions to be taken about which healthy foods should be marketed in the short term, and ways in which manufacturers and retailers could better distribute healthy snack foods. This led to the formation of partnerships with suppliers of healthy foods, along with formative research among five- to twelve-year-olds and subsequent social marketing plans in pilot communities (Grier and Bryant 2005).

What's the pitch? How do we present (frame) nutrition communication?

The ways health information messages are presented to people can have strong effects on the degree to which they respond to them. Messages about health can be presented in a positive light (e.g. 'If you eat fruit and vegetables every day you will feel more vital') or in a negative 'frame' (e.g. 'If you do not eat enough fruit and vegetables every day you may get bowel cancer'). Another form of presentation of health information involves numerical information, such as 'Women have a one in eight chance of contracting breast cancer during their lifetime'—which is very alarming. However, information from the same database suggests that the probability of a 50-year-old woman developing breast cancer before she reaches 85 years is 7.85 per cent (one in 12.7)—which is far more reassuring!

Unfortunately, expert opinion differs about the actual effects of the framing of information. It is believed that positive framing may be more effective for prevention behaviours and negative framing may be more important for detection behaviours—for example, estimates of one's risk of disease X (Finney and Iannotti 2002). However, moderating factors may include the person's degree of involvement with an issue, and whether they systematically work through the presented information or not. This is more than a mere academic issue, as information presentation may make the difference between action and inaction.

Miscommunication

There are some forms of communication which are contentious and regarded by many as misinformation, or miscommunication. Three examples are food advertising to children, the general confusion about nutrition caused by communications in the mass media, and nutrition quackery.

Television food advertising to children[5]

Advertising is part of marketing, which is used to communicate with consumers about the ways in which products or services might meet their wants and needs.

Television food advertising to children is a problem in many countries, including Britain and Australia. British and Australian children have long been exposed to more television food advertising than children of many other nationalities (Morton 1990; see Box 13.2; Dalmeny et al. 2003). Story and French (2004) showed that most food advertising during children's viewing hours is for confectionery and foods and beverages that contain large amounts of fat, sugar and salt. Long hours of exposure to television programs are associated with increased risk of obesity in children (Campbell et al. 2002; Dietz 1996). Current guidelines suggest children should spend no more than two hours per day viewing *all* electronic entertainment media (American Academy of Pediatrics 2001; Australian College of Pediatrics 1994), but Australian primary school-aged children watch television for an average of two and a half hours per day (ACNielsen Media International 2001).

Arguments for and against food advertising to children

The advertising industry promotes the view that food advertising is largely a matter of parental choice, and it is only about promoting brands of products, not whole categories of foods like chocolate. The line is that children have to be taught to be consumers so they should learn early and often, and anyway advertising isn't a powerful influence over food choice (which begs the question as to why so much money is spent on advertising). The key objections are that, since children are minors who have difficulties separating commercial reality from fantasy and cannot make legal contracts, they should not be asked to participate in purchasing. Furthermore, advertising gives rise to distorted views of a healthy diet and to unhealthy eating practices, and thus obesity and long-term disease risk.

Most consumers perceive TV food advertising as a negative influence on children's eating (Hardus et al. 2003). However, instead of wanting bans on food advertising, most consumers (in Australia) would prefer reductions in confectionery and fast food

---

5 Based on Worsley and Crawford (2005a).

BOX 13.2 TELEVISION ADVERTISING TO CHILDREN IN AUSTRALIA

Advertisements screened at times children are watching can occur at the rate of 30 per hour. On average, Australian children watch two hours and 30 minutes of television per day. This equates to viewing around 22 000 television advertisements per year.

Food advertisements, as a percentage of total advertisements on television, range from 25 to 48 per cent and average 34 per cent. A study of thirteen industrialised nations showed that Australia has the highest number of television food ads per hour during children's television viewing times.

The proportion of advertisements promoting non-nutritious foods ranges from 50 to 84 per cent, and averages 72 per cent. The largest categories of foods advertised tend to be chocolate and confectionery, fast food restaurants and sweetened breakfast cereals.

Parents can:

- limit the amount of commercial television to which their children are exposed;
- choose to watch the ABC, pay television or videos instead, especially in the early years;
- introduce children to a range of tasty foods that are good for them and that can be fun to eat;
- limit the consumption of foods advertised on television to once or twice a week. Play 'spot the gimmicks' in advertisements on television—encourage children to be sceptical about claims made in advertisements.

Source: YMA (1997).

advertising, and increases in the advertising of healthier foods (such as fruit and vegetables), possibly supported by government subsidies (Hardus et al. 2003).

Parents and educators face the challenge of providing attractive alternatives to compete with this commercialism. A number of initiatives have supported schools and families in dealing with this issue—for example, education courses have been developed to build scepticism among primary children about the mass media (Chandler 1997).

Major changes to television food advertising content are required to reduce children's exposure to models of unhealthy food and beverage consumption. These changes will require the development of advocacy coalitions and the encouragement of innovation in the food industry. As a starting point, raising people's awareness of the problems of food advertising is important. The current situation is unlikely to change without advocacy that can translate community unease into new policies. In the United Kingdom and Australia, the Parents' Jury (see Box 13.3) makes awards to food products and public figures. This unsubtle public pressure has resulted in a number of changes in the production and marketing of children's food products (see www.parentsjury.org.uk and www.parentsjury.org.au).

Several major reviews of the effects of television food advertising to children have shown that it does not educate children about the healthiness of food, and that brand advertising increases preferences for and sales of the *whole food category*. That is, when one brand is advertised the sales of the whole category tend to increase. A major systematic review was commissioned by the UK Food Standards Agency which reviewed over 29 946 reports and found 118 robust peer-reviewed studies (Hastings et al. 2003). It showed that food promotion is dominated by TV advertising of sugared breakfast cereals, soft drinks, confectionery, savoury snacks and fast food outlets. The review clearly showed that such food promotion can lead to confusion about nutrition (e.g. whether a product contains fruit), changes food preferences and purchasing behaviours, influences brand selection and alters the balance of food categories consumed. Often the TV advertising is tied into other marketing tactics such as point-of-sale activities.

BOX 13.3 ADVOCACY EXAMPLE: AWARDS GIVEN BY THE PARENTS' JURY (UK)

- The **More in My Lunchbox!** award for healthy foods suitable for children's lunch boxes. Winner: dried fruit
- The **Not in My Lunchbox!** award for the worst food targeted at children's lunch boxes
- The **Happy Gnashers!** award for healthy foods that do not use added sugars to entice children to purchase them
- The **Tooth Rot** award for a food or drink relying on sugar for its appeal to children
- The **Honest Food** award for a manufacturer or retailer taking steps to reduce food additives
- The **Additive Nightmare** award for the most blatant use of additives to make a product appealing to children
- The **High Five!** award for promoting the consumption of five portions of fruit and vegetables per day
- The **Pester Power** award for the most manipulative advertising or marketing techniques used to promote unhealthy food to children

Source: www.parentsjury.org.uk.

Much public concern about TV food advertising to children has been expressed and lobby groups against it are to be found in most countries (see Box 13.4). The response of governments has generally been limited, with increased calls for 'self-regulation' from industry and government alike. The frequency and timing of food adverting on TV may become less in the future, and may be replaced by new marketing techniques such as viral, guerrilla and stealth advertising which have been used to circumvent bans on mass media alcohol and tobacco advertising. Perhaps the strongest control over advertising to date is the consumer protection model in Quebec. This covers all forms of marketing and examines the nature and intended purpose of communications, as well as the manner, time and place

---

BOX 13.4 THE COALITION ON FOOD ADVERTISING TO CHILDREN (CFAC)

CFAC is calling for a marked reduction in the commercial promotion of foods and beverages to children under fourteen years of age. The vital first step is to extend the statutory regulations to prohibit all television food and beverage advertising during programs where children make up a significant proportion of the viewing audience. This does not preclude the promotion of healthy eating messages to children through non-commercial social marketing.

Examples of the coalition's activities

- New resource for primary schools: 'TV Food Ads: Educate and Advocate'
- Have your say and be part of our advocacy campaign
- 'Pull the Plug on TV Food' advertising campaign
- What you can do to help fight unhealthy TV food advertising to children
- Community education kit
- How to complain about a TV food advertisement
- CFAC briefing paper: 'Children's health or corporate wealth: the case for banning television food advertising to children'
- Junk food injunction newsletters
- CFAC position papers and press releases
- Visit our Media Room for news articles about TV food advertising to children
- Suggested student research projects relating to food marketing to children

Source: www.cfac.net.au.

of the communication—logos and mascots are banned. Companies can be fined for breaches of the code. In practice, the adverse publicity surrounding any breach is a major deterrent. Further developments in the control of advertising and marketing to minors may emerge from applications of the UN Convention on the Rights of the Child. An International Code on Marketing to Children is likely to be adopted by the World Health Organization in the coming years. This is likely to incorporate the Sydney Principles (Box 13.5).

Food marketing to children at school

Food advertising on television is far from being the only form of fast food marketing to children. Perhaps of more serious concern is the practice of marketing high-energy foods and beverages direct to children while they are at school. Cash-strapped schools often accept the presence of vending machines selling snack foods and soft drinks and contracts for the supply of fast foods in exchange for payments which they can use to support school sports teams, computer labs and other worthy causes. In effect, this is tantamount to accepting overweight and obesity and all they entail for the children's future ill-health in exchange for quick fixes for activities which should be supported directly by the state or by parents. This sort of arrangement reflects the view that health and nutrition don't really matter. School food policies can do much to prevent this risk exposure but in the end it is government which has to regulate the types of foods children are exposed to. This has happened in the United Kingdom where the government has been forced reluctantly by public and media uproar (notably, but not only, by Jamie Oliver's TV programs) to require schools to feed healthy foods to children throughout the day and to consider greater control of food marketing to children (see Box 13.6).

Sources of consumer confusion

There is little doubt that many people are confused about food and nutrition issues. In part, this is because they have many interests and goals relating to food and nutrition such as making sure children eat their food, trying to control their own weight, and looking for foods which just might prevent heart disease or skin wrinkling. Someone once summed up the situation by noting that *consumer nutrition knowledge is a centimetre deep and a mile wide!*

BOX 13.5 THE SYDNEY PRINCIPLES

International **Obesity** TaskForce

The Sydney Principles

Guiding principles for achieving a substantial level of protection for children against the commercial promotion of foods and beverages

Actions to reduce commercial promotions to children should:

1. **SUPPORT THE RIGHTS OF CHILDREN.** Regulations need to align with and support the United Nations Convention on the Rights of the Child and the Rome Declaration on World Food Security which endorse the rights of children to adequate, safe and nutritious food.

2. **AFFORD SUBSTANTIAL PROTECTION TO CHILDREN.** Children are particularly vulnerable to commercial exploitation, and regulations need to be sufficiently powerful to provide them with a high level of protection. Child protection is the responsibility of every section of society—parents, governments, civil society and the private sector.

3. **BE STATUTORY IN NATURE.** Only legally enforceable regulations have sufficient authority to ensure a high level of protection for children from targeted marketing and the negative impact that this has on their diets. Industry self-regulation is not designed to achieve this goal.

4. **TAKE A WIDE DEFINITION OF COMMERCIAL PROMOTIONS.** Regulations need to encompass all types of commercial targeting of children (e.g. television advertising, print, sponsorships, competitions, loyalty schemes, product placements, relationship marketing, internet) and be sufficiently flexible to include new marketing methods as they develop.

5. **GUARANTEE COMMERCIAL-FREE CHILDHOOD SETTINGS.** Regulations need to ensure that childhood settings such as schools, child care and early childhood education facilities are free from commercial promotions that specifically target children.

6. **INCLUDE CROSS-BORDER MEDIA.** International agreements need to regulate cross-border media such as internet, satellite and cable television and free-to-air television broadcast from neighbouring countries.

7. **BE EVALUATED, MONITORED AND ENFORCED.** The regulations need to be evaluated to ensure the expected effects are achieved, independently monitored to ensure compliance and fully enforced.

Source: www.iotf.org/sydneyprinciples.

BOX 13.6 EXAMPLE OF ADVOCACY WORK BY THE UK FOOD COMMISSION: HEALTH GROUPS WARN 'WORLD'S CHILDREN AT RISK FROM JUNK FOOD MARKETING'

The health of the children around the world is put at risk by the marketing of junk food, says a report from the Food Commission issued today.

The report, 'Broadcasting Bad Health: Why Food Marketing to Children Needs to be Controlled', shows that:

- The food industry's global advertising budget is $40 billion, a figure greater than the Gross Domestic Product (GDP) of 70 per cent of the world's nations.
- For every $1 spent by the World Health Organization on preventing the diseases caused by Western diets, more than $500 is spent by the food industry promoting these diets.
- In industrialised countries, food advertising accounts for around half of all advertising broadcast during children's TV viewing times. Three-quarters of such food advertisements promote high-calorie, low-nutrient foods.
- For countries with transitional economies (such as in Eastern Europe), typically 60 per cent of Foreign Direct Investment in food production is for sugar, confectionery and soft drinks. For every $100 invested in fruit and vegetable production, over $1000 is being invested in soft drinks and confectionery.
- Over half the world's population lives in less-industrialised countries such as Russia, China and India, and they are now suffering a rising tide of diet-related diseases as food companies export their products and their advertising practices.
- Companies such as KFC, Burger King, McDonald's, Kinder, Mars, Cadbury's, Nestlé, Coca-Cola and Pepsi are criticised in the Food Commission report for targeting children. The report calls for international controls on the marketing of high-calorie, low-nutrient food to children.

Source: www.foodcomm.org.uk/press_junk_marketing_03.htm.

Possible additional causes of confusion include:

- the release of research findings directly to the media (from domestic and overseas sources), usually under dramatic headlines ('Nutrient X Linked to Bowel cancer!') but with little attempt to provide a simple framework to understand the meaning of the findings;

- the focus on single nutrients by many nutritionists rather than presenting a broader view of the roles of foods and food patterns in the maintenance of health—that is, a failure to translate nutrition science into terms which can be used by consumers in their food purchasing and consumption;
- low levels of food and nutrition education among health and education professionals—for example, few medical schools or teacher training faculties provide education in nutrition and nutrition promotion despite their importance for health;
- too great an emphasis on declarative knowledge in children's nutrition education rather than a focus on the social and experiential skills associated with food consumption—that is, we need to teach how to enjoy shopping, preparing and consuming food;
- finally, the effects of previous successful nutrition education! Many older people (and quite a few younger people) hold outdated views of nutrition (e.g. many diabetics who remain obsessed with the evils of sugar). While nutrition science views old knowledge as valid (e.g. serum cholesterol was seen as problematic in the 1950s and remains so), failure to provide wide-ranging community food and nutrition education inhibits people from assimilating new findings.

Nutrition quackery

While it is encouraging that health professionals are seen to be more credible than other sources, it is certainly true that unscrupulous people can be seen by the public as credible nutritionists. These nutrition quacks often promote ineffective or dangerous diets and remedies (Herbert 2006; Herbert and Barrett 1986; Barrett and Herbert 1994). A quack is a person who makes promises of benefits for monetary gain while denying or not mentioning the difficulties or risks that may be involved in taking their advice. Nutrition quackery or misinformation is a huge business worldwide (Center for Science in the Public Interest, www.cspinet.org). It is a major problem for nutrition promotion because it takes people's resources and distracts them from adopting inexpensive but healthy behaviours. More education to develop healthy scepticism among the public is required. Box 13.7 provides examples of questions people should ask about any nutrition claims or advice.

BOX 13.7 ASSESSING SOURCES OF NUTRITION INFORMATION

You can critically review nutrition information by asking yourself the following set of questions:

- Where and when was the information produced? Is the information Australian, or at least relevant to the Australian lifestyle or population? Is it from a credible publication or journal? Is it recent?
- Why was the information produced? Was it to inform readers about sound dietary principles or to sell a product?
- Who wrote or released the information? Is the writer or person who is named as the expert trained in the area of nutrition or medicine? Does he or she represent a recognised professional body?
- Who funded the collection of information and its publication? Were research funds supplied by an organisation with a vested interest in the results?
- Is the information supported by reliable evidence? Statements must be supported with reliable evidence (which means that similar results would be found in any repeated trials) so you can see how the writer came to the conclusions. The writer should tell you how key terms like 'adequate diet' or 'overweight' were measured and what data the conclusions were based on.
- Are the conclusions valid? Does the evidence support the conclusions or statements made? Validity means that the evidence is related to the topic or question being addressed.

Source: Carey et al. (2003).

Conclusions

Communication and mediated communication are essential for the success of much nutrition promotion. To date, powerful marketing and communication techniques have not often been used in the pursuit of health and nutrition. However, as the Verb campaign shows, if they are sufficiently funded, mass communications combined with community-based approaches can be most effective. Unfortunately, the marketing of high-energy, less-nutritious food products has dwarfed the marketing of healthy food products. Fortunately, there are increasing signs that the situation is changing as food manufacturers acknowledge that many of their products are less than healthy, and as the public and the media increasingly question the unhealthy status quo. It is very likely that food marketing will continue, but under tighter, healthier guidelines.

Discussion questions

13.1 Outline the main mass communication theories and explain how they might be applied to nutrition promotion.

13.2 What are the main sources from which consumers gain their nutritional knowledge? How can nutrition promoters influence these sources?

13.3 What is nutrition quackery? Outline the ways in which claims and advice can be checked or verified.

13.4 What are the advantages and disadvantages of mass communication programs for nutrition promotion?

13.5 What is risk communication? Outline some of the ways in which nutrition promoters can help manage risk communication (with some examples).

13.6 Outline the problems inherent in food marketing and advertising to children.

References

ACNielsen Media International 2001, *Australian pay TV trends 2001*, ACNielsen Media International, Sydney.

Adams, C.J. 1990, *The sexual politics of meat*, Polity Press, Cambridge.

Ader, M., Berensson, K., Carlsson, P., Granath, M. and Urwitz, V. 2001, Quality indicators for health promotion programs, *Health Promotion International*, 16(2): 187–95.

Ajzen, I. 1991, The theory of planned behaviour, *Organizational Behaviour and Human Decision Processes*, 50: 179–211.

Albright, C.L., Flora, J.A. and Fortmann, S.P. 1990, Restaurant menu labeling: Impact of nutrition information on entree sales and patron attitudes, *Health Education Quarterly*, 17: 157–67.

Alexander, S. 2003, The kitchen garden at Collingwood College, www.stephaniealexander.com.au/kgoverview.pdf.

Allen, M.W. and Baines, S. 2002, Manipulating the symbolic meaning of meat to encourage greater acceptance of fruits and vegetables and less proclivity for red and white meat, *Appetite*, 38: 118–30.

Allen, M. and Wilson, M. 2005, Materialism and food security, *Appetite*, 45(3); 314–23.

American Academy of Pediatrics 2001, Children, adolescents, and television, *Pediatrics*, 107(2): 423–6.

Anderson, J.R. 1983, A spreading activation theory of memory, *Journal of Verbal Learning and Verbal Behavior*, 22: 261–95.

Anderson, J.R. and Bower, G.H. 1973, *Human associative memory*, Wiley, New York.

Anderson, J.W., Johnstone, B.M. and Remley, D.T. 1999, Breast feeding and cognitive development: A meta analysis, *American Journal of Clinical Nutrition*, 70: 525–35.

Andreassen, A. 1995, *Marketing social change: Changing behaviour to promote health, social development and the environment*, Jossey Bass, San Francisco.

Appel, L.J., Champagne, C.M., Harsha, D.W., Cooper, L.S., Obarzanek, E. et al. 2003, Effects of comprehensive lifestyle modification on blood pressure control: Main results of the PREMIER clinical trial, *Journal of the American Medical Association*, 289: 2083–93.

Appel, L.J., Moore, T.J., Obarzanek, E., Vollmer, W.M., Svetkey, L.P., Sacks, F.M., Bray, G.A., Vogt, T.M., Cutler, J.A., Windhauser, M.W., Lin Pao-Hwa and Karanja, N.A. 1997, Clinical trial of the effects of dietary patterns on blood pressure, *The New England Journal of Medicine*, 336(16): 1117–24.

Armstrong, C.A., Sallis, J.F., Alcaraz, J.E., Kolody, B., McKenzie, T.L. and Hovell, M.F. 1998, Children's television viewing, body fat, and physical fitness, *American Journal of Health Promotion*, 12(6): 363–8.

Auspos, P. and Kubisch, A. 2004, *Building knowledge about community change: Moving beyond evaluations*, The Aspen Institute, Washington, DC.

Australian Bureau of Statistics (ABS) 1999, *National nutrition survey 1995: Foods eaten*, ABS, Canberra.

—— 2001, *National Health Survey 2001: Data reference package*, ABS www.abs.gov.au.

—— 2002, *Census of population and housing: Selected social and housing characteristics*, ABS, Canberra, cat. No. 2015.0.

—— 2003 Disability, Australia: Preliminary, ABS, Canberra, cat. no. 4446.0.

—— 2004, *Disability, ageing and carers survey 2003: Summary of findings*, ABS, Canberra.

Australian College of Paediatrics 1994, Policy statement: Children's television, *Journal of Paediatric Child Health*, 30(1): 6–8.

Australian Institute of Health and Welfare (AIHW) 2003, *A growing problem: Trends and patterns in overweight and obesity among adults in Australia, 1980 to 2001*, Bulletin no. 8, AIHW cat. no. AUS36, Canberra.

—— 2006, *Australia's Health 2004*, AIHW, Canberra, 2003, www.aihw.gov.au/publications/index.cfm/title/10014.

Baker, A.H. and Wardle, J. 2003, Sex differences in fruit and vegetable intake in older adults, *Appetite*, 40: 269–75.

Bandura, A. 1986, *Social foundations of thought and action: A social cognitive theory*, Prentice Hall, Englewood Cliffs, NJ.

Bao, K.Q., Mori, T.A., Burke, V., Puddey, I.B. and Beilin, L. 1998, Effects of dietary fish and weight reduction on ambulatory blood pressure in overweight hypertensives, *Hypertension*, 32: 710–17.

Baranowski, T., Weber Cullen, K. and Baranowski, J. 1999, Psychosocial correlates of dietary intake: Advancing dietary intervention, *Annual Review of Nutrition*, 19: 17–40.

Barker, D.J.P 1994, *Mothers, babies and diseases in later life*, BMJ Publishing Group, London.

Barrett, S. and Herbert, V. 1994, *The vitamin pushers: How the 'health food' industry is selling America a bill of goods*, Prometheus Books, Amherst, NY.

Baum, F. 2007, Cracking the nut of health equity: Top down and bottom up pressure for action, *Promotion & Education*, 14(2): 90–5.

Baum, F., Palmer, C., Modra, C., Murray, C. and Bush, R. 2000, Families, social capital and health, in I. Winter (ed.), *Social capital and public policy in Australia*, Australian Institute of Family Studies, Melbourne.

Baur, L.A. 2001, Obesity: Definitely a growing concern. Time to implement Australia's strategy for preventing overweight and obesity, *Medical Journal of Australia*, 174: 553–4.

Beard, T. 1997, The bread of the 21st century, *Australian Journal of Nutrition and Dietetics*, 54(4): 198–203.

Beard, T., Woodward D.R., Ball P.J., Hornsby H., von Witt R.J. and Dwyer T., 1997, The Hobart Salt Study 1995: Few meet national sodium intake target, *Medical Journal of Australia*, 166(8): 404–7.

Beard, T.C., Nowson, C.A. and Riley, M.D. 2007, Traffic-light food labels, *Medical Journal of Australia*, 186(1): 19, www.mja.com.au/public/issues/186_01_010107/bea10962_letter_fm.html.

Beck, A.M., Balknas, U.N., Furst, P., Hasunen, K., Jones, L., Keller, U., Melchor, J.-C., Mikkelsen, B.E., Schauder, P., Sivonen, L., Zinck, O., Oieb, H. and Ovesen, L. 2001, Food and nutritional care in hospitals: How to prevent undernutrition—report and guidelines from the Council of Europe, *Clinical Nutrition*, 20: 455–60.

Beck, U. 1992, *Risk society: Towards a new modernity*, Sage, New Delhi.

Belk, R. 1983, Worldly possessions: Issues and criticisms, in R. Bagozzi and A.M. Tybout (eds), *Advances in Consumer Research*, 10: 514–19.

Bell, A.C., Kremer, P.J. and Swinburn, B.A. 2004, Everything in my lunchbox is healthy—except for the spoon and the chocolate, *Asia Pacific Journal of Clinical Nutrition* 13, Supplement: S38.

Bensberg, M. 1998, *Improving health promotion*, Department of Human Services, Victoria State Government, Melbourne.

Benton, D. 2002, Carbohydrate ingestion, blood glucose and mood, *Neuroscience and Biobehavioral Reviews*, 26: 293–308.

Benton, D. and Nabb, S. 2003, Carbohydrate, memory, and mood, *Nutrition Reviews*, 61, Supplement 1: 61–7.

Benton, D. and Parker, P.Y. 1998, Breakfast, blood glucose, and cognition, *American Journal of Clinical Nutrition*, 67, Supplement: 772S–8S.

Beresford, S.A.A., Thompson, B., Feng, Z., Christianson, A., McLerran, D. and Patrick, D.L. 2001, Seattle 5 a Day worksite program to increase fruit and vegetable consumption, *Preventive Medicine*, 32: 230–8.

Berkman, L.F. and Kawachi, I. (eds) 2002, *Social epidemiology*, Oxford University Press, New York.

Bettman, J.R. 1979, *An information processing theory of consumer choice*, Addison Wesley, Reading, MA.

Bettman, J.R., Payne, J.W. and Staelin, R. 1986, Cognitive considerations in designing effective labels for presenting risk information, *Journal of Public Policy and Marketing*, 5: 1–28.

Biener, L., Glanz, K., McLerran, D., Sorensen, G., Thompson, B., Basen-Engquist, K., Linnan, L. and Varnes, J. 1999, Impact of the Working Well trial on the worksite smoking and nutrition environment, *Health Education and Behavior*, 26(4): 478–94.

Bingham, S. 2006, The fibre–folate debate in colo-rectal cancer, *Proceedings of the Nutrition Society*, 65: 19–23.

Bingham, S.A., Day, N.E., Luben, R., Ferrari, P., Slimani, N., Norat, T., Clavel-Chapelon, F., Kesse, E., Nieters, A., Boeing, H., Tjonneland, A., Overvad, K., Martinez, C., Dorronsoro, M., Gonzalez, C.A., Key, T.J., Trichopoulou, A., Naska, A., Vineis, P., Tumino, R., Krogh, V., Bueno-de-Mesquita, H.B., Peeters, P.H.M., Berglund, G., Hallmans, G., Lund, E., Skeie, G., Kaaks, R. and Riboli, E. 2003, Dietary fibre in food and protection against colorectal cancer in European Prospective Investigation into Cancer and Nutrition (EPIC): An observational study, *The Lancet*, 361: 1496–501.

Birch, L.L. 1999, Development of food preference, *Annual Review of Nutrition*, 19: 41–62.

Birch, L.L. and Fisher, J.O. 1998, Development of eating behaviors among children and adolescents, *Pediatrics*, 101: 539–49.

Birch, L.L., Johnson, S.L. and Fisher, J.A. 1995, Children's eating: The development of food acceptance patterns, *Young Children*, 50: 71–8.

Blainey, G. 1975, *Triumph of the nomads: A history of ancient Australia*, Macmillan, Melbourne.

Bloom, B.S. 1956, *Taxonomy of educational objectives, handbook I: The cognitive domain*, David McKay, New York.

Blundell, J.E. 1999, The control of appetite: Basic concepts and practical implications, *Schweiz Med Wochenschr*, 129(5): 182–8.

Blundell, J.E., Lawton, C.L., Cotton, J.R. and Macdiarmid, J.I. 1996, Control of human appetite: Implications for the intake of dietary fat, *Annual Review of Nutrition*, 16: 285–319.

Booth, M.L., Wake, M., Armstrong, T., Chey, T., Hesketh, K. and Mathur, S. 2001, The epidemiology of overweight and obesity among Australian children and adolescents, 1995–1997, *Australian and New Zealand Journal of Public Health*, 25: 162–9.

Booth, S. and Smith, A. 2001, Food insecurity in Australia: Challenges for dietitians, *Australian Journal of Nutrition and Dietetics* 57: 150–6.

Boyd, M. 2008, *People, places, processes: Finding the right balance in health promotion approaches to tackle health inequalities*, Victorian Health Promotion Foundation, Maelbourne, March.

Bradford Hill, A. The environment and disease: Association or causation? Proceedings of the Royal Society of Medicine, 58(1965), 295–300.

Brookfield, S.D. 1994, *Understanding and facilitating adult learning: A comprehensive analysis of principles and effective practices*, Jossey-Bass, San Francisco.

Brug, J., Campbell, M. and van Assema, P. 1999, The application and impact of computer-generated personalized nutrition education: A review of the literature, *Patient Education and Counselling*, 36: 145–56.

Brug, J. and Klep, K.I. 2007, Children and adolescents, in M.L. Lawrence and A. Worsley (eds), *Public health nutrition: From principles to practice*, Allen & Unwin, Sydney.

Brug, J., Steenhuis, I., van Assema, P., Glanz, K. and De Vries, H. 1999, Computer-tailored nutrition education: Differences between two interventions, *Health Education Research*, 1: 249–56.

Bryan, J. 2004, Nutrients for cognitive development in school-aged children, *Nutrition Reviews*, 62: 295–306.

Bryant, C.A., Kent, E., Brown, C., Bustillo, M., Blair, C. et al. 1998, A social marketing approach to increase consumer satisfaction with the Texas WIC program, *Marketing and Health Services*, Winter: 5–17.

Bryant, C.A., Lindenberger, J.H., Brown, C., Kent, E., Schreiber, J.M., et al. 2001, A social marketing approach to increasing enrolment in a public health program: Case study of the Texas WIC Program, *Human Organisations*, 60: 234–46.

Buckingham, J., Fisher, B., and Saunders, D. 2008, Evidence-based medicine toolkit, http://www.ebm.med.ualberta.ca/ (sourced 2 March 2008).

Buckman, D.R. 2004, *Fruits and vegetables and physical activity at the worksite: Business leaders and working women speak out on access and environment*, California Department of Health Services, Public Health Institute, Los Angeles.

Buijsse, B., Feskens, E.J., Schlettwein-Gsell, D., Ferry, M., Kok, F.J., Kromhout, D. and de Groot, L.C. 2005, Plasma carotene and alpha-tocopherol in relation to 10-y all-cause and cause-specific mortality in European elderly: The Survey in

Europe on Nutrition and the Elderly, a Concerted Action (SENECA), *American Journal of Clinical Nutrition*, 82(4): 879–86.

Bundy, D. 2005, School health and nutrition: Policy and programs — Proceedings of the International Workshop on Articulating the Impact of Nutritional Deficits on the Education for All Agenda, *Food and Nutrition Bulletin*, 26, Supplement 2: S186–202.

Burke, V., Giangiulio, N., Gillam, H.F., Beilin, L.J. and Houghton, S. 2003, Physical activity and nutrition programs for couples: A randomized controlled trial, *Journal of Clinical Epidemiology*, 56: 421–32.

Burns, C. 2004, *A review of the literature describing the link between poverty, food insecurity and obesity with specific reference to Australia*, Vichealth, Melbourne.

The Cabinet Office Strategy Unit, 2008. Food: An analysis of the issues, www.cabinet office.gov.uk/strategy/work_areas/food_policy.aspx.

Callahan, S.T. and Mansfield, M.J. 2000, Type 2 Diabetes Mellitus in adolescents, *Current Opinion in Pediatrics*, 12: 310–12.

Calvo, A. and Rahrig, K. 1997, Diffusion of Innovations overview, University of South Florida, http://hsc.usf.edu/~kmbrown/Diffusionof_Innovations_Overview.htm.

Campbell, K. and Crawford, D. 2001, Family food environments as determinants of preschool-aged children's eating behaviours: Implications for obesity prevention policy — a review, *Australian Journal of Nutrition and Dietetics*, 58: 19–25.

Campbell, K., Crawford, D., Jackson, M., Cashel, K., Worsley, A., Gibbons, K. and Birch, L. 2002, Family food environments of 5–6-year-old children: Does socio-economic status make a difference?, *Asia-Pacific Journal of Clinical Nutrition*, 11: S553–61.

Campbell, K., Crawford, D. and Worsley, A. 2002, Promoting food preferences for obesity prevention: Insights from the Children's and Family Eating Study, *Proceedings of the Australian Health and Medical Congress*, 24–29 November, Melbourne.

Campbell, M.K., Tessaro, I., DeVellis, B., Benedict, S., Kelsey, K., Belton, L. and Sanhueja, A. 2002, Effects of a tailored health promotion program for female blu-collar workers: health works for women, *Preventive Medicine*, 34(3): 313–23.

Canadian Association of Food Banks. Hunger Count, p. 12, Toronto, Canada, www.cafbnacba.ca/documents/hungercount2007/pdf.

Canadian Heart Health Surveys Research Group 1992, The federal-provincial Canadian Heart Health Initiative, *Canadian Medical Association Journal* 146: 1915–16.

Cannon, G. and Leitzmann, C. 2005, The new nutrition science project, *Public Health Nutrition*, 8(6A): 673–94.

Caraher, M., Dixon, P., Lang, T. and Carr-Hill, R. 1998, Access to healthy foods: Part I — Barriers to accessing healthy foods: Differentials by gender, social class, income and mode of transport, *Health Education Journal*, 57: 191–201.

Caraher, M., Lang, T., Dixon, P. and Carr-Hill, R. 1999, The state of cooking in England: The relationship of cooking skills to food choice, *British Food Journal*, 101(8): 590–609.

Carey, D., Weston, K., Perriton, G. and Worsley, A. 2003, *People, food and health*, Heinemann Education, Melbourne.

Carleton, R.A., Lasater, T.M., Assaf, A.R., Feldman, H.A. and McKinlay, S. 1995, The Pawtucket Heart Health Program: Community changes in cardiovascular risk factors and projected disease risk, *American Journal of Public Health*, 85: 777–85.

Carver, C.S. and Scheier, M.F. 1998, *On the self-regulation of behavior*, Cambridge University Press, New York.

Casey, R. and Rozin, P. 1989, Changing children's food preferences: Parent opinions, *Appetite*, 12: 171–82.

Chadwick, E.L. 1842, *Report on the sanitary condition of the labouring population of Great Britain*, presented to both Houses of Parliament, London.

Chalmers, B. 2004, The Baby Friendly Hospital initiative: Where next?, *BJOG: An International Journal of Obstetrics and Gynecology*, 111: 198–9.

Chalmers, B. and Levin, A. 2001, *Humane perinatal care*, TEA Publishers, Tallinn, Estonia.

Chandler, D. 1997, Children's understanding of what is 'real' on television: A review of the literature, *Journal of Educational Media*, 23: 65–80.

Chapman, L.S. 2004, Guidelines for health promotion in worksite settings, *American Journal of Health Promotion*, 18(4): 6–9.

Cheadle, A., Psaty, B., Curry, S., Wagner, E., Diehr, P., Koepsell, T., Kristal, A. 1993, Can measures of the grocery store environment be used to track community-level dietary change?, *Preventive Medicine*, 222: 361–72.

Cheadle, A., Psaty, B., Diehr, P., Koepsell, T., Wagner, E., Curry, S. and Kristal, A. 1995, Evaluating community-based nutrition programs: Comparing grocery store and individual-level survey measures of program impact, *Preventive Medicine*, 24: 71–9.

Cleland, V., Worsley, A. and Crawford, D. 2004, What are ten-year-old Victorian children buying from school canteens and what do parents and teachers think about it?, *Nutrition and Dietetics*, 61(3): 145–50.

Cliska, D., Miles, E., O'Brien, M.A., Turl, C., Tomasik, H.H., Donovan, U. and Beyers, J. 2000, Effectiveness of community-based interventions to increase fruit and vegetable consumption, *Society for Nutrition Education*, 32: 341–52.

Cochrane Collaboration 2005, Policy on the granting of endorsements by the Cochrane Collaboration, www.cochrane.org/resources/policyonthegrantingof endorsements-v218apr05.doc.

Cohen, S. and Syme, S.L. 1985, Issues in the study and application of social support, in S. Cohen and S.L. Syme (eds), *Social support and health*, Academic Press, Orlando, FL, pp. 3–22.

Colby, J.J., Elder, J.P., Peterson, G., Kinsley, P.M. and Carleton, R.A. 1987, Promoting the selection of healthy food through menu item description in a family-style restaurant, *American Journal of Preventive Medicine*, 3: 171–7.

Cole-Hamilton, I. and Lang, T. 1986, *Tightening belts: A report on the impact of poverty on food*, London Food Commission, London.

Colombo, M., Mosso, C. and De Piccoli, N. 1991, Sense of community and participation in urban contexts, *Journal of Community and Applied Social Psychology*, 11: 457–64.

Conference of Principal Investigators of Heart Health, Canadian Heart Health Initiative 2002, *Process evaluation of the demonstration phase*, Health Canada, Ottawa.

Connors, M., Bisogni, C., Sobal, J. and Devine, C. 2001, Managing values in personal food systems, *Appetite*, 36: 189–200.

Conway, G.R. 1998, *The doubly green revolution: Food for all in the 21st century*, Cornell University Press, Ithaca, New York.

Cook, T.D. and Campbell, D.T. 1979, *Quasi-experimentation, design and analysis for field settings*, Chicago, Illinois: Rand McNally.

Cooke, P. 1990, *Back to the future*, Unwin Hyman, London.

Coonan, W., Worsley, A. and Maynard, E.J. 1984, *The body owner's manual*, Life Be In It, Melbourne.

Corey, S. 1953, *Action research to improve school practices*, Teachers' College Press, New York.

Craig, P.L. and Truswell, A.S. 1994, Dynamics of food habits of newly married couples: Who makes changes in the foods consumed?, *Journal of Human Nutrition and Dietetics*, 7, 347–61.

Crawford, D. and Baghurst, K.I. 1990, Diet and health: A national survey of beliefs, behaviours and barriers to change in the community, *Australian Journal of Nutrition and Dietetics*, 47, 97–104.

—— 1991, Nutrition information in Australia—the public's view, *Australian Journal of Nutrition and Dietetics*, 48, 44–54.

Crotty, P.A., Rutishauser, I.H. and Cahill, M. 1992, Food in low-income families, *Australian Journal of Public Health*, 16(2): 168–74.

Crowe, M., Harris, S., Maggiore, P. and Binns, C. 1992, Consumer understanding of food additive labels, *Australian Journal of Dietetics*, 49: 19–22.

Curhan, R.C. 1974, The effects of merchandising and temporary promotional activities on the sales of fresh fruit and vegetables in supermarkets, *Journal of Marmara University Dentistry Faculty*, 11: 286–94.

Cutter, G.R., Burke, G.L, Dyer, A.R. Friedman, G.D., Hilner, J.E., Hodges, G.H., Holley, S.B., Jacobs, D.R.Jr., Liu, K. and Manolio, T.A. 1991, Cardiovascular risk factors in young adults: The CARDIA baseline monograph. Controlled Clinical Trials, 12(1), Supplement 1S–77S.

Dalmeny, K., Hanna, E. and Lobstein, T. 2003, *Broadcasting bad news: Why food marketing to children needs to be controlled*, International Association of Consumer Food Organizations, London.

Davies, M. 2000, The role of commonsense understandings in social inequalities in health: An investigation in the context of dental health, PhD dissertation, Faculty of Medicine, University of Adelaide.

Davison, K.K. and Birch, L.L. 2001, Childhood overweight: A contextual model and recommendations for future research, *Obesity Reviews*, an official journal of the International Association for the Study of Obesity, 2: 159–171.

De Garine, I. 1972, The socio-cultural aspects of nutrition, *Ecology of Food and Nutrition*, 1: 143–63.

de Groot, L.C., Verheijden, M.W., de Henauw, S., Schroll, M. and van Staveren, W.A. 2004, SENECA Investigators: Lifestyle, nutritional status, health, and mortality in elderly people across Europe: A review of the longitudinal results of the SENECA study, *Journal of Gerontology Series A: Biological Sciences and Medical Sciences*, 59(12): 1277–84.

de Lorgeril, M., Salen, P., Martin, J.-L., Monjaud, I., Delaye, J. and Mamelle, N. 1999, Mediterranean diet, traditional risk factors, and the rate of cardiovascular complications after myocardial infarction—final report of the Lyon Diet Heart Study, *Circulation*, 99: 779–85.

deFleur, M.L. and Ball-Rokeach, S. 1989, *Theories of mass communication* (5th ed.), Longman, White Plains, New York.

Department for Environment, Food and Rural Affairs (DEFRA) (UK) 2005, *The validity of food miles as an indicator of sustainable development: Final report*, http://statistics.defra.gov.uk/esg/reports/foodmiles.

Department of Education and Training, Victoria 2003, *Guidelines for school canteens and other school food services*, Executive memorandum no. 2003/017, Melbourne, www.sofweb.vic.edu.au/scln/docs/ExecMemo017.doc.

Department of Human Services, Victoria 2001, *What's there to eat? The practical guide to feeding families*, Melbourne.

——— 2003a, *Integrated health promotion resource kit*, Melbourne.

——— 2003b, *Evaluation of the six child nutrition fact sheets*, research report, Melbourne.

——— 2005, Give Breastfeeding a Boost: Community-based approaches to improving breastfeeding rates, Department of Human Services Victoria, Australia.

Devince, C.M. and Edstrom, E. 2001, Continuity in women's food and nutrition trajectories through mid-life and older age: A ten-year follow-up, *Journal of Nutrition Education*, 33: 215–23.

Diabetes Prevention Program Research Group 2002, Reduction in the incidence of type 2 diabetes with lifestyle intervention or metformin, *New England Journal of Medicine*, 346: 393–403.

Diamond, J. 1997, *Guns, germs, and steel: The fates of human societies*, W.W. Norton, New York.

——— 2005, *Collapse: How societies choose to fail or succeed*, Viking, New York.

Dietz, W.H. 1996, The role of lifestyle in health: The epidemiology and consequences of inactivity, *Proceedings of the Nutrition Society*, 55(3): 829–40.

Dodgson, J.E., Chee, Y.O. and Yap, T.S. 2004, Workplace breastfeeding support for hospital employees, *Journal of Advanced Nursing*, 47(1): 91–100.

Donovan, R.J., Egger, G. and Francas, M. 1999, TARPARE: A method for selecting target audiences for public health interventions, *Australian and New Zealand Journal of Public Health*, 23(3): 280–4.

Dooris, M. 2005, Healthy settings: Challenges to generating evidence of effectiveness, *Health Promotion International*, 21: 55–65.

Douglas, M. and Nicod, M. 1974, Taking the biscuit: The structure of British meals, *New Society*, 19: 744–7.

Dowey, A.J. 1996, Psychological determinants of children's food preferences, unpublished doctoral dissertation, University of Wales, Bangor.

Dowler, E. and Finer, C. (eds) 2003, *The welfare of food: Rights and responsibilities in a changing world*, Oxford, Blackwell.

Eat Well Queensland 2002, Smarter eating for a healthier state, www.health.qld.gov.au/QPHF/FoodNutrition.htm.

Egger, G. and Swinburn, B. 1997, An 'ecological' approach to the obesity pandemic, *British Medical Journal*, 315(7106): 477–80.

Ehrlich, P.R. 1971, *The population bomb*, Ballantyne Books, New York.

Erlich, R. and Murkies, A. 2001, *Color me healthy*, Nutrition Australia, Melbourne.

Eldridge, A.L., Snyder, P.A., Faus, N.G. and Kotz, K. 1997, Development and evaluation of a labeling program for low-fat foods in a discount department store foodservice area, *Journal of Nutrition Education*, 29: 159–61.

Elmer, P.J., Obarzanek, E., Vollmer, W.M., Simons-Morton, D., Stevens, V.J., Young, D.R., Lin, P.H., Champagne, C., Harsha, D.W., Svetkey, L.P., Ard, J., Brantley, P.J., Proschan, M.A., Erlinger, T.P. and Appel, L.J. 2006, PREMIER Collaborative Research Group: Effects of comprehensive lifestyle modification on diet, weight, physical fitness, and blood pressure control: 18-month results of a randomized trial, *Annals of Internal Medicine*, 144(7): 485–95.

Epstein, S. 1994, Integration of the cognitive and the psychodynamic unconscious, *American Psychologist*, 49(8): 709–24.

Euromonitor 2000, *Consumer lifestyles UK: Integrated market information system*, Euromonitor, London.

Farquhar, J.W., Fortmann, S.P., Flora, J.A., Taylor, C.B., Haskell, W.L., Williams, P.T., Maccoby, N. and Wood, P.D. 1990, Effects of community wide education on cardiovascular disease risk factors: The Stanford Five-City Project, *Journal of American Medical Association*, vol. 264, no. 3, pp. 359–65.

Feather, N.T., 1982, *Expectations and Actions: Expectancy–Value models in psychology*, Hillsdale, New Jersey: Lawrence Erlbaum Associates.

Fell, Christine. 1984, *Women in Anglo-Saxon England and the impact of 1066*, British Museum Publications, London.

Festinger, L., Riecken, H.W. and Schachter, S. 1956, *When prophecy fails: A social and psychological study of a modern group that predicted the destruction of the world*, University of Minnesota Press, Minneapolis.

Finney, L.J. and Iannotti, R.J. 2002, Message framing and mammography: A theory driven intervention, *Behavioral Medicine*, 28: 5–14.

Flegal, K.M., Graubard, B.I., Williamson, D.F. and Gail, M.H. 2005, Excess deaths associated with underweight, overweight and obesity, *Journal of the American Medical Association*, 293: 1861–67.

Foerster, S.B., Kizer, K.W., Disogra, L.K., Bal, D.G., Krieg, B.F. and Bunch, K.L. 1995, California's 5 a day—for better health campaign: An innovative population-based effort to effect large-scale dietary change, *Journal of Preventive Medicine*, 11: 124–31.

Foley, R.M. 1998, The Food Cent$ project: A practical application of behaviour change theory, *Australian Journal of Nutrition and Dietetics*, 55: 1.

Foley, R.M. and Pollard, C.M. 1998, Food Cent$—implementing and evaluating a nutrition education project focusing on value for money, *Australian and New Zealand Journal of Public Health*, 22: 494–501.

Food Marketing Institute 1992, *Trends in Australia 1992: Survey on consumer shopping*, Australian Supermarket Institute, Sydney.

Food Standards Australia and New Zealand (FSANZ) 2003, *Food labelling issues: Quantitative research with consumers*, Evaluation Report Series No. 4, Canberra, June.

French, S.A. 2003, Pricing effects on food choices, *Journal of Nutrition*, 133(3): 841S–43S.

French, S.A., Jeffery, R.W., Forster, J.L., McGovern, P.G., Kelder, S.H. and Baxter, J.E. 1994, Predictors of weight change over two years among a population of working adults: The Healthy Worker Project, *International Journal of Obesity and Related Metabolic Disorders*, 18(3): 145–54.

French, S.A., Jeffery, R.W., Story, M., Breitlow, K.K., Baxter, J.S., Hannan, P. and Snyder, M.P. 2001, Pricing and promotion effects on low-fat vending snack purchases: The CHIPS study, *American Journal of Public Health*, 9(1): 112–17.

Frewer, L. and Miles, S. 2001, Risk perception, communication and trust: How might consumer confidence in the food supply be maintained?, in L. Frewer, E. Risvik and H. Schifferstein (eds), *Food, people and society: A European perspective of consumers' food choices*, Springer Verlag, Berlin Heidelberg, pp. 401–14.

Friends of the Earth 2006, *Towards a community supported agriculture*, Friends of the Earth, Brisbane.

Future Foundation, 2005, Convenience Food Sector 2015, London: Future Foundation.

Furst, T., Connors, M., Bisogni, C.A., Sobal, J. and Falk, L.W. 1996, Food choice: A conceptual model of the process, *Appetite*, 23(3): 247–66.

Gabriel, Y. and Lang, T. 1995, *The unmanageable consumer: Contemporary consumption and its fragmentation*, Sage, London.

Galal, O., Neumann, C. and Hulot, J. (eds) 2005, Proceedings of the International Workshop on Articulating the Impact of Nutritional Deficits on the Education for All Agenda, *Food and Nutrition Bulletin*, 26, Supplement.

Geissen Declaration 2005, International Union of Nutrition Sciences Public Health Nutrition, 8(6A), 783–6.

Gerbner, G., Holsti, O.R., Krippendorff, K., Païsleg, W.J. and Stone, P.J. (eds) 1969, *The analysis of communication content: Developments in scientific theories and computer techniques*, Wiley, New York.

Gerstman, B.B. 1998, *Epidemiology kept simple: An introduction to classic and modern epidemiology*, Wiley-Liss, New York.

Gibbons, K.L. 2002, The primary school years, *Medical Journal of Australia*, 176: S115–16.

Giddens, A. 1993, *New rules of sociological method*, Polity Press, Cambridge.

Glanz, K., Hewitt, A. and Rudd, J. 1992, Consumer behaviour and nutrition education: An integrative review, *Journal of Nutrition Education*, 24: 267–77.

Glanz, K. and Mullis, R.M. 1988, Environmental interventions to promote healthy eating: A review of models, programs, and evidence, *Health Education Queensland*, 15: 395–415.

Glanz, K., Sallis, J.F., Saelens, B.E. and Frank, L.D. 2005, Healthy nutrition environments: Concepts and measures, *American Journal of Health Promotion*, 19(5): 330–3.

Glanz, K. and Yaroch, A.L. 2004, Strategies for increasing fruit and vegetable intake in grocery stores and communities: Policy, pricing, and environmental change, *Preventive Medicine*, 39: S75–80.

Glasgow, R.E., McCaul, K.D. and Fisher, K.J. 1993, Participation in worksite health promotion: A critique of the source, *Health Education Queensland*, 20(3): 391–408.

Glasgow, R.E., Vogt, T.M. and Boles, S.M. 1999, Evaluating the public health impact of health promotion interventions: The RE-AIM framework, *American Journal of Public Health*, 89: 1322–7.

Glass, T.A. and McAtee, M.J. 2006, Behavioral science at the crossroads in public health: Extending the horizons, envisioning the future, *Social Science and Medicine*, 62: 1650–71.

Godfrey, C., Caraher, M. and Angus, K. 2003, *Review of research on the effects of food promotion to children: Final report*, Food Standards Agency, London.

Gopaldas, T. 2005, Improved effect of school meals with micronutrient supplementation and deworming: Proceedings of the International Workshop on Articulating the Impact of Nutritional Deficits on the Education for All Agenda, *Food and Nutrition Bulletin*, 26, Supplement 2: S220–9.

Gorder, D.D., Dolecek, T.A. and Coleman, G.G. et al. 1986, Dietary intake in the Multiple Risk Factor Intervention Trial (MRFIT): Nutrient and food group changes over 6 years, *Journal of the American Dietary Association*, 86(6): 744–51.

Grantham-McGregor, S. 2005, Can the provision of breakfast benefit school performance?, *Food and Nutrition Bulletin*, 26: S144–58.

Green, L. and Kreuter, M. 2005, *Health promotion planning: An educational and environmental approach* (4th ed.), McGraw-Hill, New York.

Greenwald, A.G., Banaji, M.R., Rudman, L.A., Farnham, S.D., Nosek, B.A. and Mellott, D.S. 2002, A unified theory of implicit attitudes, stereotypes, self-esteem, and self-concept, *Psychological Review*, 109, 3–25.

Grier, S. and Bryant, C.A. 2005, Social marketing in public health, *Annual Review of Public Health*, 26: 319–39.

Griffiths, A.E. 2003, Social Change and Family Stress in Australia; The perpectives of Family Support Workers in Victoria, Honours Thesis, School of Health Sciences, Deakin University, Melbourne.

Grossman, J. and Webb, K. 1991, Local food and nutrition policy, *Australian Journal of Public Health*, 15, 271–6.

Grunert, K. 2005, Food quality and safety: Consumer perception and demand, *European Review of Agricultural Economics*, 32(3): 369–91.

Grunert, K.G., Brunso, K. and Bisp, S. 1997, Food-related lifestyle: Development of a cross-culturally valid instrument for market surveillance, in L.R. Kahle and L. Chiagouris (eds), *Values, lifestyles and psychographics*, Lawrence Erlbaum, Mahwah, New Jersey, pp. 337–54.

Grunert, K.G. and Wills, J.M. 2007, A review of European research on consumer response to nutrition information on food labels, *Journal of Public Health Nutrition*, in press.

Gussow, J.D. and Contento, I. 1984, Nutrition education in a changing world, *World Reviews of Nutrition and Dietetics*, 44: 1–56.

Hamilton, W.L., Cook, J.T., Thompson, W.W., Buron, L.F., Frongillo, E.A. Jr, Olson, C.M. and Wehler, C.A. 1997, Household food security in the United States in 1995: Summary Report of the Food Security Measurement Project, report prepared for the USDA Food and Consumer Service, Alexandria, VA.

Hancock, T. and Perkins, F. 1985, The mandala of health: A conceptual model and teaching tool, *Health Education*, 24: 8–10.

Hardus, P.M., van Vuuren, C.L., Crawford, D. and Worsley, A. 2003, Public perceptions of the causes of obesity among primary school children and views regarding its prevention, *International Journal of Obesity*, 27: 1465–71.

Harris, M. and Ross, E.B. (eds) 1987, *Food and evolution: Toward a theory of human food habits*, Temple University Press, Philadelphia, Pennsylvania, pp. 181–205.

Hastings, G., Stead, M., McDermott, L., Forsyth, A., MacKintosh, A.M., Rayner, M., Godfrey, C., Caraher, M. and Angus, K. 2003, *Review of research on the effects of food promotion to children: Final report*, Food Standards Agency, London.

Haveman-Nies, A., de Groot, C.P.G.M. and van Staverern, W.A. 2003, Dietary quality, lifestyle factors and healthy ageing in Europe: The SENECA study, *Age and Ageing*, 32: 427–34.

He, F.J., Nowson, C.A. and MacGregor, G.A. 2006, Fruit and vegetable consumption and stroke: Meta-analysis of cohort studies, *The Lancet*, 367: 320–6.

Heinig, M.J. and Dewey, K.G. 1997, Health effects of breast feeding for mothers: A critical review, *Nutrition Research Reviews*, 10: 35–56.

Herbert, V. 2006, Unproven (questionable) dietary and nutritional methods in cancer prevention and treatment, *American Cancer Society Second National Conference on Diet, Nutrition, and Cancer*, 58(S8): 1930–41.

Herbert, V. and Barrett, S. 1986, Twenty-one ways to spot a quack, *Nutrition Forum Newsletter*, 65–8, September.

Hetzel, B.S. and Pandav, C.S. 1997, *SOS for a Billion*, Oxford University Press, New Delhi.

Heywood, P. and Lund-Adams, M. 1991, The Australian food and nutrition system: A basis for policy formulation and analysis, *Australian Journal of Public Health*, 15: 258–70.

Hibbeln, J.R. and Salem, N. 1995, Dietary polyunsaturated fatty acids and depression: When cholesterol does not satisfy, *American Journal of Clinical Nutrition*, 62: 1–9.

Hjermann, I., Velve Byre, K., Holme, I. and Leren, P. 1981, Effect of diet and smoking intervention on the incidence of coronary heart disease: Report from the Oslo Study Group of a randomised trial in healthy men, *The Lancet*, 12(8259): 1303–10.

Hofstede, G. 2005, *Cultures and organizations: Software of the mind* (2nd ed.), McGraw-Hill, New York.

Holliday, R. 1999, Ageing in the 21st century, *The Lancet*, 354, Supplement 4.

Holt, S.H., Miller, J.C., Petocz, P. and Farmakalidis, E. 1995, A satiety index of common foods, *European Journal of Clinical Nutrition*, 49: 675–90.

Horne, P.J., Lowe, C.F., Flemming, P.F. and Dowey, A.J. 1995, An effective procedure for changing food preferences in 5–7-year-old children, *Proceedings of the Nutrition Society*, 54: 441–52.

Horwath, C.C. 1999, Applying the transtheoretical model to eating behaviour change: Challenges and opportunities, *Nutrition Research Reviews*, 12, 281–317.

Hovland, C.L. 1959, Reconciling conflicting results derived from experimental and survey studies of attitude change, *American Psychologist*, 14: 8–17.

Howson C.P., Kennedy E.T., and Abraham Horwitz, A. (eds) 1998, *Prevention of micronutrient deficiencies: Tools for policymakers and public health workers*, National Academies Press, Institute of Medicine, Washington DC.

Huhman, M., Potter, L.D., Wong, F.L., Duke, L.D. and Heitzler, C.D. 2005, Effects of a mass media campaign to increase physical activity among children: Year 1 Results of the VERB campaign, *Pediatrics*, 116: 277–84.

Hunink, M.G., Goldman, L., Tosteson, A.N., Mittleman, M.A., Goldman, P.A., Williams, L.W., Tserat, J. and Weinstein, M.C., 1997, The recent decline in mortality from coronary heart disease, 1980–1990: The effect of secular trends in risk factors and treatment, *Journal of the American Medical Association*, 277: 535–42.

Hunt, M.K., Lederman, R., Potter, S., Stoddard, A. and Sorensen, G. 2000, Results of employee involvement in planning and implementing the Treatwell 5-a-Day worksite study, *Health Education Behaviour*, 27(2): 223–31.

Hunt, M.K., Lederman, R., Stoddard, A., Potter, S., Phillips, J. and Sorensen, G. 2000, Process tracking results from the Treatwell 5-a-Day Worksite Study, *American Journal of Health Promotion*, 14(3): 179–87.

Hunt, M.K., Lefebre, C., Hixson, M.L., Banspach, S.W., Assaf, A.R. and Carleton, R.A. 1990, Pawtucket Heart Health Program Point-of-Purchase Nutrition Education Program in Supermarkets, *American Journal of Public Health*, 80: 730–1.

Huon, G., Wardle, J. and Szabo, M. 1996, Improving children's eating patterns: Intervention programs and underlying principles, *Australian Journal of Nutrition and Diet*, 53: 156–65.

Huot, I., Paradis, G. and Ledoux, M. 2004, Effects of the Quebec Heart Health Demonstration Project on adult dietary behaviours, *Preventive Medicine*, 38: 137–48.

International Obesity Taskforce with the European Childhood Obesity Group 2002, Obesity in Europe: Copenhagen: IOTF, 11 September, http://www.iotf.org/media/euobesity.pdf.

International Union of Nutrition Sciences 2005, The Giessen Declaration, *Public Health Nutrition*, 8(6A): 117–120, http://www.iuns.org/features/05-09%20 NNS%20Declaration.pdf.

Ippolito, R.A. 2003, *Education versus savings as explanation for better health: Evidence from the Health and Retirement Survey*, Working Paper in Law and Economics No. 03–04, School of Law, George Mason University, Arlington, VA.

James, P.T., Leach, R., Kalamara, E. and Shayeghi, M. 2001, The worldwide obesity epidemic, *Obesity Research*, 9, Supplement: 228S–33S.

Janer, G., Sala, M. and Kogevinas, M. 2002, Health promotion trials at worksites and risk factors for cancer, *Scandinavian Journal of Work, Environment and Health*, 28(3): 141–57.

Jeffery, R.W. 2004, How can health behavior theory be made more useful for intervention research?, *International Journal of Behavioural Nutrition and Physical Activity*, 1: 10.

Jeffery, R.W., Adlis, S.A. and Forster, J.L. 1991, Prevalence of dieting among working men and women: The healthy worker project, *Health Psychology*, 10(4): 274–81.

Jenkins, D.J.A., Jenkins, A.L., Wolever, T.M.S., Vuksan, V., Rao, A.V., Thompson, L.U. and Josse, R.D., 1994, Low glycemic index: Lente carbohydrates and physiological effects of altered food frequency, *American Journal of Clinical Nutrition*, 59, Supplement: 706S–9S.

Johnson, A. and Baum, F. 2001, Health promotion hospitals: A typology of different organizational approaches to health promotion, *Health Promotion International*, 16: 281–7.

Johnson, D.W. and Johnson, R. 1985a, Nutrition education: A model for effectiveness; a synthesis of research, *Journal of Nutrition Education*, 17(2), Supplement: S1–44.

—— 1985b, Nutrition education's future, *Journal of Nutrition Education*, 17: S20–7.

Jussame, R.A. and Judson, D.H. 1992, Public perceptions about food safety in the United States and Japan, *Rural Sociology*, 57: 235–49.

Kaati, G., Bygren, L.O., Vester, M., Karlsson, A. and Sjostrom, M. 2005, Outcomes of comprehensive lifestyle modification in inpatient setting, *Patient Education and Counseling*, 31 August.

Kawachi, I. and Berkman, L.F. 2000, Social cohesion, social capital and health, in L.F. Berkman and L. Kawachi (eds), *Social epidemiology*, Oxford University Press, New York.

—— (eds) 2003, *Neighborhoods and health*, Oxford University Press, New York.

Kazdin, A.E. 1982, Single-case research designs: Methods for clinical and applied settings, New York: Oxford Press.

Kelsey, J.L., Whittemore, A.S., Evans, A.S. and Thompson, D.W. 1996, *Methods in observational epidemiology*, Oxford University Press, Oxford.

Kemmis, S. 1988, Action research in retrospect and prospect, in S. Kemmis and R. McTaggart (eds), *The action research reader* (3rd ed.), Deakin University Press, Geelong, pp. 27–39.

Kent, L. and Worsley, A. 2006, *Dietary predictors of Body Mass Index*, AGM, British Psychological Society, Sussex.

Kickbusch, I. 1986, Lifestyles and health, *Social Sciente & Medicine*, 2: 117–24.

Kirk, M.C. and Gillespie, A.H. 1990, Factors affecting the food choices of working mothers with young families, *Journal of Nutrition Education*, 2: 161–8.

Klapper, J.T. 1960, *The effects of mass communication*, The Free Press, Glencoe.

Klein, B.W. 1996, Food security and hunger measures: Promising future for state and local household surveys, *Family Economics and Nutrition Reviews*, 9: 31–7.

Klepp, K.I., Pereze-Rodrigo, C., De Bourdeaudhuij, I., Due, P., Elmadfa, I., Haraldsdottir, J., Konig, J., Sjostrom, M., Thorsdottir, I., Vaz de Almeida, M.D., Yngve, A. and Brug, J. 2005, Promoting fruit and vegetable consumption among European schoolchildren: Rationale, conceptualization and design of the Pro Children project, *Annals of Nutritional Metabolism*, 49(4): 212–20.

Knowles, M.S. 1984, *Andragogy in action: Applying modern principles of adult learning*, Jossey Bass, San Francisco.

Kok, G., Schaalma, H., Ruiter, R.A.C. and Van Empelen, P. 2004, Intervention mapping: Protocol for applying health psychology theory to prevention programmes, *Journal of Health Psychology*, 9(1): 85–98.

Kotler, P. and Keller, K.L. 2006, *Marketing management* (12th ed.), Pearson, Upper Saddle River, New Jersey.

Kramer, M.S., Chalmers, B., Hodnett, E., Sevkovskaya, Z., Dzikovich, I. and Shapiro, S., Collet, J.P., Vanilovich, I., Mezen, I., Ducruet, T., Shishko, G., Zubovich, V., Mknuik, D., Gluchanina, E., Dombrovskiy, V., Ustinovitch, A., Kot, T., Bogdanovich, N., Ovchinikova, L. and Helsing, E., for the Probit Study Group 2001, Promotion of breastfeeding intervention trial (PROBIT): A randomized trial in the Republic of Belarus, *Journal of the American Medical Association*, 285: 413–20.

Kramer, M.S. and Kakuma, R. 2002, *Optimal duration of exclusive breastfeeding* (Cochrane Review), The Cochrane Library, Issue 1, Update Software, Oxford.

Krammer, D., Anderson, A. and Marshall, D. 1999, Living together and eating together: Changes in food choice and eating habits during the transition from single to married/cohabiting, *Sociological Review*, 46: 48–72.

Kristal, A.R., Glanz, K., Tilley, B.C. and Li, S. 2000, Mediating factors in dietary change: Understanding the impact of a worksite nutrition intervention, *Health Education and Behavior*, 27(1): 112–25.

Kristal, A., Goldenhar, L., Muldoon, J. and Morton, R.F. 1997, Evaluation of a supermarket intervention to increase consumption of fruits and vegetables, *American Journal of Health Promotion*, 11: 422–5.

Kuhn, T. 1962, *The structure of scientific revolutions*, University of Chicago Press, Chicago.

Lang, T. and Heasman, M.A. 2004, *Food wars: The global battle for minds, mouths, and markets*, Earthscan, London.

Laslett, P. 2000, *The world we have lost: Further explored*, Routledge, London.

Lauterborn, R. 1990, New marketing litany: 4Ps passé: C-words take over, *Advertising Age*, 26, October.

Lea, E., Crawford, D. and Worsley, A. 2006, Consumers' readiness to eat a plant-based diet, *European Journal of Clinical Nutrition*, 60(3): 342–51.

Lea, E. and Worsley, A. 2003, Benefits and barriers to the consumption of a vegetarian diet, *Public Health Nutrition*, 6(5): 505–11.

—— 2004, What proportion of South Australian adult non-vegetarians hold similar beliefs to non-vegetarians?, *Nutrition and Dietetics*, 61: 11–21.

Lehmann, V. 1998, Patent on seed sterility threatens seed saving, *Biotechnology and Development Monitor*, 35: 6–8.

Levesque, L., Guilbault, G., Delormier, T. and Potvin, L. 2005, Unpacking the black box: A deconstruction of the programming approach and physical activity interventions implemented in the Kahnawake Schools Diabetes Prevention Project, *Health Promotion Practice*, 6(1): 64–71.

Levy, A.S., Schucker, R.E., Tenney, J.E. and Mathews, O. 1985, The impact of a nutrition information program on food purchases, *Journal of Public Policy*, 4: 1–16.

Lewin, K. 1947a, Frontiers in Group Dynamics, *Human Relations*, 1(1): 5–41.

—— 1947b, Frontiers in Group Dynamics II, *Human Relations*, 1(2): 143–53.

Lewis, C.J. and Yetley, E.A. 1992, Focus group sessions on formats of nutrition labels, *Journal of the American Dietetic Association*, 92(1): 62–6.

Lewis, J.M. and Pollard, C.M. 2002, Use of vocational education and training to increase the capacity of industry to improve nutritional health, *Health Promotion Journal Australia*, 13: 3.

Lewis, M.K. and Hill, A.J. 1998, Food advertising on British children's television: A content analysis and experimental study with nine-year-olds, *International Journal of Obesity*, 22(3): 206–14.

Ley, P. 1991, *Guidelines for warning labels: A report to the preventative strategies panel of the public health committee*, DHH, Canberra.

Lieb, S. 1991, Principles of adult learning: Vision, Fall, http://honolulu.hawaii.edu/ intranet/committees/FacDevCom/guidebk/teachtip/adults-2.htm.

Light, L., Portnoy, B., Blair, J., Smith, J.M., Brown Rogers, A., Tuckermanty, E., Tenney, J. and Odonna, M. 1989, Nutrition education in supermarkets, *Family and Community Health*, 12(1): 43–52.

Lowe, C.F., Dowey, A.J. and Horne, P.J. 1998, Changing what children eat, in A. Murcott (ed.), *The nation's diet: The social science of food choice*, Longman, London, pp. 57–80.

Lowe, C., Horne, P., Bowsery, M., Egerton, C. and Tapper, K. 2001, Increasing children's consumption of fruit and vegetables, *Public Health Nutrition*, 4: 387.

Luepker, R.V., Perry, C.L., McKinlay, S.M., Nader, P.R., Parcel, G.S., Stone, E.J., Webber, L.S., Elder, J.P., Feldman, H.A., Johnson, C.C., Kelder, S.H. and Wu, M. (for the CATCH Collaborative Group) 1996, Outcomes of a field trial to improve children's dietary patterns and physical activity, *Journal of the American Medical Association*, 275(10): 768–76.

Maccoby, N. and Farquhar, J.W. 1975, Communication for health: Unselling heart disease, *Journal of Communication*, 25(3): 114–26.

Maccoby, N. and Wood, P.D. 1990, Effects of community-wide education on cardiovascular disease risk factors: The Stanford Five City Project, *Journal of the American Medical Association*, 264(3): 359–65.

Maddock, B., Warren, C. and Worsley, A. 2005, A survey of canteens and food services in Victorian schools, *Nutrition and Dietetics*, 62: 76–81.

Magarey, A. 2000, What are Australian children eating? Implications for obesity, *Proceedings of the 2000 Pre-Olympic Congress*, September, Brisbane.

Magarey, A., Daniels, L.A. and Boulton, T.J.C. 2001, Prevalence of overweight and obesity in Australian children and adolescents: Reassessment of 1985 and 1995 data against new standard international definitions, *Medical Journal of Australia*, 174: 561–4.

Magarey, A., Daniels, L.A. and Smith, A. 2001, Fruit and vegetable intakes of Australians aged 2–18 years: An evaluation of the 1995 National Nutrition Survey data, *Australia and New Zealand Journal of Public Health*, 25(2): 155–61.

Margetts, B.M., Little, P. and Warm, D. 1999, Interaction between physical activity and diet: Implications for blood pressure management in primary care, *Public Health and Nutrition*, 3A: 377–82.

Marmot, M. and Wilkinson, R. (eds) 1999, *Social determinants of health*, Oxford University Press, Oxford.

Maslow, A. 1971, *The farther reaches of human nature*, The Viking Press, New York.

McClelland, J.W., Palmer Keenan, D., Lewis, J., Foerster, S., Sugerman, S., Mara, P., Wu, S., Lee, S., Keller, K., Hersey, J. and Lindquist, C. 2001, Review of evaluation tools used to assess the impact of nutrition education on dietary intake and quality, weight management practices, and physical activity of low-income audiences, *Journal of Nutrition Education*, 33: S35–48.

McCron, R. 1978, *A case study in the promotion of social action through television*, Centre for Mass Communication Research, University of Leicester.

McCron, R. and Budd, J. 1979, Mass communication and health education, in *Health education: Perspectives and choices*, Allen & Unwin, London.

—— 1981, The role of mass media in health education: An analysis, in M. Meyer (ed.), *Health education by television and radio*, Saur, Munich, pp. 118–37.

McIntyre, E., Hiller, J.E. and Turnbull, D. 1999, Determinants of infant feeding practices in a low socio-economic area: Identifying environmental barriers to breastfeeding, *Australia and New Zealand Journal of Public Health*, 23(2): 207–9.

McIntyre, E., Pisaniello, D., Gun, R., Sanders, C. and Frith, D. 2002, Balancing breastfeeding and paid employment: A project targeting employers, women and workplaces, *Health Promotion International*, 17(3): 215–22.

McIntyre, E., Turnbull, D. and Hiller, J.E. 1999, Breastfeeding in public places, *Journal of Human Lactation*, 15(2): 131–5.

McIntyre, S., McKay, L., Cummins, S. and Burns, C. 2005, Out-of-home food outlets and area deprivation: Case study in Glasgow, UK, *International Journal of Behavioral Nutrition and Physical Activity*, 2(16) http://eprints.ncl.ac.uk/file_store/nclep_241200062319.pdf.

McLaughlin, E.W. 2004, The dynamics of fresh fruit and vegetable pricing in the supermarket channel, *Preventive Medicine*, 39: S81–7.

McMichael, A.J., Powles, J., Butler, C.D. and Uauy, R. 2007, Food, livestock production, energy, climate change and health, *Lancet*, 370: 1253–63.

McMichael, P. 2000, *Development and social change: A global perspective*, Pine Forge Press, Thousand Oaks, California.

—— 2007, Global developments in the food system, in M. Lawrence and A. Worsley (eds), *Public health nutrition: From principles to practice*, Allen & Unwin, Sydney.

McNeill, W.H. 1998, *Plagues and peoples*, Anchor, New York.

Mendelsohn, H. 1968, Which shall it be: Mass education or mass persuasion for health?, *American Journal of Public Health*, 58: 131–7.

—— 1973, Some reasons why information campaigns can succeed, *Public Opinion Quarterly*, 37(1): 50–61.

Miaback, E.W., Rothschild, M.L. and Novelli, W.D. 2002, Social marketing, in K. Glenz (ed.), *Health behavior and health education theory: Research and practice* (3rd ed.), Jossey-Bass, San Francisco.

Miller, G.A. 1957, The magical number seven, plus or minus two: Some limits on our capacity for processing information, *Psychological Review*, 63: 81–97.

Miller, W. and Lennie J. 2005, Empowerment evaluation: A practical method for evaluating a school breakfast program, *Proceedings of the Australian Evaluation Society 2005 International Conference*, Brisbane, 10–12 October, www.aes.asn.au.

Montague, M. 2004, *Healthy eating and physical activity: Enhancing policy and practice in Victorian family day care and long day care*, Eat Well Victoria Partnership and the Victorian Health Promotion Foundation, Melbourne.

Moon, L., Meyer, P. and Grau, J. 1999, Australia's young people: Their health and wellbeing, The first report on the health of young people aged 12–24 years by the Australian Institute of Health and Welfare, AIHW, cat. no. PHE19.

Moore, G.F, Tapper, K., Murphy, S., Lynch, R., Pimm, C., Raisanen, L. and Moore, L. 2007, Associations between deprivation, attitudes towards eating breakfast and breakfast eating behaviours in 9–11-year-olds, *Public Health Nutrition*, 10(6): 582–9.

Moore, L., Moore, G.F., Tapper, K., Lynch, R., Desousa, C., Hale, J., Roberts, C. and Murphy, S. 2007, Free breakfasts in schools: Design and conduct of a cluster randomised controlled trial of the Primary School Free Breakfast Initiative in Wales, BioMed Central Public Health, 7: 258.

Moore, V. and Davies, M. 2001, Lifecycle nutrition and cardiovascular health, *Asia Pacific Journal of Clinical Nutrition*, 10: 113–17.

Morley, J., Swinburn, B.A., Metcalf, P.A. and Raza, F. 2002, Fat content of chips, quality of frying fat and deep-frying practices in New Zealand fast food outlets, *Australian and New Zealand Journal of Public Health*, 26: 101–6.

Morton, H. 1990, Television food advertising: A challenge for the new public health in Australia, *Community Health Studies*, 14(2): 153–61.

Muller, S. 2003, *Eating all together: Five times better—Children's nutrition in the West*, Report to the National Child Nutrition Program, Melbourne.

Must, A., Jacques, P.F., Dullal, G.E., Bajema, C.J. and Dietz, W.J. 1992, Long-term morbidity and mortality of overweight adolescents: A follow-up of the Harvard Growth Study of 1922 to 1935, *New England Journal of Medicine*, 327: 1350–5.

Nader, P.R., Stone, E.J., Lytle, L.A., Perry, C.L., Osganian, S.K., Kelder, S., Webber, L.S., Elder, J.P., Montgomery, D., Feldman, H.A., Wu, M., Johnson, C., Parcel, G.S. and Luepker, R.V. 1999, Three-year maintenance of improved diet and physical activity: The CATCH cohort—child and adolescent trial for cardiovascular health, *Archives of Pediatric and Adolescent Medicine*, 153(7): 695–704.

National Health and Medical Research Council (NHMRC) 1998, *The Australian guide to healthy eating*, NHMRC, Canberra.

—— 1999, *A guide to the development, implementation and evaluation of clinical practice guidelines*, NHMRC, Canberra.

—— 2003a, *Dietary guidelines for Australians*, NHMRC, Canberra.

—— 2003b, *The Australian dietary guidelines for children and adolescents*, NHMRC, Canberra.

—— 2005, *Nutrient reference values for Australia and New Zealand including recommended dietary intakes*, NHMRC, Canberra.

National Heart Foundation (NHF) 2003, *Eat smart, play smart: A manual for out of school hours care*, NHF, Melbourne.

—— 2006, *Summary of evidence statement on the relationships between dietary electrolytes and cardiovascular disease*, NHF, Melbourne.

Noelle Newmann, E. 1973, Return to the concept of powerful media, *Studies of Broadcasting*, 9: 67–112.

Nowson, C., Booth, A., Worsley, A., Margerison, C. and Jorna, M. 2004, Approaches to weight loss with increased fruit, vegetables and dairy,

Proceedings Dietitians Association of Australia 22nd National Conference, Melbourne, p. 137.

Nowson, C.A. and Margerison, C. 2002, Vitamin D intake and vitamin D status of Australians, *Medical Journal of Australia*, 177(3): 149–52.

Nowson, C.A., Worsley, A., Margerison, C., Jorna, M.K., Frame, A.G., Torres, S.J. and Godfrey, S. 2004, Blood pressure response to dietary modification in free-living individuals, *Journal of Nutrition*, 134: 2322–9.

Nowson, C.A., Worsley, A., Margerison, C., Jorna, M.K., Godfrey, S.J. and Booth, A.O. 2005, Blood pressure change with weight loss is affected by diet type in men, *American Journal of Clinical Nutrition*, 81(5): 983–9.

Nuffield Foundation Report on Public Health Ethics 2007, www.nuffield bioethics.org (sourced 2 March 2008).

Nutbeam, D. 2000, Health literacy as a public health goal: a challenge for contemporary health education and communication strategies into the 21st century, *Health Promotion International*, 15: 259–67.

Nutbeam, D. and Harris, E. 2003, *Theory in a nutshell: A guide to health promotion theory* (2nd ed.), McGraw Hill, Sydney.

Nutbeam, D. and St Ledger, L. 1997, *Priorities for research into health promoting schools in Australia*, Australian Health Promoting Schools Association, University of Sydney, Sydney.

Nutrition Society of Australia 2007, Specialist Competencies in Nutrition Science, http://www.nsa.asn.au/documents/SpecialistCompetenciesinNutritionScience.pdf.

O'Dea, J. 2003, Why do kids eat healthful food? Perceived benefits of and barriers to healthful eating and physical activity among children and adolescents, *Journal of the American Dental Association*, 103: 497–501.

O'Loughlin, J., Renaud, L., Richard, L., Gomez, L.S. and Paradis, G. 1998, Correlates of the sustainability of community-based heart health promotion interventions, *Preventive Medicine*, 27: 702–12.

Oldenburg, B., Sallis, J.F., Harris, D. and Owen, N. 2002, Checklist of health promotions environments at worksites (CHEW), *American Journal of Health Promotion*, 16(5): 288–99.

Olson, C.M. and Holben, D.H. 2002, Position of the American Dietetic Association—Domestic Food and Nutrition Security, *Journal of the American Dietetic Association*, 102(12): 1840–7(8).

Ounpuu, S. 1996, *Methodological considerations for application of the transtheoretical model to dietary fat reduction*, Nutrition Department, University of Guelph, Canada.

Paradis, G., Levesque, L., Macaulay, A.C., Cargo, M., McComber, A., Kirby, R., Receveur, O., Kishchuk, N. and Potvin, L. 2005, Impact of a diabetes prevention program on body size, physical activity, and diet among Kanien'keha:ka (Mohawk) children 6 to 11 years old: 8-year results from the Kahnawake Schools Diabetes Prevention Project, *Pediatrics*, 115(2) 333–9.

Parke, R.D. 2004, Development in the family, *Annual Review of Psychology*, 55: 365–99.

Patterson, R.E., Kristal, A.R., Bienerl Varnes, J., Feng, Z., Glanz, K., Stables, G., Chamberlain, R.M. and Probart, C. 1998, Durability and diffusion of the nutrition intervention in the Working Well trial, *Preventive Medicine*, 27: 668–73.

Patterson, T.E. and McClure, R.D. 1976, *The unseeing eye: The myth of television power in national politics*, Putnam, New York.

Payne, J.R., Bettman, J.R. and Johnson, E.J. 1992, Behavioural decision research: A constructive processing perspective, *Annual Review of Psychology*, 43: 87–132.

Payne, J.W., Bettman, J.R. and Schkade, D.A. 1999, Measuring constructed preferences: Towards a building code, *Journal of Risk and Uncertainty*, 19: 243–70.

Pennington, J.A.T., Wisniowski, L.A. and Logan, G.B. 1988, In-store nutrition information programs, *Journal of Nutrition Education*, 20: 5–10.

Pereira, M.A., Kartashov, A.I., Ebbeling, C.B., Van Horn, L., Slattery, M.L., Jacobs Jr, D.R. and Ludwig, D.S. 2005, Fast-food habits, weight gain, and insulin resistance (the CARDIA study): 15-year prospective analysis, *The Lancet*, 365(9453): 36–42.

Pilgrim, F.J. and Kamen, M.K. 1959, Patterns of food preferences through factor analysis, *Journal of Marketing*, 24: 68–72.

Pirog, R. 2001, Institutional food markets: The Iowa experience, www.ag.iastate.edu/centers/Leopold.

—— 2003. The food odometer: Comparing food miles for local versus conventional produce sales to Iowa institutions, www.leopold.iastate.edu/pubs/staff/files/food_travel072103.pdf.

Pirog, Rich and Schuh, Patrick 2002, The load less traveled: Examining the potential of using food miles in ecolabels, *Proceedings from Ecolabels and the Greening of the Food Market Conference*, November, p. 69.

Pollard, C. 2001, Healthy eating bears fruit for WA schools, *Healthview* (publication of Department of Health, Western Australia), Summer 2001–02: 1.

Pollard, C.M., Lewis, J.M., Barkess, J. and Toquero, A. 2002, The development and preliminary testing of an invoice-based menu assessment tool for long day childcare centres, *Nutrition and Diet Journal*, 59: 3.

Pollard, C.M, Lewis, J.M. and Miller, M.R. 1999, Food service in long day care centres—an opportunity for public health intervention, *Australian and New Zealand Journal of Public Health*, 23: 6.

—— 2001, Start right—eat right award scheme: Implementing food and nutrition policy in child care centers, *Health Education Behaviour*, 28(3): 320–30.

Pollay, R. 1986, The distorted mirror: Reflections on the unintended consequences of advertising, *Journal of Marketing*, 50: 18–36.

—— 1987, On the value of reflections on the values in 'the distorted mirror', *Journal of Marketing*, 51: 104–9.

Pollitt, E. 1993, Iron deficiency and cognitive function, *Annual Review of Nutrition*, 13: 521–37.

—— 2001, The developmental and probabilistic nature of the functional consequences of iron-deficiency anemia in children, *Journal of Nutrition*, 131(2S–2): 669S–75S.

Pollitt, E. and Mathews, R. 1998, Breakfast and cognition: An integrative summary, *American Journal of Clinical Nutrition*, 67, Supplement: S805–13.

Pomerleau, J., Lock, K., Knai, C. and McKee, M. 2005, Interventions designed to increase adult fruit and vegetable intake can be effective: A systematic review of the literature, *Journal of Nutrition*, 135: 2486–95.

Pomeroy, S. 2007, General Practitioners' nutrition promotion among cardiac patients. PhD thesis, Deakin University, Melbourne.

Popkin, B.M. 2001, The Nutrition transition and obesity in the developing world, *Journal of Nutrition*, 131: 871S–3S.

Powers, W.T. 1979, Quantitative analysis of purposive systems: Some spadework at the foundations of scientific psychology, *Psychological Review*, 85(5): 417–35.

Preamble to the Constitution of the World Health Organization as adopted by the International Health Conference, New York, 19–22 June 1946.

Price, G., MacKay, S. and Swinburn, B. 2000, The Heartbeat Challenge Program: Promoting healthy changes in New Zealand workplaces, *Health Promotion International*, 15(1): 49–55.

Prochaska, J. and DiClemente, C. 1984, *The transtheoretical approach: Crossing traditional boundaries of therapy*, Dow Jones-Irwin, Homewood, Illinois.

Pusey, M. 1991, *Economic rationalism in Canberra: A nation-building state changes its mind*, Cambridge University Press, Cambridge.

—— 2003, *The experience of middle Australia: The dark side of economic reform*, Cambridge University Press, Melbourne.

Puska, P., Nissinen, A., Tuomilehto, J. et al. 1985, The community-based strategy to prevent coronary heart disease: Conclusions from the 10 years of the North Karelia Project, *Annual Review of Public Health*, 6: 147–93.

Puska, P., Salonen, J.T., Nissinen, A., Tuomilehto, J. and Vartianinen, E. et al. 1983, Change in risk factors for coronary heart disease during 10 years of a community intervention program (North Karelia project), *British Medicine Journal*, 267: 1840–4.

Quinn, F. 1990, *Crowning the Customer: How to become customer driven*, Dublin O'Brien Press.

Ransley, J.K., Donnelly, J.K., Khara, T.N., Botham, H., Arnot, H., Greenwood, D.C. and Cade, J.E. 2001, The use of supermarket till receipts to determine the fat and energy intake in a UK population, *Public Health Nutrition*, 4(6): 1279–86.

Rasmussen, M., Krølner, R., Klepp, K.I., Lytle, L., Brug, J., Bere, E. and Due, P. 2006, Determinants of fruit and vegetable consumption among children and adolescents: A review of the literature. Part I: Quantitative studies, *International Journal of Behavioral Nutrition and Physical Activity*, 3: 22.

Raynor, M. 1992, *Just read the label*, Coronary Prevention Group, London.

Raynor, M., Scarborough, P., Stockley, L. and Boxer, A. 2005, *Nutrient profiles: Further refinement and testing of model SSCg3d*, FSA, London, www.food. gov.uk/multimedia/pdfs/npreportsept05.pdf.

Reidpath, D., Burns, C., Garrard, J., Mahoney, M. and Townsend, M. 2002, An ecological study of the relationship between social and environmental determinants of obesity, *Health and Place*, 8: 141–5.

Resnicow, K., Cross, D. and Wynder, E. 1993, The Know Your Body program: A review of evaluation studies, *Bulletin of the New York Academy of Medicine*, 70(3): 188–207.

Resnicow, K., Wallace, D.C., Jackson, A., Digirolamo, A. and Odom, E., Wang, D.T., Dudley, W.N., Davis, M. and Baranowski, T. 2000, Dietary change through African American churches: Baseline results and program description of the Eat for Life trial, *Journal of Cancer Education*, 15: 156–63.

Reynolds, J. 2000a, A rationale for an empowerment approach to school-based nutrition education, *Journal of the Home Economics Institute of Australia*, 7(4): 14–26.

—— 2000b, Learning and teaching considerations for school-based, adolescent, food and nutrition education, *Journal of the Home Economics Institute of Australia*, 7(1): 14–26.

—— 2000c, The construction of food-related behaviour and implications for school-based adolescent food and nutrition education, *Journal of the Home Economics Institute of Australia*, 7(1): 2–13.

Rice, C. 1993, *Consumer behaviour: Behavioural aspects of marketing*, Butterworth Heinemann, London.

Rodgers, A.B., Kessler, L.G., Portnoy, B., Potosky, A.L., Patterson, B., Tenney, J., Thompson, F.E., Krebs-Smith, S.M., Breen, N. and Mathews, O. 1994 Eat for Health: A supermarket intervention for nutrition and cancer risk reduction, *American Journal of Public Health*, 84: 72–6.

Rogers, E.M. 1995, *The diffusion of innovations*, Free Press, New York.

Rogers, E.M. and Shoemaker, F.F. 1971, *Communication of innovations: A cross cultural approach*, Free Press, New York.

Rogers, I.S., Emmett, P.M. and Golding, J. 1997, The incidence and duration of breast feeding, *Early Human Development*, 49, Supplement: S45–74.

Rolls, B.J., Castellanos, V.H., Halford, J.C., Kilara, A., Panyam, D. and Pelkman, C.L. 1998, Volume of food consumed affects satiety in men, *American Clinical Nutrition*, 67: 1170–7.

Rose, G. 1985, Sick individuals and sick population, *International Journal of Epidemiology*, 14: 32–8.

Rosenthal, R., Rosnow, R.L. and Rubin, D.B. 2000, *Contrasts and effect sizes in behavioral research: A correlational approach*, Cambridge University Press, Cambridge.

Rothschild, M.L. 1999, Carrots, sticks, and promises: A conceptual framework for the management of public health and social issue behaviors, *Journal of Marketing*, 63: 24–37.

Rozin, P. 1987, Psychobiological perspectives on food preferences and avoidances, *Annual Review of Nutrition*, 6: 433–56.

Rozin, P. and Vollmecke, T. 1986, Food likes and dislikes, *Annual Review of Nutrition*, 6: 433–56.

Royal Australian College of General Practitioners (RACGP) 2004, *Smoking, nutrition, alcohol and physical activity (SNAP): A population health guide to behavioural risk factors in general practice*, Royal Australian College of General Practitioners, Melbourne.

Rubinelli, S. and Haes, J. (eds) 2005, Tailoring health messages: Bridging the gap between social and humanistic perspectives on health communication, *Proceedings of the International Conference*, Monte Verità, 6–10 July, www.theme.usilu.net/pages/THEME_proceedings.pdf.

Rudder, G. 1999, Food in focus, *Retail World*, 52(20): 20; 52(21): 7; 52(22): 7; 52(23): 6–7; 52(24): 6–7.

Russell, G. 2007, Contributions of parent socialisation to children's food preferences, PhD thesis, Deakin University.

Russell, G. and Worsley, A. 2007a, Children's food preferences, *Public Health Nutrition*, in press.

—— 2007b, Do children's food preferences align with dietary recommendations?, *European Journal of Clinical Nutrition*, in press.

Sahay, T.B., Ashbury, F.D., Roberts, M. and Rootman, I. 2006, Effective components for nutrition interventions: A review and application of the literature, *Health Promotion Practice*, 7(4): 418–27.

Sandman, P.M. 1987, Risk communication: Facing public outrage, *EPA Journal*, 21–22 November.

—— 1989, Hazard versus outrage in the public perception of risk, in V.T. Covello, D.B. McCallum and M. Pavlov (eds), *Effective risk communication*, Plenum Press, New York.

—— 1993, *Responding to community outrage: Strategies for effective risk communication*, American Industrial Hygiene Association, Fairfax, Virginia.

Sandman, P.M. and Lanard, J. 2003, *Risk communication recommendations for infectious disease outbreaks*, World Health Organization, Geneva, www.psandman.com/articles/who-srac.htm.

Savige, G., Wahlqvist, M.L., Lee, D. and Snelson, B. 2001, *Agefit: Fitness and nutrition for an independent future*, Pan Macmillan, Melbourne.

Scarborough, P., Rayner, M. and Stockley, L. 2007, Developing nutrient profile models: A systematic approach, *Public Health Nutrition*, 10: 330–6.

Schramm, W. 1971, The nature of communication between humans, in *The process and effects of mass communication*, Free Press, Glencoe, Illinois.

Schucker, R.E., Levy, A.S., Tenney, J.E. and Mathews, O. 1992, Nutrition shelf labeling and consumer purchase behavior, *Journal of Nutrition Education*, 24: 75–81.

Schwartz, S.H. 1992, Universals in the content and structure of values: Theoretical advances and empirical tests in 20 countries, *Advances in Experimental Social Psychology*, 25: 1–65.

Scientific Advisory Committee on Nutrition 2003, *Salt and health*, The Stationery Office, London.

Scott, V. and Worsley, A. 1994, Ticks, claims, tables and food groups: A comparison for nutritional labelling, *Health Promotion International*, 9: 27–38.

—— 1997, Consumer views on nutrition labels in New Zealand, *Australian Journal of Nutrition and Dietetics*, 54(1): 6–13.

Seattle 5 a Day Worksite Program to Increase Fruit and Vegetable Consumption 2001, *Preventive Medicine*, 32: 230–8.

Setter, T., Kouris-Blazos, A. and Wahlqvist, M. 2000, *School based healthy eating initiatives: Recommendations for success*, Healthy Eating Healthy Living Program (Monash University) and VicHealth, Melbourne.

Seymour, J.D., Yaroch, A.L., Serdula, M., Blanck, M. and Khan, L.K. 2004, Impact of nutrition environmental interventions on point-of-purchase behavior in adults: A review, *Preventive Medicine*, 39: S108–36.

Shaddish, W.R, Cook, T. and Campbell, D.T. 2002, *Experimental and quasi-experimental designs for generalized causal inference*, Houghton Mifflin, New York.

Shah, M., French, S.A., Jeffery, R.W., McGovern, P.G., Forster, J.L. and Lando, H.A. 1993, Correlates of high fat/calorie food intake in a worksite population: the Healthy Worker Project, *Addictive Behavior*, 18(5): 583–94.

Shannon, C.E. 1949, Synthesis of two-terminal switching circuits, *Bell Systems Technical Journal*, 28: 656–715.

Sharp, D. 2004, Labelling salt in food: If yes, how?, *The Lancet*, 364(9451): 2079–81.

Shea, S. and Basch, C.E. 1990, A review of five major community-based cardio-vascular disease prevention programs. Pt. II: Intervention strategies, evaluation methods, and results, *American Journal of Health Promotion* 4: 279–87.

Shine, A., O'Reilly, S. and O'Sullivan, K. 1997, Consumer attitudes to nutrition labelling, *British Food Journal*, 99: 283–9.

Sikorski, J., Renfrew, M.J., Pindoria, S. and Wade, A. 2003, Support for breast-feeding mothers: A systematic review, *Paediatric and Perinatal Epidemiology*, 17: 407–17.

Sims, L.S. 1978, Food-related value-orientations, attitudes and beliefs of vegetarians and non vegetarians, *Ecology of Food and Nutrition*, 7: 23–35.

Skemp, R.R. 1979, *Intelligence, learning and action: A foundation for theory and practice in education*, John Wiley, London.

Sloane, D.C., Diamant, A.L., Lewis, L.B., Yancey, A.K., Flynn, G., Nascimento, L.M., McCarthy, W.J., Guinyard, J.J. and Cousineau, M.R. 2003, REACH Coalition of the African American Building a Legacy of Health Project: Improving the nutritional resource environment for healthy living through community-based participatory research, *Journal of General Internal Medicine*, 18(7): 568–75.

Slovic, P. 1987, Perception of risk, *Science*, 236(4799): 280–5.

Smith, A., Kellett, E. and Schmerlaib, Y. 1998, *The Australian guide to healthy eating: Background information for nutrition educators*, NHMRC, Canberra.

Smith, G.P. and Thorwart, M.L. 1998, Volume of food consumed affects satiety in men, *American Journal of Clinical Nutrition*, 67: 1170–7.

Soloman, M.R. 1994, *Consumer behavior*, Allyn and Bacon, Needham Heights, MA.

Sorensen, G., Emmons, K., Hunt, M.K. and Johnson, D. 1998, Implications of the results of community intervention trials, *Annual Review of Public Health*, 19: 379–416.

Sorensen, G., Hunt, M.K., Cohen, N., Stoddard, A. and Stein, E. 1998, Worksite and family education for dietary change: The Treatwell 5-a-Day Program, *Health Education Research*, 13(4): 577–91.

Sorensen, G., Linnan, L. and Hunt, M.K. 2004, Worksite-based research and initiatives to increase fruit and vegetable consumption, *Preventive Medicine*, 39: S94–100.

Sorensen, G., Stoddard, A., Peterson, K., Cohen, N., Hunt, M.K., Stein, E., Palombo, R. and Lederman, R. 1999, Increasing fruit and vegetable consumption through worksites and families in the Treatwell 5-a-Day Study, *American Journal of Public Health*, 89(1): 54–60.

St Jeor, S.T., Krenkel, J.A., Plodkowski, R.A., Veach, T.L., Tolles, R.L. and Kimmel, J.H. 2006, Medical nutrition: A comprehensive, school-wide curriculum review, *American Journal of Clinical Nutrition*, 83(4): 963S–7S.

Stanton, R. 1999, What's happening in nutrition?, *Proceedings of the Home Economics Society of Australia*, January: 47–53.

Steenhuis, I.H., van Assema, P. and Glanz, K. 2001, Strengthening environmental and educational nutrition programmes in worksite cafeterias and supermarkets in The Netherlands, *Health Promotion International*, 16(1): 21–33.

Stephen, L. 1990, *The English utilitarians—volume 2*, Duckworth, London.

Stephenson, L.S., Latham, M.C. and Ottesen, E.A. 2000, Global malnutrition, *Parasitology*, 121: S5–22.

Steptoe, A., Pollard, T.M. and Wardle, J. 1995, Development of a measure of the motives underlying the selection of food: The Food Choice Questionnaire, *Appetite*, 25: 267–84.

Story, M. and French, S. 2004, Food advertising and marketing directed at children and adolescents in the US, *International Journal of Behavioural Nutrition and Physical Activity*, 1: 3, 1–17.

Story, M., Lytle, L.A., Birnbaum, A.S. and Perry, C.L. 2002, Peer-led, school-based nutrition education for young adolescents: Feasibility and process evaluation of the TEENS study, *Journal of Scholastic Health*, (3): 121–7.

Strand, B.H. and Tverdal, A. 2004, Can cardiovascular risk factors and lifestyle explain the educational inequalities in mortality from ischaemic heart disease and from other heart diseases? 26 year follow up of 50,000 Norwegian men and women, *Journal of Epidemiology and Community Health*, 58(8): 705–9.

Swinburn, B., Egger, G. and Raza, F. 1999, Dissecting obesogenic environments: The development and application of a framework for identifying and prioritizing environmental interventions for obesity, *Preventive Medicine*, 29: 563–70.

Synovate 2005, *Qualitative evaluation of alternative food signposting concepts: Report of Findings for UK Food Standards Agency*, November.

Tansey, G. and Worsley, A. 1995, *The food system: A guide*, Earthscan, London.

Tapper, K., Horne, P.J. and Lowe, C.F. 2003, Food dudes to the rescue, *The Psychologist*, 16: 18–21.

Tapper, K., Murphy, S., Moore, L., Lynch, R. and Clark, R. 2007, Evaluating the Free School Breakfast Initiative in Wales: Methodological Issues, *British Food Journal*, 109: 206–15.

Temple, E.C., Hutchinson, I., Laing, D.G. and Jinks, A.L. 2002, Taste development: Differential growth rates of tongue regions in humans, *Developmental Brain Research*, 135: 65–70.

Thomas, J., Sutcliffe, K., Harden, A., Oakley, A., Oliver, S., Rees, R., Brunton, G. and Kavanagh, J. 2003, *Children and healthy eating: A systematic review of barriers and facilitators*, EPPI-Centre, Social Science Research Unit, Institute of Education, University of London.

Thompson, B., Coronado, G., Snipes, S.A. and Puschel, K. 2003, Methodologic advances and ongoing challenges in designing community-based health promotion programs, *Annual Review of Public Health*, 24: 315–49.

Thompson, B. and Kinne, S. 1999, Social change theory: Applications to community health, in N. Bracht (ed.), *Health promotion at the community level: New advances*, Sage, Thousand Oaks, California, pp. 45–65.

Thompson, B., Shannon, J., Beresford, S.A.A., Jacobson, P.E. and Ewings, J.A. 1995, Implementation aspects of the Seattle '5 a Day' intervention project: Strategies to help employees make dietary changes, *Topics in Clinical Nutrition*, 11: 58–75.

Thompson, B., Wallack, I., Lichtenstein, E. and Pedchacek, T. 1991, Principles of community organisation and partnership for smoking cessation in the Community Intervention Trial for Smoking Cessation (COMMIT), *International Quarterly of Community Health Education*, 11: 187–200.

Tilley, B.C., Glanz, K., Kristal, A.R., Hirst, K., Li, S.S., Vernon, S.W. and Myers, R. 1999, Nutrition intervention for high-risk auto workers: Results of the Next Step Trial, *Preventive Medicine*, 28: 284–92.

Timperio, A., Ball, K., Roberts, R. and Salmon, J. 2005, *The built environment and children's eating, physical activity and weight status*, Report to VicHealth, Melbourne.

Tones, K. and Tilford, S. 2001, *Health education: Effectiveness, efficiency and equity*, Chapman Hall, London.

Tooty Fruity Vegie in Preschools: Preliminary Evaluation Report 2008, North Coast Area Health Service, NSW Health.

Trichopoulou, A., Costacou, T., Bamia, C. and Trichopoulos, D. 2003, Adherence to a Mediterranean diet and survival, *New England Journal of Medicine*, 348: 2599–608.

Tudor-Smith, C., Nutbeam, D., Moore, L. and Catford, J. 1998, Effects of the Heartbeat Wales programme over five years on behavioural risks for cardiovascular disease: Quasi-experimental comparison of results from Wales and a matched reference area, *British Medical Journal*, 316: 818–22.

Tugwell, P., de Sabigny, D., Hawker, G. and Robinson, V. 2006, Applying clinical epidemiological methods to health equity: The equity effectiveness loop, *British Medical Journal*, 332: 358–61.

Turk, T., Ewing, M. and Newton, F.J. 2006, Using ambient media to promote HIV/AIDS protective behaviour change, *International Journal of Advertising*, 25(3): 333–59.

Turrell, G., Hewitt, B., Patterson, C., Oldenburg, B. and Gould, T. 2002, Socioeconomic differences in food purchasing behaviour and suggested implications for diet-related health promotion, *Journal of Human Nutrition and Dietetics*, 15(5): 355–64.

Turrell, G., Stanley, L., de Looper, M. and Oldenburg, B. 2006, *Health inequalities in Australia: Morbidity, health behaviours, risk factors and health service use*, Health Inequalities Monitoring Series No. 2, AIHW cat. no. PHE 72, Queensland University of Technology and the Australian Institute of Health and Welfare, Canberra, available at www.aihw.gov.au/publications/index.cfm/ title/10041 (sourced 2 March 2008).

Twigg, J. 1979, Food for thought: Purity and vegetarianism, *Religion*, 9: 13–35.

Underhill, P. 2000, *Why we buy: The science of shopping*, Simon and Schuster, New York.

UNICEF 1997, *The state of the world's children*, Oxford University Press and UNICEF, Oxford.

US Census Bureau 2002, Americans with Disabilities: (P70-107), http://www.census. gov/hhes/www/disability/sipp/disable02.html (sourced 2 March 2008).

—— 2003, *Disability status: 2000*, www.census.gov/prov/2003pubs/c2bkr.pdf.

Van Binsbergen, J.P. and Drenthen, A.J.M. 2003, Patient information letters on nutrition: Development and implementation, *American Journal of Clinical Nutrition*, 77, Supplement: 1035S–1038S.

van der Maesen, L.J. and Nijhuis, H.G. 2000, Continuing the debate on the philosophy of modern public health: Social quality as a point of reference, *Journal of Epidemiological Community Health*, 54(2): 134–42.

Verheijden, M.W. and Kok, F.J. 2005, Public health impact of community-based nutrition and lifestyle interventions, *European Journal of Clinical Nutrition*, 39 (supplement1) S66–S76.

Victorian Population Health Survey 2006, Department of Human Services, Victoria, Australia 2006, http://www.health.vic.gov.au/healthstatus/downloads /vphs.2006/vphs2006.pdf.

Wahlqvist, M.L. (ed.) 1995, *Food habits in later life: A cross cultural study*, CD ROM, United Nations University Press, Tokyo.

Wahlqvist, M.L., Darmadi-Blackberry, I., Kouris-Blazos, A., Jolley, D., Steen, B., Lukito, W. and Horie, Y. 2005, Does diet matter for survival in long-lived cultures?, *Asia Pacific Journal of Clinical Nutrition*, 14(1): 2–6.

Wallack, L.M. 1981, Mass media campaigns: The odds against finding behavior change, *Health Education Quarterly*, 8(3): 209–60.

—— 1983, Mass media campaigns in a hostile environment, *Journal of Drug Education*, 28(2): 51–63.

Wandel, M. 1994, Understanding consumer concern about food-related health risks, *British Food Journal*, 96(7): 35–40.

—— 1997, Food labelling from a consumer perspective, *British Food Journal*, 99: 212–19.

Wanless, D. 2002, Securing Our Future Health: Taking a Long-Term View Final Report, London: HM Treasury, http://www.hm-treasury.gov.uk/Consultations_ and_legislation/wanless/consult_wanless_final.cfm.

Ward, M. 2006, Nutrient Reference Values (NRV), *Pabulum*, 62: 13.

Wardle, J., Parmenter, K. and Waller, J. 2000, Nutrition knowledge and food intake, *Appetite*, 34: 269–75.

Waring, Marilyn 1998, *Counting for nothing: What men value and what women are worth*, Bridget Williams Books, Wellington, New Zealand.

Warren, R. 1958, Toward a reformulation of community theory, *Community Development Review*, 9: 41–8.

Wass, A. 1994, *Promoting health: The primary health care approach*, WB Saunders Bailliere Tindall, Sydney.

Watkins, J. 1997, Top notch warts and all: Obituary of Karl Popper, in *The British Academy, Obituary of Karl Raimund Popper: 1902–1994, Proceedings of the British Academy*, 94: 645–84, www.britac.ac.uk/pubs/src/popper.

Watt, G.C.M. 1996, All together now: Why social deprivation matters to everyone, *British Medical Journal*, 312: 1026–9.

Weil, P. and Harmata, R. 2002, Rekindling the flame: Routine practices that promote hospital community leadership, *Journal of Healthcare Management*, 47: 98–109.

Weinstein, N.D. and Rothman, A.J. 2005, Commentary: Revitalizing research on health behaviour theories, *Health Education Research*, 20(3): 294–7.

Wheelock, V. 1992, Healthy eating: The food issue of the 1990s, *British Food Journal*, 94(2): 3–8.

—— (ed.) 2006, *Healthy eating in schools: A handbook of practical case studies*, Verner Wheelock Associates, Skipton, Yorks, UK.

Whelton, P.K., Appel, L.J., Espeland, M.A., Applegate, W.B., Ettinger Jr, W.H., Kostis, J.B., Kumanyika, S., Lacy, C.R., Johnson, K.C., Folmar, S. and Cutler, J.A. for the TONE Collaborative Skipton Yorks 1998, Sodium reduction and weight loss in the treatment of hypertension in older persons: A randomized controlled trial of nonpharmacologic interventions in the elderly (TONE), *Journal of the American Medical Association*, 279: 839–46.

Whitehead, D. 2004, The European Health Promoting Hospitals (HPH) project: How far on?, *Health Promotion International*, 19: 259–67.

Whitehead, F. 1973, Nutrition education research, *World Review of Nutrition and Dietetics*, 17: 91–149.

Wilcox, B.J., Suzuki, M., Willcox, C.D. and Weil, A. 2001, *The Okinawa way*, Mermaid Books, Williamsberg, Virginia.

Wilkinson, R. and Marmot, M. (eds) 2003, Social determinants of health: The solid facts, 2nd edition, World Health Organization, Copenhagen.

Williams, P.L., McIntyre, L., Dayle, J.B. and Raine, K. 2003, The 'wonderfulness' of children's feeding programs, *Health Promotion International*, 18, (2): 163–70.

Wilson, A.C., Forsyth, J.S., Greene, S.A., Irvine, L., Hau, C. and Howie, P.W. 1998, Relation of infant diet to childhood health: Seven year follow up of cohort of children in Dundee infant feeding study, *British Medical Journal*, 316: 21–5.

Wilson, D.H., Wakefield, M.A., Steven, I.D., Rohrsheim, R.A., Esterman, A.J. and Graham, N.M.H. 1990, 'Sick of smoking': Evaluation of a targeted minimal smoking cessation intervention in general practice, *Medical Journal of Australia*, 152: 518–21.

Wilson, M.G., Holman, P.B. and Hammock, A. 1996, A comprehensive review of the effects of worksite health promotion on health-related outcomes, *American Journal of Health Promotion*, 10(6): 429–35.

Willett, W.C., Dietz, W.H. and Colditz, G.A. 1999, Guidelines for Healthy Weight, *New England Journal of Medicine*, 341: 427–34.

Winkleby, M.A., Taylor, B., Jatulis, D. and Fortmann, S.P. 1996, The long-term effects of a cardiovascular disease prevention trial: The Stanford Five-City Project, *American Journal of Public Health*, 86: 1773–9.

Winkler, E., Turrell, G. and Paterson, C. 2006, Does living in a disadvantaged area entail limited opportunities to purchase fresh fruit and vegetables in terms of price, availability and variety?, *Health and Place*, 12: 306–19.

World Cancer Research Fund and the American Institute for Cancer Research 1997, *Food, nutrition and the prevention of cancer: A global perspective*, American Institute for Cancer Research, New York.

—— 2007, *Food, nutrition, physical activity and the prevention of cancer*: A global perspective, AICR, Washington DC, pp. 48–62.

World Health Organization 1946, Preamble to the Consitution of the World Health Organization, International Health Conference, New York, 19–22 June, (in official Records of the World Health Organization, no. 2, p. 100).

World Health Organization (WHO)1981, *International code of marketing of breast-milk substitutes*, Geneva.

—— 1986, *Ottawa Charter for health promotion*, WHO, Geneva.

—— 1990, *International code of marketing of breast-milk substitutes: Synthesis of reports on action taken (1981–90)*, WHO/MCH/NUT/90.1, Geneva.

—— 1991, *The Budapest Declaration of Health Promoting Hospitals*, WHO, Copenhagen.

—— 1996, *Ljubjana Charter on Reforming Health Care*, WHO, Copenhagen.

—— 1997, *The Vienna Recommendations on Health Promoting Hospitals*, WHO, Copenhagen.

—— 1998, *Health promotion evaluation: Recommendations to policymakers: Report of the WHO European Working Group on Health Promotion Evaluation*, WHO, Copenhagen.

—— 2000, *Obesity: Preventing and managing the global epidemic*, WHO Obesity Technical Report Series 894, WHO, Geneva.

—— 2002, *The world health report 2002: Reducing risks, promoting healthy life*, WHO, Copenhagen.

WHO Collaborative Study Team on the Role of Breastfeeding on the Prevention of Infant Mortality 2000, Effect of breastfeeding on infant and child mortality due to infectious disease in less developed countries: A polled analysis, *The Lancet*, 355: 451–5.

Worsley, A. 1988, Cohabitation–gender effects on human food consumption, *International Journal of Biosocial Sciences*, 10: 107–22.

—— 1991, Mother's work and food consumption: Going out to work changes mother's food diets?, *Ecology of Food and Nutrition*, 25: 59–69.

—— 1996, Which nutrition information do shoppers want on food labels?, *Asia Pacific Journal of Clinical Nutrition*, 5: 70–8.

—— 2006a, Lay people's views of school food policy options: Associations with confidence, personal values and demographics, *Health Education Research*, 22 November 2006, info:doi/10.1093/her/cyl138.

—— 2006b, Lay people's views of school food policy options: Associations with confidence, personal values and demographics, submitted to *Appetite*.

—— 2007, Lay people's views of school children's food services: Demographic associations, *British Food Journal*, 109(6): 429–42.

Worsley, A. and Adams, J. 1992, Point of purchase nutrition promotion: Literature review, unpublished manuscript, Food Policy Research Unit, CSIRO Division of Human Nutrition, Adelaide.

Worsley, A., Baghurst, K.I. and Skrzypiec, G. 1995, Meat consumption and young people: Final report to the Meat Research Corporation, CSIRO, Adelaide.

Worsley, A., Blasche, R., Ball, K. and Crawford, D. 2004, The relationship between education and food consumption in the 1995 Australian National Nutrition Survey, *Public Health Nutrition*, 7(5), 649–63.

Worsley, A., Coonan, W. and Worsley, A.J. 1987, The first Body Owner's Programme: An integrated school-based physical and nutrition education programme, *Health Promotion*, 2, 39–49.

Worsley, A. and Crawford, D. 2005a, *Children's healthy eating: What works*, Report of the Review of Children's Healthy Eating Interventions to the Department of Human Services, Victoria.

—— 2005b, *Promoting healthy eating for children: A planning guide for practitioners*, Department of Human Services, Victoria.

Worsley, A. and Lea, E. 2003, Personal values and consumers' trust in sources of nutrition information, *Ecology of Food and Nutrition*, 42: 129–51.

Worsley, A. and Murphy, S. 1994, Sectorial attitudes to the food system: Sources of agreement and disagreement, *Health Promotion International*, 231–40.

Worsley, A. and Scott, V. 2000, Exploratory studies of consumers' concerns about food and health in Australia and New Zealand, *Asia Pacific Journal of Clinical Nutrition*, 9: 24–32.

Worsley, A. and Shannon, P. 1991, *Interim report of the evaluation of the South Australian Health and Social Welfare Councils*, Social Health Branch, South Australian Health Commission, Adelaide.

Worsley, A. and Skrzypiec, G. 1997, Teenage vegetarianism: Beauty or the beast?, *Nutrition Research*, (3): 391–404.

—— 1998a, Teenage vegetarianism: Prevalence, social and cognitive contexts, *Appetite*, 30: 151–70.

—— 1998b, Personal predictors of consumers' food and health concerns, *Asia-Pacific Journal of Clinical Nutrition*, 7: 15–23.

Worsley, A. and Worsley, A.J. 1991, New Zealand general practitioners nutrition opinions, *Australian Journal of Nutrition and Dietetics*, 48: 7–10.

—— 1992, Dietary supplementation, naturalistic values and attitudes to the food supply, *Ecology of Food and Nutrition*, 28: 211–17.

Wright, J., Franks, A., Ayres, P., Jones, K., Roberts, T. and Whitty, P. 2002, Public health in hospitals: The missing link in health improvement, *Journal of Public Health Medicine*, 24: 152–5.

Yancey, A.K., Kumanyika, S.K., Ponce, N.A., McCarthy, W.J., Fielding, J.K., Leslie, J.P. and Akbar, J. 2004, Population-based interventions engaging people of color in healthy eating and active living: A review, *Preventing Chronic Disease* (serial online), www.cdc.gov/pcd/issues/2004/jan/03_0012.htm.

Yanovitsky, I. and Stryker, J. 2001, Mass media social norms and health promotion efforts: A longitudinal study of media effects on youth binge drinking, *Communication Research*, 28(2): 208–39.

Young, L. and Nestle, M. 2002, The contribution of expanding portion sizes to the US obesity epidemic, *American Journal of Public Health*, 92(2): 246–9.

Young, L. and Swinburn, B. 2002, Impact of the Pick the Tick food information programme on the salt content of food in New Zealand, *Health Promotion International*, 17: 13–19.

Young Media Australia (YMA) 1997, *Sugar shows and fast food frenzies*, Report of the 'Good to eat or good for you?' Project, YMA, Adelaide.

Index